BARRON'S

Regents Exams and Answers

Living Environment
Fourth Edition

Gabrielle I. Edwards
Former Science Consultant
Board of Education
City of New York

Former Assistant Principal
Supervision
Science Department
Franklin D. Roosevelt High School
Brooklyn, New York

Marion Cimmino
Former Teacher, Biology and
Laboratory Techniques
Franklin D. Roosevelt High School
Brooklyn, New York

Frank J. Foder
Teacher, Advanced Placement
Biology
Franklin D. Roosevelt High School
Brooklyn, New York

G. Scott Hunter
Former Teacher, Biology and
Advanced Biology
Former Consultant, State Education
Department Bureaus of Science and
Testing Former School Business
Administrator, Schodack, NY
Former Superintendent of Schools
Mexico and Chatham, New York

T0274169

Published by Kaplan North America, LLC d/b/a Barron's Educational Series
1515 West Cypress Creek Road
Fort Lauderdale, Florida 33309
www.barronseduc.com

ISBN: 978-1-5062-9133-8

10 9 8 7 6 5 4 3 2 1

Kaplan North America, LLC d/b/a Barron's Educational Series print books are available at special quantity discounts to use for sales promotions, employee premiums, or educational purposes. For more information or to purchase books, please call the Simon & Schuster special sales department at 866-506-1949.

Contents

Glossary   132

Regents Examinations, Answers, and Student Self-Appraisal Guides   167

How to Use This Book

Organization of the Book

Study Questions

Section 1 of the book consists of 173 questions taken from past Regents examinations in biology. Most of the questions require that you select a correct response from four choices given. A few questions provide you with a list of words or phrases from which to select the one that best matches a given description. Others are constructed-response, graphical analysis, or reading comprehension questions. You should become familiar with the formats of the questions that appear on the Living Environment Regents examination.

Each question in Section 1 of this book has a well-developed answer. Each answer provides the number of the correct response, the reason why the response is correct, and explanations of why the other choices are incorrect.

A useful feature of Section 1 is the student Self-Appraisal Guide. This device allows you to determine where your learning strengths and weaknesses lie in the major topics of each unit. For specific topics within the units, the numbers of related questions are given. As you attempt to answer each question in Section 1, you may wish to circle on the appraisal form the numbers of the questions that you are unable to answer. The circled items then help you to identify at a glance subject matter areas in which you need additional study.

New York State Learning Standards

Commencement Standards

There are several commencement standards required of students in New York State public schools regarding their performance in math, science, and technology. The Core Curriculum for the Living Environment addresses two of these standards:

Standard 1: **Students will use mathematical analysis, scientific inquiry, and engineering design, as appropriate, to pose questions, seek answers, and develop solutions.**

Standard 4: **Students will understand and apply scientific concepts, principles, and theories pertaining to the physical setting and living environment and recognize the historical development of ideas in science.**

The Core Curriculum for the Living Environment was built from these two commencement standards. It is important to recognize that the Core Curriculum is not a syllabus. It does not prescribe what will be taught and learned in any particular classroom. Instead, it defines the skills and understandings that you must master in order to achieve the commencement standards for life science.

Instead of memorizing a large number of details at the commencement level, then, you are expected to develop the skills needed to deal with science on the investigatory level, generating new knowledge from experimentation and sharpening your abilities in data analysis. You are also expected to read and understand scientific literature, taking from it the facts and concepts necessary for a real understanding of issues in science. You are required to pull many facts together from different sources to develop your own opinions about the moral and ethical problems facing modern society concerning technological advances. These are thinking skills that do not respond to simple memorization of facts and scientific vocabulary.

Key Ideas, Performance Indicators, and Major Understandings

Each commencement standard is subdivided into a number of Key Ideas. Key Ideas are broad, unifying statements about what you need to know. Within Standard 1, three Key Ideas are concerned with laboratory investigation and data analysis. Together, these unifying principles develop your ability to deal with data and understand how professional science is carried out in biology. Within Standard 4, seven Key Ideas present a set of concepts that are central to the science of biology. These Key Ideas develop your understanding of the essential characteristics of living things that allow them to be successful in diverse habitats.

Within each Key Idea, Performance Indicators are presented that indicate which skills you should be able to demonstrate through your mastery of the Key Idea. These Performance Indicators give guidance to both you and your teacher about what is expected of you as a student of biology.

Performance Indicators are further subdivided into Major Understandings. Major Understandings give specific concepts that you must master in order to achieve each Performance Indicator. It is from these Major Understandings that the Regents assessment material will be drawn.

Laboratory Component

A meaningful laboratory experience is essential to the success of this or any other science course. You are expected to develop a good sense of how scientific inquiry is carried out by the professional scientist and how these same techniques can assist in the full understanding of concepts in science. The Regents requirement of 1,200 minutes of successful laboratory experience, coupled with satisfactory written reports of your findings, should be considered a minimum.

Students are required to complete four laboratory experiences required by the New York State Education Department and tested on the Regents Examination. See Barron's *Let's Review: The Living Environment* for a complete treatment of this requirement.

Regents Examinations

Section 2 of the book consists of actual Biology Regents examinations and answers. These Regents examinations are based on the New York State Core Curriculum for the Living Environment.

Assessments: Format and Scoring

The format of the Regents assessment for the Living Environment is as follows, based on actual Regents examinations.

Part A: Variable number (usually 30) of multiple-choice questions that test the student's knowledge of specific factual information. **All** questions must be answered on Part A.

Part B: Variable number (usually 25) of questions, representing a mixture of multiple-choice and constructed-response items. Questions may be based on the student's direct knowledge of biology, interpretation of experimental data, analysis of readings in science, and ability to deal with representations of biological phenomena. **All** questions must be answered on Part B1–B2.

Part C: Variable number (usually 17) of multiple-choice and constructed-response questions. Questions may be based on the student's direct knowledge of biology, interpretation of experimental data, analysis of readings in science, and ability to deal with representations of biological phenomena. **All** questions must be answered on Part C.

Part D—Laboratory component: Variable number (usually 13) of multiple-choice and constructed-response items. This component of the examination aims to assess student knowledge of and skills on any of four required laboratory experiences supplied to schools by the New York State Education Department. The content of these questions will reflect specific laboratory experiences. You are strongly encouraged to include review of these laboratory experiences as part of your year-end Regents preparation activity.

The following chart summarizes the current laboratory requirement for New York State public schools:

Laboratory Requirements

Laboratory Title	Description
The Beaks of Finches	Explores the adaptive advantages of beaks with different physical characteristics
Relationships and Biodiversity	Explores the relationship between DNA structure and the biochemistry of inheritance
Making Connections	Explores the effects of physical activity on human metabolic activities
Diffusion Through a Membrane	Explores the nature of cross-membrane transport in living cells
Adaptations for Reproductive Success in Flowering Plants*	Not yet available
DNA Technology*	Not yet available
Environmental Conditions and Seed Germination*	Not yet available

*Not yet available as of this writing.

Studying questions from past Regents examinations is an invaluable aid in developing a mindset that will enable you to approach questions with understanding. Although exact questions are not repeated, question types are repeated. If you practice questions that require interpretation, problem solving, and graph construction, you will do well on the entire exam. During the school year, the 30 required laboratory lessons teach you certain manipulative skills. Questions involving identification, measurement, and other laboratory procedures are based on the laboratory exercises. Review of past materials gives you insight as to the types of questions that you may be asked to answer. Study the questions in the Regents exams in this book diligently.

How to Study

General Suggestions for Study

You've spent all year learning many different facts and concepts about biology—far more than you could ever hope to remember the "first time around." Your teacher has drilled you on these facts and concepts; you've done homework, taken quizzes and tests, performed laboratory experiments, and reviewed the material at intervals throughout the year.

Now it's time to put everything together. The Regents exam is only a few weeks away. If you and your teacher have planned properly, you will have finished all new information about 3 weeks before the exam. Now you have to make efficient use of the days and weeks ahead to review all that you've learned in order to score high on the Regents.

The task ahead probably seems impossible, but it doesn't have to be! You've actually retained much more of the year's material than you realize! The review process should be one that helps you to recall the many facts and concepts you've stored away in your memory. Your Barron's resources, including Barron's *Let's Review: The Living Environment,* will help you to review this material efficiently.

You also have to get yourself into the right frame of mind. It won't help to be nervous and stressed during the review process. The best way to avoid being stressed during any exam is to be well rested, prepared, and confident. We're here to help you prepare and to build your confidence. So let's get started on the road to a successful exam experience!

To begin, carefully read and follow the steps outlined below.

1. Start your review early; don't wait until the last minute. Allow at least 2 weeks to prepare for the Regents exam. Set aside an hour or two a day over the next few weeks for your review. Less than an hour a day is insufficient time for you to concentrate on the material meaningfully; more than 2 hours daily will yield diminishing returns on your investment of time.

2. Find a quiet, comfortable place to study. You should seat yourself at a well-lighted work surface, free of clutter, in a room without distractions of any kind. You may enjoy watching TV or listening to music curled up in a soft chair, but these and other diversions should be avoided when doing intense studying.

3. Make sure you have the tools you need, including this book, a pen and pencil, and some scratch paper for taking notes and doing calculations. Keep your class notebook at hand for looking up information between test-taking sessions. It will also help to have a good review text, such as Barron's *Let's Review: The Living Environment*, available for reviewing important concepts quickly and efficiently.

4. Concentrate on the material in the "Study Questions and Answers" section of this book. Read carefully and thoughtfully. Think about the questions that you review,

and try to make sense out of them. Choose the answers carefully. (See the following section, "Using This Book for Study," for additional tips on question-answering techniques.)

5. Use available resources, including a dictionary and the glossary in this book, to look up the meanings of unfamiliar words in the practice questions. Remember that the same terms can appear on the Regents exam you will take, so take the opportunity to learn them now.

6. Remember: study requires time and effort. Your investment in study now will pay off when you take the Regents exam.

Using This Book for Study

This book is an invaluable tool if used properly. Read carefully and try to answer *all* the questions in Section 1 and on the practice exams. The more you study and practice, the more you will increase your knowledge of biology and the likelihood that you will earn a high grade on the Regents exam. To maximize your chances, use this book in the following way:

1. Answer all of the questions in the section entitled "Study Questions and Answers." Check your responses by using "Answers to Topic Questions," including "Wrong Choices Explained," following each question set. Record the number of correct responses on each topic in the "Self-Appraisal Guide" at the end of the section to identify your areas of strength and weakness. Use a good review text, such as Barron's *Let's Review Regents: Living Environment,* to study each area on which you did poorly. Finally, go back to the questions you missed on the first round and be sure that you fully understand what each question asks and why the correct answer is what it is.

2. When you have completed the questions in "Study Questions and Answers," go on to the examination section. Select the first complete examination and take it under test conditions.

3. Interpret the term *test conditions* as follows:

   ■ Be well rested; get a good night's sleep before attempting *any* exam.

   ■ Find a quiet, comfortable room in which to work.

   ■ Allow no distractions of any kind.

   ■ Select a well-lighted work surface free of clutter.

   ■ Have your copy of this book with you.

   ■ Bring to the room a pen, a pencil, some scratch paper, and a watch or alarm clock set for the 3-hour exam limit.

4. Take a deep breath, close your eyes for a moment, and RELAX! Tell yourself, "I know this stuff!" You have lots of time to take the Regents exam; use it to your advantage by reducing your stress level. Forget about your plans for later. For the present, your number 1 priority is to do your best, whether you're taking a practice exam in this book, or the real thing.

5. Read all test directions carefully. Note how many questions you must answer to complete each part of the exam. If test questions relate to a reading passage, diagram, chart, or graph, be sure you fully understand the given information before you attempt to answer the questions that relate to it.

6. When answering multiple-choice questions on the Regents exam, TAKE YOUR TIME! Pay careful attention to the "stem" of the question; read it over several times. These questions are painstakingly written by the test preparers, and every word is chosen to convey a specific meaning. If you read the question carelessly, you may answer a question that was never asked! Then read each of the four multiple-choice answers carefully, using a *pencil* to mark in the test booklet the answer you think is correct.

7. Remember that three of the multiple-choice answers are *incorrect*; these incorrect choices are called "distracters" because they seem like plausible answers to poorly prepared or careless students. To avoid being fooled by these distracters, you must think clearly, using everything you have learned about biology since the beginning of the year. This elimination process is just as important to your success on the Living Environment Regents exam as knowing the correct answer! If more than one answer seems to be correct, reread the question to find the words that will help you to distinguish between the correct answer and the distracters. When you have made your best judgment about the correct answer, circle the number in *pencil* in your test booklet.

8. Constructed-response questions appear in a number of different forms. You may be asked to select a term from a list, write the term on the answer sheet, and define the term. You may be asked to describe some biological phenomenon or state a biological fact using a complete sentence. You may be asked to read a value from a diagram of a measuring instrument and write that value in a blank on the answer paper. When answering this type of question, care should be exercised to follow directions precisely. If a complete sentence is called for, it must contain a subject and a verb, must be punctuated, and must be written understandably in addition to answering the scientific part of the question accurately. Values must be written clearly and accurately and include a unit of measure, if appropriate. Failure to follow the directions for a question may result in a loss of credit for that question.

9. A special type of constructed-response question is the essay or paragraph question. Typically, essay or paragraph questions provide an opportunity to earn multiple credits for answering the question correctly. As in the constructed-response questions described above, it is important that you follow the directions given if you hope to earn the maximum number of credits for the question. Typically, the question outlines exactly what must be included in your essay to gain full credit. Follow these directions step by step, double-checking to be certain that all question components are addressed in your answer. In addition, your essay or paragraph should follow the rules of good grammar and good communication so that it is readable and understandable. And, of course, it should contain correct information that answers all the parts of the question asked.

10. Graphs and charts are a special type of question that requires you to organize and represent data in graphical format. Typically for such questions, you are expected to place unorganized data in ascending order in a data chart or table. You may also be asked to plot organized data on a graph grid, connect the plotted points, and label the graph axes appropriately. Finally, questions regarding data trends and extrapolated projections may be asked, requiring you to analyze the data in the graph and draw inferences from it. As with all examination questions, always follow all directions for the question. Credit can be granted only for correctly following directions and accurately interpreting the data.

11. When you have completed the exam, relax for a moment. Check your time; have you used the entire 3 hours? Probably not. Resist the urge to quit. Go back to the beginning of the exam, and, in the time remaining, *retake the exam in its entirety*. Try to ignore the penciled notations you made the first time. If you come up with a different answer the second time through, read over the question with extreme care before deciding which response is correct. Once you have decided on the correct answer, mark your choice in ink in the answer booklet.

12. Score the exam using the Answer Key at the end of the exam. Review the "Answers Explained" section for each question to aid your understanding of the exam and the material. Remember that it's just as important to understand why the incorrect responses are incorrect as it is to understand why the correct responses are correct!

13. Finally, focus your between-exam study on your areas of weakness in order to improve your performance on the next practice exam. Complete all the practice exams in this book using the techniques outlined above.

14. Be sure to sign the declaration on your answer sheet. Unless this declaration is signed, your paper cannot be scored.

Test-Taking Tips—
A Summary

The following pages contain seven tips to help you achieve a good grade on the Living Environment Regents exam.

TIP 1

Be Confident and Prepared

Suggestions

- Review previous tests.
- Use a clock or watch, and take previous exams at home under examination conditions (i.e., don't have the radio or television on).
- Get a review book. (The preferred book is Barron's *Let's Review Regents: Living Environment.*)
- Talk over the answers to questions on these tests with someone else, such as another student in your class or someone at home.
- Finish all your homework assignments.
- Look over classroom exams that your teacher gave during the term.
- Take class notes carefully.
- Practice good study habits.
- Know that there are answers for every question.
- Be aware that the people who made up the Regents exam want you to pass.
- Remember that thousands of students over the last few years have taken and passed a Biology Regents. You can pass, too!
- Complete your study and review at least one day before taking the examination. Last-minute "cramming" may hurt, rather than enhance, your performance on the exam.

- Be well rested when you enter the exam room. A good night's sleep is essential preparation for any examination.

- On the night before the exam day: lay out all the things you will need, such as clothing, pens, and admission cards.

- Bring with you two pens, two pencils, an eraser, and, if your school requires it, an identification card. Decide before you enter the room that you will remain for the entire 3-hour examination period, and either bring a wristwatch or sit where you can see a clock.

- Once you are in the exam room, arrange things, get comfortable, and attend to personal needs (the bathroom).

- Before beginning the exam, take a deep breath, close your eyes for a moment, and RELAX! Repeat this technique any time you feel yourself "tensing up" during the exam.

- Keep your eyes on your own paper; do not let them wander over to anyone else's paper.

- Be polite in making any reasonable demands of the exam room proctor, such as changing your seat or having window shades raised or lowered.

TIP 2

Read Test Instructions Carefully

Suggestions

- Be familiar with the format of the examination.

- Know how the test will be graded.

- Read all directions carefully. Be sure you fully understand supplemental information (reading passages, charts, diagrams, graphs) before you attempt to answer the questions that relate to it.

- Underline important words and phrases.

- Ask for assistance from the exam room proctor if you do not understand the directions.

TIP 3
Read Each Question Carefully and Read Each Choice Before Recording Your Answer

Suggestions

- When answering the questions, TAKE YOUR TIME! Be sure to read the "stem" of the question and each of the four multiple-choice answers very carefully.

- If you are momentarily "stumped" by a question, put a check mark next to it and go on; come back to the question later if you have time.

- Remember that three of the multiple-choice answers (known as "distracters") are incorrect. If more than one answer seems to be correct, reread the question to find the words that will help you to distinguish between the correct answer and the distractors.

- When you have made your best judgment about the correct answer, circle the appropriate number in pencil on your answer sheet.

TIP 4
Budget Your Test Time (3 Hours)

Suggestions

- Bring a watch or clock to the test.
- The Regents examination is designed to be completed in 1½ to 2 hours.
- If you are absolutely uncertain of the answer to a question, mark your question booklet and move on to the next question.
- If you persist in trying to answer every difficult question *immediately*, you may find yourself rushing or unable to finish the remainder of the examination.
- When you have completed the exam, relax for a moment. Then go back to the beginning, and, in the time remaining, *retake the exam in its entirety*. Pay particular attention to questions you skipped the first time. Once you have decided on a correct response for multiple-choice questions, mark an "X" in ink through the penciled circle on the answer sheet.
- Plan to stay in the room for the entire 3 hours. If you finish early, read over your work—there may be some things that you omitted or that you may wish to add. You also may wish to refine your grammar, spelling, and penmanship.

TIP 5

Use Your Reasoning Skills

Suggestions

- Answer *all* questions.
- Relate (connect) the question to anything that you studied, wrote in your notebook, or heard your teacher say in class.
- Relate (connect) the question to any film, demonstration, or experiment you saw in class, any project you did, or to anything you may have learned from newspapers, magazines, or television.
- Look over the entire test to see whether one part of it can help you answer another part.

TIP 6

Don't Be Afraid to Guess

Suggestions

- In general, go with your first answer choice.
- Eliminate obvious incorrect choices.
- If still unsure of an answer, make an educated guess.
- There is no penalty for guessing; therefore, answer ALL questions. An omitted answer gets no credit.

TIP 7

Sign the Declaration

Suggestions

- Be sure to sign the declaration on your answer sheet.
- Unless this declaration is signed, your paper cannot be scored.

Tips for Teachers

Classroom Use

All teachers will be able to use this book with their students as a companion to their regular textbooks and will find that their students gain considerable self-confidence and ability in test taking through its consistent use.

The Living Environment Core Curriculum defines the skills and abilities students should have at the point of commencement at the upper-secondary level. It is assumed that science concepts have been taught and assessed at an age-appropriate level throughout their career, so that little additional detail needs to be presented at the upper-secondary level.

An excellent companion to this book (and any comprehensive biology text) is Barron's *Let's Review: The Living Environment* (Hunter). The factual material and organization of this book lend themselves well to the development of Standards-Based Learning Units (SBLUs) and Essential Questions. The level of detail is consistent with what students really have to know in order to do well on the New York State Regents Examination on the Living Environment.

Application-Based Curriculum

The curriculum focus can be characterized as application-based—one that is less concerned about content and more concerned about thinking. It is less about *how much* students know and more about *what they can do* with what they know. The latter, after all, is what real learning is all about; these are the abilities that will last a lifetime, not facts and scientific terminology.

This being said, it is acknowledged that students will have a difficult time expressing their views and making moral and ethical judgments about science if they lack a working knowledge of scientific principles and do not have at least a passing understanding of the terms used by biologists. For this reason, teachers and administrators will need to develop local curricula that complement the Core Curriculum. It is up to the teacher or administrator to decide what examples and factual knowledge best illustrate the concepts presented in the Core Curriculum, what concepts need to be reinforced and enhanced, what experiences will add measurably to students' understanding of science, and what examples of local interest should be included.

The teacher will immediately recognize the need to go beyond this level in the classroom, with examples, specific content, and laboratory experiences that complement and illuminate these Major Understandings. It is at this level that the locally developed curriculum is essential. Each school system is challenged to develop an articulated K-12 curriculum in mathematics, science, and technology that will position students to achieve a passing standard at the elementary and intermediate levels, such that success is maximized at the commencement level.

The addition of factual content must be accomplished without contradicting the central philosophy of the learning standards. If local curricula merely revert to the fact-filled syllabi of the past, then little will have been accomplished in the standards movement other than to add yet another layer of content and requirements on the heads of students. A balance must be struck between the desire to build students' ability to think and analyze and the desire to add to the content they are expected to master.

Laboratory Experience

The reduction of factual detail in the Core Curriculum (1982–1999) should allow a more in-depth treatment of laboratory investigations to be planned and carried out than was possible under the previous syllabus. Laboratory experiences should be designed to address Standard 1 (inquiry techniques) but should also take into account Standards 2 (information systems), 6 (interconnectedness of content), and 7 (problem-solving approaches). They should also address the laboratory skills listed in Appendix A of the Core Curriculum.

Part D of the examination assesses student knowledge of and skills on any of four required laboratory experiences supplied to schools by the New York State Education Department. The specific laboratory experiences required in any year will vary according to a preset schedule (see chart on next page).

Questions on this section can be a combination of multiple-choice and constructed-response questions similar to those found in Parts A, B, and C of the Living Environment Regents Examination. The content of these questions reflect the four specific laboratory experiences required for a particular year. Teachers are strongly encouraged to include review of these laboratory experiences as part of their year-end Regents preparation activity.

The following chart summarizes the current laboratory requirement for New York State public schools:

Laboratory Title	Description
The Beaks of Finches	Explores the adaptive advantages of beaks with different physical characteristics
Relationships and Biodiversity	Explores the relationship between DNA structure and the biochemistry of inheritance
Making Connections	Explores the effects of physical activity on human metabolic activities
Diffusion Through a Membrane	Explores the nature of cross-membrane transport in living cells
Adaptations for Reproductive Success in Flowering Plants*	Not yet available
DNA Technology*	Not yet available
Environmental Conditions and Seed Germination*	Not yet available

*Not yet available as of this writing.

Study Questions and Answers

Questions on Standard 1

Students will use mathematical analysis, scientific inquiry, and engineering design, as appropriate, to pose questions, seek answers, and develop solutions.

Key Idea 1—Purpose of Scientific Inquiry

The central purpose of scientific inquiry is to develop explanations of natural phenomena in a continuing and creative process.

Performance Indicator	Description
1.1	The student should be able to elaborate on basic scientific and personal explanations of natural phenomena and develop extended visual models and mathematical formulations to represent one's thinking.
1.2	The student should be able to hone ideas through reasoning, library research, and discussion with others, including experts.
1.3	The student should be able to work toward reconciling competing explanations and clarify points of agreement and disagreement.
1.4	The student should be able to coordinate explanations at different levels of scale, points of focus, and degrees of complexity and specificity, and recognize the need for such alternative representations of the natural world.

Base your answers to questions 1 through 4 on the passage below and on your knowledge of biology.

To Tan or Not to Tan

Around 1870, scientists discovered that sunshine could kill bacteria. In 1903, Niels Finsen, an Icelandic researcher, won the Nobel Prize for his use of sunlight therapy against infectious diseases. Sunbathing then came into wide use as a treatment for tuberculosis, Hodgkin's disease (a form of cancer), and common wounds. The discovery of vitamin D, the "sunshine vitamin," reinforced the healthful image of the Sun. People learned that it was better to live in a sun-filled home than a dark dwelling. At that time, the relationship between skin cancer and exposure to the Sun was not known.

In the early twentieth century, many light-skinned people believed that a deep tan was a sign of good health. However, in the 1940s, the rate of skin cancer began to increase and reached significant proportions by the 1970s. At this time, scientists began to realize how damaging deep tans could really be.

Tanning occurs when ultraviolet radiation is absorbed by the skin, causing an increase in the activity of melanocytes, cells that produce the pigment melanin. As melanin is produced, it is absorbed by cells in the upper region of the skin, resulting in the formation of a tan. In reality, the skin is building up protection against damage caused by the ultraviolet radiation. It is interesting to note that people with naturally dark skin also produce additional melanin when their skin is exposed to sunlight.

Exposure to more sunlight means more damage to the cells of the skin. Research has shown that, although people usually do not get skin cancer as children, each time a child is exposed to the Sun without protection, the chance of that child getting skin cancer as an adult increases.

Knowledge connecting the Sun to skin cancer has greatly increased since the late 1800s. Currently, it is estimated that ultraviolet radiation is responsible for more than 90% of skin cancers. Yet, even with this knowledge, about 2 million Americans use tanning parlors that expose patrons to high doses of ultraviolet radiation. A recent survey showed that at least 10% of these people would continue to do so even if they knew for certain that it would give them skin cancer.

Many of the deaths due to this type of cancer can be prevented. The cure rate for skin cancer is almost 100% when it is treated early. Reducing exposure to harmful ultraviolet radiation helps to prevent it. During the past 15 years, scientists have tried to undo the tanning myth. If the word "healthy" is separated from the word "tan," maybe the occurrence of skin cancer will be reduced.

1. State *one* known benefit of daily exposure to the Sun.   [1]

_____

2. Explain what is meant by the phrase "the tanning myth."   [1]

_____

3. Which statement concerning tanning is correct?
   (1) Tanning causes a decrease in the ability of the skin to regulate body temperature.
   (2) Radiation from the Sun is the only radiation that causes tanning.
   (3) The production of melanin, which causes tanning, increases when skin cells are exposed to the Sun.
   (4) Melanocytes decrease their activity as exposure to the Sun increases, causing a protective coloration on the skin.   3 _____

4. Which statement concerning ultraviolet radiation is *not* correct?
   (1) It may damage the skin.
   (2) It stimulates the skin to produce antibodies.
   (3) It is absorbed by the skin.
   (4) It may stimulate the skin to produce excess pigment.   4 _____

5. Current knowledge concerning cells is a result of the investigations and observations of many scientists. The work of these scientists forms a well-accepted body of knowledge about cells. This body of knowledge is an example of a
   (1) hypothesis   (3) theory
   (2) controlled experiment   (4) research plan   5 _____

6. In his theory of evolution, Lamarck suggested that organisms will develop and pass on to offspring variations that they need in order to survive in a particular environment. In a later theory of evolution, Darwin proposed that changing environmental conditions favor certain variations that promote the survival of organisms. Which statement is best illustrated by this information?
   (1) Scientific theories that have been changed are the only ones supported by scientists.
   (2) All scientific theories are subject to change and improvement.
   (3) Most scientific theories are the outcome of a single hypothesis.
   (4) Scientific theories are not subject to change.   6 _____

Base your answers to questions 7 and 8 on the passage below and on your knowledge of biology.

The number in the parentheses ( ) at the end of a sentence is used to identify that sentence.

They Sure Do Look Like Dinosaurs

When making movies about dinosaurs, film producers often use ordinary lizards and enlarge their images many times (1). We all know, however, that although they look like dinosaurs and are related to dinosaurs, lizards are not actually dinosaurs (2).

Recently, some scientists have developed a hypothesis that challenges this view (3). These scientists believe that some dinosaurs were actually the same species as some modern lizard that had grown to unbelievable sizes (4). They think that such growth might be due to a special type of DNA called repetitive DNA, often referred to as "junk" DNA because scientists do not understand its functions (5). These scientists studied pumpkins that can reach sizes of nearly 1,000 pounds and found them to contain large amounts of repetitive DNA (6). Other pumpkins that grow to only a few ounces in weight have very little of this kind of DNA (7). In addition, cells that reproduce uncontrollably have almost always been found to contain large amounts of this DNA (8).

7. State *one* reason why scientists formerly thought of repetitive DNA as "junk." [1]

_____

_____

8. Write the number of the sentence that provides evidence supporting the hypothesis that increasing amounts of repetitive DNA are responsible for increased sizes of organisms. [1]

_____

Answers Explained

Key Idea 1—Purpose of Scientific Inquiry

1. One response is required. Acceptable responses include:
    - *Kills bacteria*
    - *Produces vitamin D*
    - *Treats diseases and/or wounds*

2. One response is required. Acceptable responses include:
    - *The "tanning myth" involves people believing that a tan is a sign of good health.*
    - *The "tanning myth" says that a good tan is good for people.*

3. **3** *The production of melanin, which causes tanning, increases when skin cells are exposed to the Sun* is the correct statement concerning tanning. Melanin is a dark pigment that is produced in specialized skin cells in response to ultraviolet radiation in sunlight or an artificial source. This information is found in the third paragraph of the passage.

    WRONG CHOICES EXPLAINED:
    (1) *Tanning causes a decrease in the ability of the skin to regulate body temperature* is not a correct statement concerning tanning. There is no known relationship between tanning and body temperature regulation.

    (2) *Radiation from the Sun is the only radiation that causes tanning* is not a correct statement concerning tanning. Tanning can also occur when the skin is exposed to artificial sources of ultraviolet radiation. A reference is made in the fourth paragraph of the passage to "tanning parlors" where people can be exposed to artificial doses of ultraviolet radiation.

    (4) *Melanocytes decrease their activity as exposure to the Sun increases, causing a protective coloration on the skin* is not a correct statement concerning tanning. Melanin is produced as a protective pigment that helps prevent deep penetration of ultraviolet radiation into the deep layers of the skin. When ultraviolet radiation is absorbed by melanocytes, their activity increases, not decreases.

4. 2 *It stimulates the skin to produce antibodies* is not a correct statement concerning ultraviolet radiation. There is no information in the passage relating to the production of antibodies as a result of absorption of ultraviolet radiation, and no known research indicates this type of relationship.

WRONG CHOICES EXPLAINED:
(1), (3), (4) *It may damage the skin, it is absorbed by the skin,* and *it may stimulate the skin to produce excess pigment* are all correct statements concerning ultraviolet radiation. Ultraviolet radiation is an invisible but extremely powerful form of electromagnetic radiation. It can penetrate unshielded living tissues and alter the genetic makeup of the cells it encounters. In humans, this radiation can cause the production of melanin from melanocytes; in extreme cases, it can stimulate the growth of skin cancer.

5. 3 The body of knowledge described in this question is an example of a *theory*. When scientists begin to study a phenomenon in nature, their first step is normally to investigate it through repeated observation and experimentation. As a result of the analysis of the large quantity of data gathered during this process, the scientists then formulate a theory ("well-accepted body of knowledge") that describes the phenomenon in a way that is consistent with the data.

WRONG CHOICES EXPLAINED:
(1) A *hypothesis* is not the body of knowledge described in this question. Scientists develop a hypothesis ("educated guess") around their preliminary observations concerning a natural phenomenon. The hypothesis may be proven accurate or inaccurate as a result of the experimentation used to test it. For this reason, a hypothesis cannot be considered a "well-accepted body of knowledge."

(2) A *controlled experiment* is not the body of knowledge described in this question. A controlled experiment is a scientific method used to test an experimental hypothesis. The data that results from a controlled experiment can be used to support the development of a "well-accepted body of knowledge," but it does not constitute that body of knowledge.

(4) A *research plan* is not the body of knowledge described in this question. A research plan may be a series of controlled experiments designed to test various aspects of a natural phenomenon. The data that results from the research plan can be used to support the development of a "well-accepted body of knowledge," but it does not constitute that body of knowledge.

6. **2** *All scientific theories are subject to change and improvement* is the statement best illustrated by the information given. Both Lamarck and Darwin developed their theories of evolution based on observations made and inferences drawn before there was a good understanding of the genetic basis of variation. Lamarck's earlier theory of "use and disuse" was disproven by later experiments of other scientists. Darwin's later theory of "natural selection," though much closer to the currently accepted scientific theory of evolution, has been modified and improved on by the work of later scientists who have had the benefit of modern-day research in genetics, paleontology, and other sciences.

WRONG CHOICES EXPLAINED:

(1) *Scientific theories that have been changed are the only ones supported by scientists* is not the statement best illustrated by the information given. Scientists generally support theories that have stood the test of good scientific research. A theory that has not changed, as long as it is still supported by such research, is generally supported by most scientists.

(3) *Most scientific theories are the outcome of a single hypothesis* is not the statement best illustrated by the information given. In fact, scientific theories are based on the results of many experiments that each contain their own independent hypotheses.

(4) *Scientific theories are not subject to change* is not the statement best illustrated by the information given. Scientists are constantly questioning and reevaluating scientific theories. It is likely that a vast majority of all scientific theories undergo at least some modification.

7. One response is required that indicates a reason why scientists formerly thought of repetitive DNA as "junk." Acceptable responses include: [1]

- *Scientists did not understand the function of repetitive DNA.*
- *They didn't know what it did, and so they thought it was junk.*

8. One credit is allowed for indicating that either sentence 6 or sentence 7 provides evidence supporting the hypothesis that increased amounts of repetitive DNA are responsible for increased sizes of organisms. These sentences give information about the results of scientific investigations that measured the amount of repetitive DNA in the cells of giant pumpkins and miniature pumpkins and found that giant pumpkins contain more of this kind of DNA than miniature pumpkins.

Key Idea 2—Methods of Scientific Inquiry

Beyond the use of reasoning and consensus, scientific inquiry involves the testing of proposed explanations involving the use of conventional techniques and procedures and usually requiring considerable ingenuity.

Performance Indicator	Description
2.1	The student should be able to devise ways of making observations to test proposed explanations.
2.2	The student should be able to refine research ideas through library investigations, including electronic information retrieval and reviews of literature, and through peer feedback obtained from review and discussion.
2.3	The student should be able to develop and present proposals including formal hypotheses to test explanations (i.e., predict what should be observed under specific conditions if the experiment is true).
2.4	The student should be able to carry out research for testing explanations, including selecting and developing techniques, acquiring and building apparatus, and recording observations as necessary.

Base your answers to questions 9 and 10 on the diagram below of the field of view of a light compound microscope and on your knowledge of microscopes.

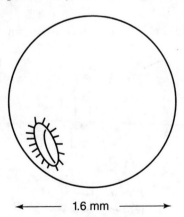

← 1.6 mm →

9. In order to center the organism in the field of view, the slide should be moved

(1)   down and to the right   (3)   up and to the right
(2)   down and to the left   (4)   up and to the left   9 _____

10. The approximate length of the organism is

(1)   500 μm   (3)   50 μm
(2)   1,600 μm   (4)   1.6 μm   10 _____

11. After viewing an organism under low power, a student switches to high power. The student should first

(1)   adjust the mirror
(2)   center the organism
(3)   raise the objective and switch to high power
(4)   close the diaphragm   11 _____

12. Using one or more complete sentences, explain why a specimen viewed under the high-power objective of a microscope appears darker than when it is viewed under low power.

_____

_____

    Base your answers to questions 13 and 14 on the diagram below of a compound light microscope.

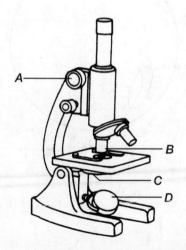

13. The letter *C* represents

    (1) the mirror

    (2) the diaphragm

    (3) the eyepiece

    (4) the high-power objective   13 _____

14. Select and name *one* of the labeled parts, and in one or more complete sentences describe its function.

_____

_____

15. The letter "p" as it normally appears in print is placed on the stage of a compound light microscope. Which best represents the image observed when a student looks through the microscope?

    (1) p   (3) b

    (2) q   (4) d   15 _____

16. To separate the parts of a cell by differences in density, a biologist would probably use

    (1) a microdissection instrument

    (2) an ultracentrifuge

    (3) a compound light microscope

    (4) an electron microscope   16 _____

17. The diagram below represents the field of view of a microscope. What is the approximate diameter, in micrometers, of the cell shown in the field?

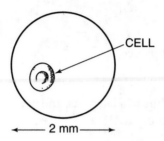

(1)   50 μm   (3)   1,000 μm
(2)   500 μm   (4)   2,000 μm   17 _____

Base your answer to the question on the diagram below.

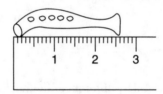

18. How many millimeters long is the organism resting on the metric ruler?

_____

Questions 19 through 21 are based on the experiment described below.

A test tube was filled with a molasses solution, sealed with a membrane, and inverted into a beaker containing 200 mL of distilled water. A second test tube was filled with a starch solution, sealed with a membrane, and inverted into a beaker containing 200 mL of distilled water. After several hours, the water in each beaker was tested for the presence of molasses and starch.

The diagrams show the setup of the experiment.

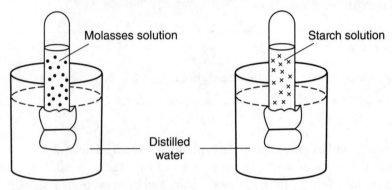

At the Start of the Experiment

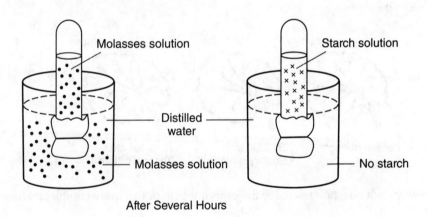

After Several Hours

Answer each question related to the experiment in one or more complete sentences.

19. What principle was being tested in the experiment?

_____

_____

20. What reagents were used in the experiment to test for the presence of molasses and starch?

_____

21. Draw one conclusion from this experiment.

_____

Questions 22 and 23 are based on the experiment described below.

An opaque disk was placed on several leaves of a geranium plant. The remaining leaves of the plant were untreated. After the plant had been exposed to sunlight, a leaf on which a disk had been placed was removed and tested as shown in parts *B* and *C* of the diagram below.

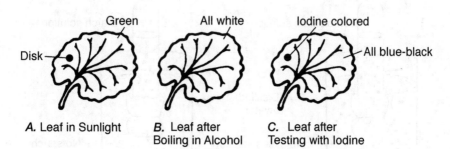

Green   All white   Iodine colored

Disk   All blue-black

A. Leaf in Sunlight   *B.* Leaf after   *C.* Leaf after
                              Boiling in Alcohol   Testing with Iodine

Answer each question related to the experiment in one or more complete sentences.

22. What conclusion can be drawn from the result of the experiment?

_____

23. What process was being investigated by the experiment?

_____

Questions 24 and 25 are based on the experiment described below.

A student added 15 mL of water to each of three test tubes, labeled A, B, and C. A 1-cc piece of raw potato was added to tube B. A 1-cc piece of cooked potato was added to tube C. Five drops of hydrogen peroxide (H_2O_2) were added to each test tube. The results are shown in the following diagram.

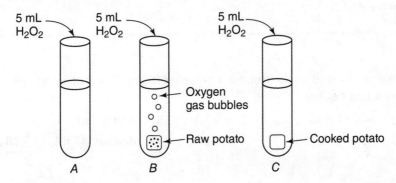

24. What conclusion can be drawn from the experiment?

_____

25. Which test tube is the control? Explain the reason for your choice.

_____

Base your answers to questions 26 through 28 on the diagram of the measuring device shown below.

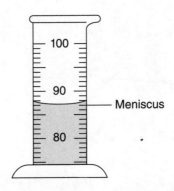

26. What is the name of this measuring device?

_____

27. In one complete sentence describe the procedure that you would follow to read the meniscus.

_____

28. What must a student do to obtain a volume of 85 milliliters of liquid in this measuring device?

(1)   Add 2.0 mL.   (3)   Add 2.5 mL.

(2)   Remove 2.0 mL.   (4)   Remove 8.7 mL.   28 _____

Answers Explained

Key Idea 2—Methods of Scientific Inquiry

9. **2** Specimens viewed under the microscope appear upside-down, backward, and reversed.

   WRONG CHOICES EXPLAINED:
   (1), (3), (4) With any of these choices, the specimen would be moved out of the field of view.

10. **1** The field of view is given as 1.6 mm. 1 mm = 1000 μm. 1.6 mm × 1000 μm = 1600 μm. The diagram shows that three specimens would fit across the field of view. One-third of 1600 μm = 533 μm. Of the choices given, *500 μm* (choice 1) is closest to this value.

    WRONG CHOICES EXPLAINED:
    (2), (3), (4) Each of these choices is mathematically incorrect.

11. **2** The student should first *center the organism*. The field of view is smaller under high power; therefore, less of a specimen can be seen. If the organism is not centered, it may fall out of the field of view under high power.

    WRONG CHOICES EXPLAINED:
    (1) The *mirror is adjusted* for maximum light under low power. Because the diameter of the high-power objective is very small, it is impossible to adjust the light under high power.

    (3) A compound light microscope is parfocal; that is, it is not necessary to *lift the high-power objective* to focus under high power. The specimen remains in focus when switching from low power to high power.

    (4) *Closing the diaphragm* reduces the amount of light entering the objective. Therefore, the specimen would appear very dark and would be difficult to see.

12. *The diameter of the high-power objective is smaller than the diameter of the low-power objective. Less light enters through the high-power objective, and therefore the specimen appears darker.*

13. 2 The letter *C* represents *the diaphragm.*

WRONG CHOICES EXPLAINED:
(1) *The mirror* is represented by *D.*

(3), (4) *The eyepiece* and *the high-power objective* are not labeled on the diagram.

14. *Coarse adjustment* (A)—*used to focus a specimen under the low-power objective.*

or

Low-power objective (B)—*along with the standard eyepiece, magnifies a specimen 100×.*

or

Diaphragm (C)—*regulates the amount of light entering the objectives.*

or

Mirror (D)—*provides a source of light that illuminates the specimen.*

15. 4 The image of a specimen as seen under a microscope is upside-down (*d*). The right side is on the left side, and the top is on the bottom.

WRONG CHOICES EXPLAINED:
(1) In this choice (*p*) there is no change in the way the image of the letter appears.

(2) In this choice (*q*) the image of the letter is reversed in only one direction: The right and left sides are reversed.

(3) In this choice (*b*) the image of the letter is reversed in only one direction: The top and bottom are reversed.

16. 2 The *ultracentrifuge* is a machine that spins at a very high speed. A test tube of a liquid containing the parts of ruptured cells is placed in the machine. Each cell part has its own density (mass per unit volume). When the machine rotates, the cell parts fall to different levels in the test tube depending on their density.

WRONG CHOICES EXPLAINED:
(1) *A microdissection instrument* enables a biologist to remove a cell part from a single living cell. A micromanipulator is an example of such an instrument.

(3) A cell is transparent under a light microscope. Its structures cannot be seen unless the cell is stained. *A compound light microscope* can be used to view, but not to separate, cell parts.

(4) *An electron microscope* uses beams of electrons to view freeze-dried specimens; it cannot be used to separate cell parts for study.

17. **2** Study the information given in the diagram. Notice that the diameter of the circle is 2 mm. Since 1 mm is equal to 1,000 μm, 2 mm are equal to 2,000 μm. In relation to the entire circle, how large is the cell? Is the cell one-half as large or one-fourth as large? Dividing the circle into four parts shows us that the diameter of the cell is about one-quarter the diameter of the circle. Dividing 4 into 2,000 results in *500 μm*.

    WRONG CHOICES EXPLAINED:
    (1) *50 μm* is too small. The cell is ten times larger than 50.

    (3) *1,000 μm* is too large. The cell is not one-half the diameter.

    (4) *2,000 μm* is the diameter of the circle. The cell is only one-fourth as large.

18. The organism is *26 millimeters* long.

19. The principle of *diffusion* was being tested in the experiment.

20. *Benedict's solution* was used to test for the presence of molasses in the beaker. *Iodine* was used to test for the presence of starch in the beaker.

21. *Molasses can diffuse through a membrane.*

    or

    Starch cannot diffuse through a membrane.

22. *No starch was produced in the area covered by the disk.*

23. *The process of photosynthesis was being investigated.*

24. *Raw potato contains an enzyme that breaks down hydrogen peroxide.*

    or

    Cooking a potato destroys the enzyme that breaks down hydrogen peroxide.

25. *Test tube A is the control.* A control is the part of the experiment that provides the basis of comparison for the variable being tested.

26. The device is known as a *graduated cylinder*.

27. *The meniscus should be read at eye level.*

28. To obtain a volume of 85 mL, *2.0 mL* must be removed. The graduated cylinder contains 87 mL of liquid.

Key Idea 3—Analysis in Scientific Inquiry

The observations made while testing proposed explanations, when analyzed using conventional and invented methods, provide new insights into natural phenomena.

Performance Indicator	Description
3.1	The student should be able to use various methods of representing and organizing observations (e.g., diagrams, tables, charts, graphs, equations, matrices) and insightfully interpret the organized data.
3.2	The student should be able to apply statistical analysis techniques, when appropriate, to test if chance alone explains the results.
3.3	The student should be able to assess correspondence between the predicted result contained in the hypothesis and actual result, and reach a conclusion as to whether the explanation on which the prediction was based is supported.
3.4	The student should be able to, based on the results of the test and through public discussion, revise the explanation and contemplate additional research.
3.5	The student should be able to develop a written report for public scrutiny that describes the proposed explanation, including a literature review, the research carried out, its result, and suggestions for further research.

29. If curve *A* in the diagram represents a population of hawks in a community, what would most likely be represented by curve *B?*

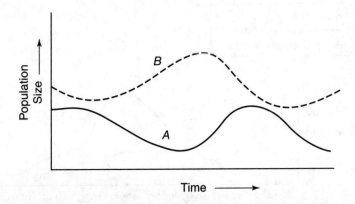

(1) the dominant trees in that community

(2) a population with which the hawks have a mutualistic relationship

(3) variations in the numbers of producers in that community

(4) a population on which the hawks prey   29 _____

Base your answers to questions 30 and 31 on this graph and on your knowledge of biology. The graph below depicts changes in the population growth rate of Kaibab deer.

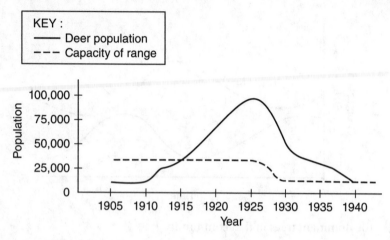

30. About how many deer could the range have supported in 1930 without some of them starving to death?

  (1)   12,000   (3)   50,000

  (2)   35,000   (4)   100,000   30 _____

31. In which year were the natural predators of the deer most likely being killed off faster than they could reproduce?

  (1)   1905   (3)   1930

  (2)   1920   (4)   1940   31 _____

32. Which process is illustrated by the diagram?

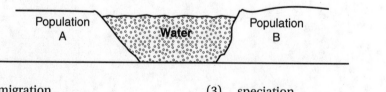

  (1)   migration   (3)   speciation

  (2)   adaptive radiation   (4)   isolation   32 _____

Base your answers to questions 33 through 36 on the information and data table below and on your knowledge of biology. The table shows the average systolic and diastolic blood pressure measured in millimeters of mercury (mm Hg) for humans between the ages of 2 and 14 years.

Data Table

Age	Average Blood Pressure (mm Hg)	
	Systolic	Diastolic
2	100	60
6	101	64
10	110	72
14	119	76

Directions **(33–36): Using the information in the data table, construct a line graph on the grid provided, following the directions below.**

33. Mark an appropriate scale on each labeled axis.

34. Plot the data for systolic blood pressure on your graph. Surround each point with a small triangle and connect the points.

35. Plot the data for diastolic blood pressure on your graph. Surround each point with a small circle and connect the points.

Example: 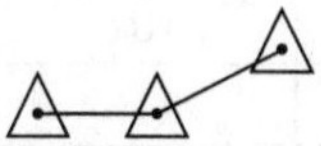

Example: 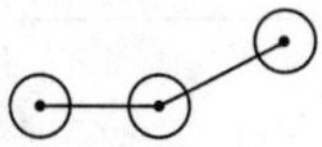

The Effect of Age on Human Blood Pressure

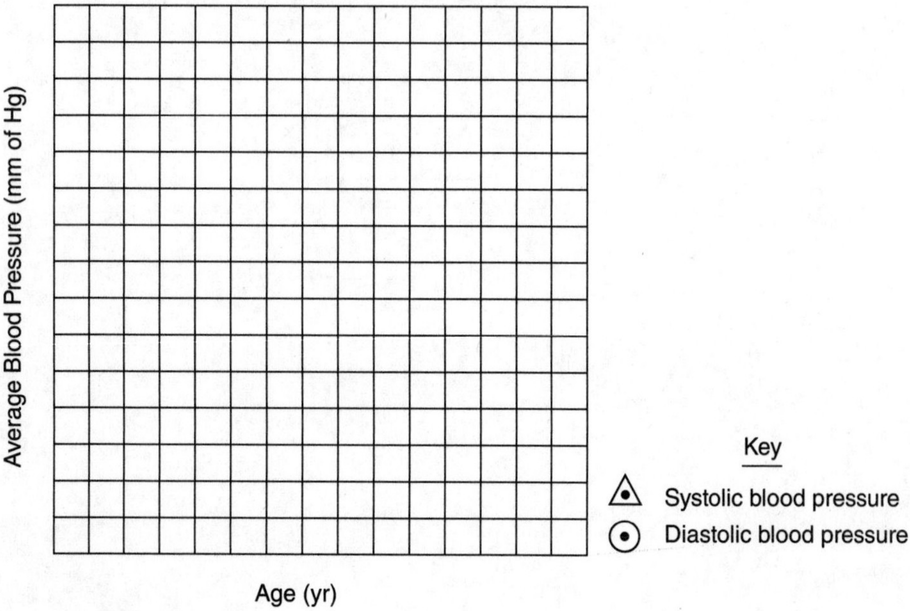

36. Using one or more complete sentences, state one conclusion that compares systolic blood pressure to diastolic blood pressure in humans between the ages of 2 and 14 years.

_____

_____

37. The graph below shows the results of an experiment.

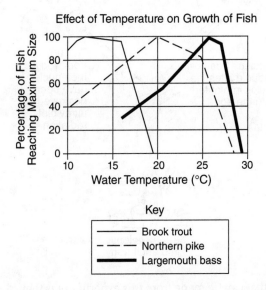

Effect of Temperature on Growth of Fish

At 16°C, what percentage of the brook trout reached maximum size?

(1) 30%   (3) 75%

(2) 55%   (4) 95%   37 _____

38. An experiment is represented in the diagram below.

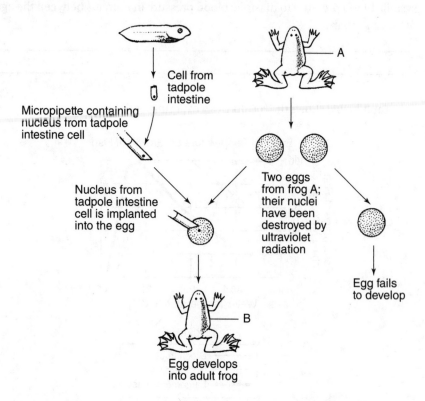

An inference that can be made from this experiment is that

(1)   adult frog *B* will have the same genetic traits as the tadpole

(2)   adult frog *A* can develop only from an egg and a sperm

(3)   fertilization must occur in order for frog eggs to develop into adult frogs

(4)   the nucleus of a body cell fails to function when transferred to other cell types

38 _____

39. The charts below show the relationship of recommended weight to height in men and women age 25–29.

Height-Weight Charts

MEN Age 25–29 Weight (lb)				WOMEN Age 25–29 Weight (lb)			
Height Feet | Inches	Small Frame	Medium Frame	Large Frame	Height Feet | Inches	Small Frame	Medium Frame	Large Frame
5 2	128–134	131–141	138–150	4 10	102–111	109–121	118–131
5 3	130–136	133–143	140–153	4 11	103–113	111–123	120–134
5 4	132–138	135–145	142–156	5 0	104–115	113–126	122–137
5 5	134–140	137–148	144–160	5 1	106–118	115–129	125–140
5 6	136–142	139–151	146–164	5 2	108–121	118–132	128–143
5 7	138–145	142–154	149–168	5 3	111–124	121–135	131–147
5 8	140–148	145–157	152–172	5 4	114–127	124–138	134–151
5 9	142–151	148–160	155–176	5 5	117–130	127–141	137–155
5 10	144–154	151–163	158–180	5 6	120–133	130–144	140–159
5 11	146–157	154–166	161–184	5 7	123–136	133–147	143–163
6 0	149–160	157–170	164–188	5 8	126–139	136–150	146–167
6 1	152–164	160–174	168–192	5 9	129–142	139–153	149–170
6 2	155–168	164–178	172–197	5 10	132–145	142–156	152–173
6 3	158–172	167–182	176–202	5 11	135–148	145–159	155–176
6 4	162–176	171–187	181–207	6 0	138–151	148–162	158–179

The recommended weight for a 6′0″ tall man with a small frame is closest to that of a

(1)   5′10″ man with a medium frame

(2)   5′9″ woman with a large frame

(3)   6′0″ man with a medium frame

(4)   6′0″ woman with a medium frame   39 _____

Base your answers to questions 40 through 43 on the information below and on your knowledge of biology.

A group of biology students extracted the photosynthetic pigments from spinach leaves using the solvent acetone. A spectrophotometer was used to measure the percent absorption of six different wavelengths of light by the extracted pigments. The wavelengths of light were measured in units known as nanometers (nm). One nanometer is equal to one-billionth of a meter. The following data were collected:

yellow light (585 nm)—25.8% absorption

blue light (457 nm)—49.8% absorption

orange light (616 nm)—32.1% absorption

violet light (412 nm)—49.8% absorption

red light (674 nm)—41.0% absorption

green light (533 nm)—17.8% absorption

40. Complete all three columns in the data table below so that the wavelength of light either increases or decreases from the top to the bottom of the data table.

Color of Light	Wavelength of Light (nm)	Percent Absorption by Spinach Extract

Directions (41–42): Using the information in the data table, construct a line graph on the grid provided, following the directions below.

41. Mark an appropriate scale on the axis labeled "Percent Absorption."

42. Plot the data from the data table on your graph. Surround each point with a small circle and connect the points.

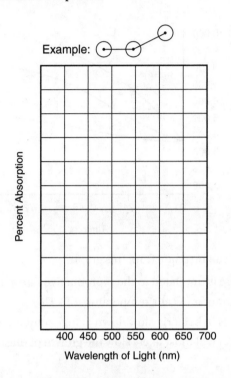

43. Which statement is a valid conclusion that can be drawn from the data obtained in this investigation?

(1)   Photosynthetic pigments in spinach plants absorb blue light and violet light more efficiently than red light.

(2)   The data would be the same for all pigments in spinach plants.

(3)   Green light and yellow light are not absorbed by spinach plants.

(4)   All plants are efficient at absorbing violet light and red light.   43 _____

44. The graph below represents the results of an investigation of the growth of three identical bacterial cultures incubated at different temperatures.

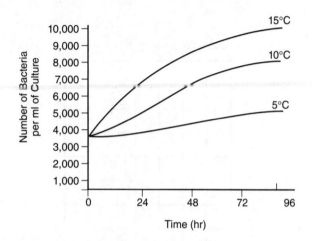

Which inference can be made from this graph?

(1) Temperature is unrelated to the reproductive rate of bacteria.

(2) Bacteria cannot grow at a temperature of 5°C.

(3) Life activities in bacteria slow down at high temperatures.

(4) Refrigeration will most likely slow the growth of these bacteria.   44 _____

45. A study was conducted using two groups of ten plants of the same species. During the study, the plants were kept under identical environmental conditions. The plants in one group were given a growth solution every 3 days. The heights of the plants in both groups were recorded at the beginning of the study and at the end of a 3-week period. The data showed that the plants given the growth solution grew faster than those not given the solution.

When other researchers conduct this study to test the accuracy of the results, they should

(1) give growth solution to both groups

(2) make sure that the conditions are identical to those in the first study

(3) give an increased amount of light to both groups of plants

(4) double the amount of growth solution given to the first group   45 _____

46. Worker bees acting as scouts are able to communicate the distance of a food supply from the hive by performing a "waggle dance." The graph below shows the relationship between the distance of a food supply from the hive and the number of turns in the waggle dance every 15 seconds.

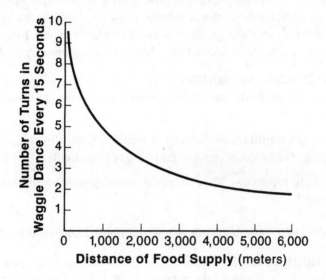

Using one or more complete sentences, state the relationship between the distance of the food supply from the hive and the number of turns a bee performs in the waggle dance every 15 seconds.

_____

47. Based on experimental results, a biologist in a laboratory reports a new discovery. If the experimental results are valid, biologists in other laboratories should be able to perform

(1) an experiment with a different variable and obtain the same results

(2) the same experiment and obtain different results

(3) the same experiment and obtain the same results

(4) an experiment under different conditions and obtain the same results

47 _____

Answers Explained

29. 4 The diagram shows the population growth cycle for two organisms. An examination of the graph shows that the population growth cycle of the hawks closely follows the cycle of population *B*. There is a slight lag in the cycles. This type of graph is used to show a predator-prey relationship. *The hawks prey on population* B.

WRONG CHOICES EXPLAINED:
(1) Hawks are *not* herbivores. They do not live off the *dominant trees* in the community.

(2) If the two populations benefited equally from the relationship (were *mutualistic*), the peaks of the two graphs would coincide.

(3) Hawks are carnivores. They do not depend directly on the *producers* in the community.

30. 1 According to the graph, the range could support *12,000* deer in 1930.

WRONG CHOICES EXPLAINED:
(2) The carrying capacity of the range was *35,000*. The *carrying capacity* is the maximum number of individuals that can be supported by the area. The number is usually constant unless severe environmental changes occur.

(3) The actual number of deer occupying the range in 1930 was *50,000*.

(4) A population of *100,000* deer was reached in 1925.

31. 2 In nature, a predator-prey relationship keeps the prey population in check. In *1920*, the population of deer increased. It was at this time that the predators were removed by human hunting.

WRONG CHOICES EXPLAINED:
(1) The deer probably entered the region in *1905*. It takes time for an organism to adjust to a new environment.

(3) In *1930*, the deer population, which had exceeded the carrying capacity, was declining. The decline was caused by starvation. The deer had consumed almost all of the available vegetation in the area.

(4) By *1940*, the deer population reached the new carrying capacity of the range. The carrying capacity had been greatly reduced by overgrazing by the previously unchecked deer population.

32. 4 As the result of *isolation*, the members of populations *A* and *B* are separated and are prevented from interbreeding. Variations that occur in one area are not transmitted to the individuals in the other area. Consequently, over a long period of time, the genetic differences become accentuated, and the new variations are maintained.

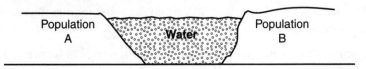

WRONG CHOICES EXPLAINED:

(1) *Migration* is not possible when populations are separated by a geographical barrier such as water.

(2) *Adaptive radiation* refers to a branching evolution and is not depicted in the diagram.

(3) *Speciation* indicates formation of new species from a parent population. This is not indicated in the diagram.

33–35.

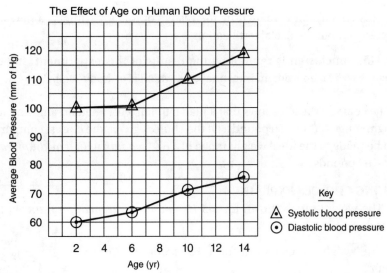

36. *Systolic pressure is higher than diastolic pressure.* Or *Both systolic pressure and diastolic pressure increase between the ages of 2 and 14.*

[*Note:* Other correct, complete-sentence responses are acceptable.]

37. **4** Brook trout growth is represented by the lighter solid line on the graph. Tracing along the horizontal axis to 16°C and then tracing up to the point at which the brook trout line is encountered, we see that the growth rate is between 80% and 100% but closer to 100%; we estimate it is about *95%*.

WRONG CHOICES EXPLAINED:
(1) Brook trout growth rate is at *30%* when the water temperature is about 18°C. Largemouth bass growth rate is about 30% at 16°C.

(2) None of the species indicated have a growth rate of *55%* at 16°C. Brook trout growth rate is at 55% at about 17.5°C.

(3) Brook trout growth rate falls to *75%* at about 17°C. The growth rate of northern pike is at 75% at about 16°C.

38. **1** The diagram represents an experiment in cloning. Because the nucleus (which contains the genetic material) used in this experiment comes from a tadpole, the egg will produce new cells that have *genetic characteristics identical to those of the tadpole.*

WRONG CHOICES EXPLAINED:
(2) It is unclear from this diagram how frog *A* came into being.

(3) This conclusion is refuted by the results of this experiment; a new frog was created without the use of fertilization.

(4) This conclusion is refuted by the results of this experiment; the tadpole nucleus was taken from an intestinal cell, which is a "body" cell.

39. **4** The chart on the left shows that a man 6′0″ tall with a small frame has an ideal weight range of 149–160 pounds. Of the choices given, the closest comparison can be made to the ideal weight range of a *6′0″ woman with a medium frame* at 148–162 pounds.

WRONG CHOICES EXPLAINED:
(1) The weight range shown for a *5′10″ man with a medium frame* is 151–163 pounds.

(2) The weight range shown for a *5′9″ woman with a large frame* is 149–170 pounds.

(3) The weight range shown for a *6′0″ man with a medium frame* is 157–170 pounds.

40.

Color of Light	Wavelength of Light (nm)	Percent Absorption by Spinach Extract
violet	412	49.8
blue	457	49.8
green	533	17.8
yellow	585	25.8
orange	616	32.1
red	674	41.0

41–42.

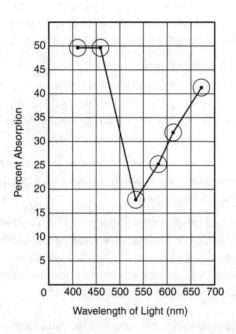

43. **1** The statement *photosynthetic pigments in spinach plants absorb blue light and violet light more efficiently than red light* is a valid conclusion that can be drawn from the data. The high point of the chart/graph data is clearly shown to be above the blue and violet wavelengths of light.

WRONG CHOICES EXPLAINED:
(2) The statement *the data would be the same for all pigments in spinach plants* is not supported by the results of this experiment. The chart/graph data show considerable variation in the experimental results as the wavelength of light varies.

(3) The statement *green light and yellow light are not absorbed by spinach plants* is not supported by the results of this experiment. Although the chart/graph data show a lower absorption rate at these wavelengths, there is still some absorption in this range.

(4) The statement *all plants are efficient at absorbing violet light and red light* is not supported by the results of this experiment. The experimental data are limited to the absorption of light by pigments found in one type of plant. These data cannot be extended to all plants unless all other types of plants are tested under the same experimental conditions and the results are found to be similar.

44. **4** Of those given, *refrigeration will most likely slow the growth of these bacteria* is the most reasonable inference that can be made from the graph data. The graph clearly shows a slower rate of growth (reproduction) at 5°C than at 10°C or 15°C.

WRONG CHOICES EXPLAINED:
(1) The inference that *temperature is unrelated to the reproductive rate of bacteria* is not supported by the data. Temperature is the independent (experimental) variable in this study. It clearly has an influence on the bacterial reproductive rate.

(2) The inference that *bacteria cannot grow at a temperature of 5°C* is not supported by the data. The graph clearly shows that growth at this temperature, while slow, occurs at a steady pace.

(3) The inference that *life activities in bacteria slow down at high temperatures* is not supported by the data. The data indicate that, if anything, bacterial activity increases with increasing temperature. No data are shown for bacterial growth at temperatures above 15°C, and so we cannot draw any inference about what happens to the rate of bacterial growth at these extremes.

45. **2** These researchers should *make sure that the conditions are identical to those in the first study.* The validity of any scientific experiment can be verified only if the same results are obtained under the same experimental conditions. Any change in these conditions invalidates the results of the verification study.

WRONG CHOICES EXPLAINED:
(1) If the researchers *give growth solution to both groups,* there will be no control group against which to compare the experimental group. The results of the verification study will be invalid because the experimental conditions will have been changed.

(3) If the researchers *give an increased amount of light to both groups of plants,* the original experimental method will be altered. The results of the verification study will be invalid because the experimental conditions will have been changed.

(4) If the researchers *double the amount of growth solution given to the first group,* the original experimental method will not be followed. The results of the verification study will be invalid because the experimental conditions will have been changed.

46. *The number of turns in the waggle dance decreases as the distance of the food supply from the hive increases.* Or *The closer to the hive the food source is located, the more turns there are in the waggle dance.*

   [*Note:* Any correct, complete-sentence answer is acceptable.]

47. **3** Other biologists in other laboratories should be able to perform *the same experiment and obtain the same results* if the experimental results are valid. Any experimental results obtained by one scientist must be validated through independent research by other scientists following the same procedures.

   WRONG CHOICES EXPLAINED:
   (1) If different scientists perform *an experiment with a different variable and obtain the same results,* the original experimental results will be invalidated.

   (2), (4) If different scientists perform *the same experiment and obtain different results,* or *an experiment under different conditions and obtain the same results,* they will neither validate nor invalidate the results of the original experiment. All variables and conditions must be kept the same if the experimental results are to be properly tested.

Questions on Standard 4

Students will understand and apply scientific concepts, principles, and theories pertaining to the physical setting and living environment and recognize the historical development of ideas in science.

Key Idea 1—Application of Scientific Principles

Living things are both similar to and different from each other and from nonliving things.

Performance Indicator	Description
1.1	The student should be able to explain how diversity of populations within ecosystems relates to the stability of ecosystems.
1.2	The student should be able to describe and explain the structures and functions of the human body at different organizational levels (e.g., systems, tissues, cells, organelles).
1.3	The student should be able to explain how a one-celled organism is able to function despite lacking the levels of organization present in more complex organisms.

48. In which life function is the potential energy of organic compounds converted to a form of stored energy that can be used by the cell?

(1)   transport   (3)   excretion

(2)   respiration   (4)   regulation   48 _____

49. Which life activity is *not* required for the survival of an individual organism?

(1)   nutrition   (3)   reproduction

(2)   respiration   (4)   synthesis   49 _____

50. Which function of human blood includes the other three?

(1)   transporting nutrients

(2)   transporting oxygen

(3)   maintaining homeostasis

(4)   collecting wastes   50 _____

51. In the human body, the blood with the greatest concentration of oxygen is found in the

(1)   left atrium of the heart

(2)   cerebrum of the brain

(3)   nephrons of the kidney

(4)   lining of the intestine   51 _____

52. Which type of vessel normally contains valves that prevent the backward flow of materials?

(1)   artery   (3)   capillary

(2)   arteriole   (4)   vein   52 _____

Directions (53–55): For each of questions 53 through 55, select the excretory structure, chosen from the list below, that best answers the question. Then record its number in the space provided at the right.

Excretory Structures
(1) Alveolus
(2) Nephron
(3) Sweat gland
(4) Liver

53. Which structure forms urine from water, urea, and salts?

_____

54. Which structure removes carbon dioxide and water from the blood?

_____

55. Which structure is involved in the breakdown of red blood cells?

_____

56. The bones of the lower arm are connected to the muscles of the upper arm by

(1)   ligaments   (3)   cartilage

(2)   tendons   (4)   skin   56 _____

57. The diagram below shows the same type of molecules in area *A* and area *B*. With the passage of time, some molecules move from area *A* to area *B*.

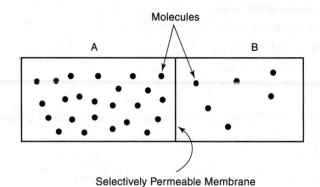

Molecules

A   B

Selectively Permeable Membrane

The movement is the result of the process of

(1)   phagocytosis   (3)   diffusion

(2)   pinocytosis   (4)   cyclosis   57 _____

58. Which is the principal inorganic compound found in cytoplasm?

(1)   lipid   (3)   water

(2)   carbohydrate   (4)   nucleic acid   58 _____

59. A specific organic compound contains only the elements carbon, hydrogen, and oxygen in a ratio of 1:2:1. This compound is most probably a

(1)   nucleic acid   (3)   protein

(2)   carbohydrate   (4)   lipid   59 _____

60. Compared to ingested food molecules, end-product molecules of digestion are usually

(1)   smaller and more soluble

(2)   larger and more soluble

(3)   smaller and less soluble

(4)   larger and less soluble   60 _____

61. The cellular function of the endoplasmic reticulum is to

    (1)   provide channels for the transport of materials

    (2)   convert urea to a form usable by the cell

    (3)   regulate all cell activities

    (4)   change light energy into chemical bond energy   61 _____

62. In which organelles are polypeptide chains synthesized?

    (1)   nuclei   (3)   ribosomes

    (2)   vacuoles   (4)   cilia   62 _____

63. Which organelle contains hereditary factors and controls most cell activities?

    (1)   nucleus

    (2)   cell membrane

    (3)   vacuole

    (4)   endoplasmic reticulum   63 _____

64. Centrioles are cell structures involved primarily in

    (1)   cell division   (3)   enzyme production

    (2)   storage of fats   (4)   cellular respiration   64 _____

65. Which cell structure contains respiratory enzymes?

    (1)   cell wall   (3)   mitochondrion

    (2)   nucleolus   (4)   vacuole   65 _____

66. Which process is represented below?

simple organic molecules $\xrightarrow{\text{enzymes}}$ complex organic molecules $+ H_2O$

    (1)   hydrolysis   (3)   digestion

    (2)   synthesis   (4)   respiration   66 _____

67. Amino acids derived from the digestion of a piece of meat are transported to living cells of an animal. In the cell they are

  (1) converted to cellulose

  (2) used to attack invading bacteria

  (3) synthesized into specific proteins

  (4) incorporated into glycogen molecules   67 _____

68. Which of the following variables has the *least* direct effect on the rate of a hydrolytic reaction regulated by enzymes?

  (1) temperature

  (2) pH

  (3) carbon dioxide concentration

  (4) enzyme concentration   68 _____

69. Which term refers to the chemical substance that aids in the transmission of the impulse through the area indicated by X?

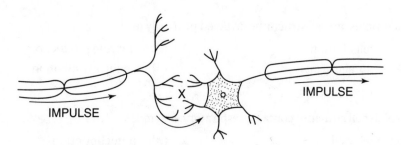

  (1) neurotransmitter   (3) neuron

  (2) synapse   (4) nerve   69 _____

70. Which lists human nervous-system structures in order of increasing size?

  (1) neuron, nerve, ganglion, receptor

  (2) nerve, ganglion, neuron, receptor

  (3) neuron, receptor, ganglion, nerve

  (4) ganglion, receptor, nerve, neuron   70 _____

71. Glands located within the digestive tube include

    (1) gastric glands and thyroid glands

    (2) gastric glands and intestinal glands

    (3) thyroid glands and intestinal glands

    (4) adrenal glands and intestinal glands   71 _____

72. In humans, which substance is directly responsible for controlling the calcium levels of the blood?

    (1) adrenalin   (3) parathormone

    (2) insulin   (4) thyroxin   72 _____

Base your answer to question 73 on the word equation below.

glucose $\rightarrow$ 2 pyruvic acid $\rightarrow$ 2 ethyl alcohol + 2 carbon dioxide + energy

73. The process represented by the word equation is known as

    (1) aerobic respiration

    (2) fermentation

    (3) chemosynthesis

    (4) dehydration synthesis   73 _____

74. The excretory organelles of some unicellular organisms are contractile vacuoles and

    (1) cell membranes   (3) ribosomes

    (2) cell walls   (4) centrioles   74 _____

75. Which is a type of asexual reproduction that commonly occurs in many species of unicellular protists?

    (1) external fertilization

    (2) tissue regeneration

    (3) binary fission

    (4) vegetative propagation   75 _____

Answers Explained

48. 2 *Respiration* is the life function by which ATP is made available to cells. Carbo-hydrate molecules are organic compounds. The breakdown of the carbohydrate molecules releases the energy stored in the bonds of the compounds. Potential energy is stored energy. The released potential energy is used to produce ATP.

WRONG CHOICES EXPLAINED:
(1) *Transport* is the life function by which materials are distributed throughout an organism.

(3) *Excretion* is the life function by which the wastes of metabolism are removed from an organism. Carbon dioxide, water, ammonia, and urea are metabolic wastes.

(4) *Regulation* is the life activity by which an organism responds to changes in its environment. The responses are controlled by the nervous system and the endo-crine system.

49. 3 *Reproduction*, the life function through which a parent organism gives rise to off-spring, is not necessary for the survival of the parent. Although reproduction is not required for the survival of an individual, it is necessary for the survival of a species. If a given species loses its potential for reproduction, it will become extinct.

WRONG CHOICES EXPLAINED:
(1) *Nutrition* is a collective term that refers to the biochemical processes by which cells extract nutrient molecules from food substances. The nutrients are used to build tissues, provide energy, and regulate the many biochemical activ-ities that occur in cells. Without nutrition, cells die and, consequently, so do organisms. Each organism is dependent on adequate nutrition for survival.

(2) *Respiration* refers to the series of chemical changes that fuel molecules undergo to release chemical energy for cells. Respiration is necessary for the survival of the individual. Tissue cells cannot live without a means of obtaining chemical energy to power cellular activities such as active transport and metab-olism. Of course, death of tissue cells means death of the individual.

(4) *Synthesis* occurs when small molecules are joined chemically to form large molecules. Enzymes, hormones, and body tissues are the results of syntheses, without which an individual organism cannot survive. Synthesis is a building-up process in which molecules vital to the life of the organism are produced.

50. 3 *Maintaining homeostasis* is the function of human blood that includes the other three. By transporting nutrients, oxygen, wastes, and other materials around the body, the blood helps to make essential materials available to every living body cell while removing potentially harmful materials from these tissues. Equal distribution of these materials helps to promote a steady state in the tissues essential to homeostatic balance.

WRONG CHOICES EXPLAINED:
(1), (2), (4) *Transporting nutrients, transporting oxygen,* and *collecting wastes* are all functions of the blood that are involved in maintaining homeostasis. Nutrients provide cells with dissolved food molecules. Oxygen is used by cells in the release of energy from these food molecules. Wastes such as urea and carbon dioxide are carried away from the cells for excretion into the environment.

51. 1 Blood that has just returned from the lungs has the greatest concentration of oxygen. The *left atrium of the heart* receives blood directly from the lungs.

WRONG CHOICES EXPLAINED:
(2) Brain tissue is one of the largest consumers of oxygen. The blood circulating in the *cerebellum* gives up most of its oxygen to the nerve cells.

(3) The largest concentration of metabolic wastes is found in the *nephrons.* The nephrons are filtering units in the kidney.

(4) The largest concentration of digested nutrients is found in the *lining of the intestine.* Absorption of nutrients occurs through the villi in the small intestine.

52. 4 *Veins* are blood vessels that carry blood to the heart. They contain valves that prevent the backflow of blood. The blood in veins is usually deoxygenated; the exception is the pulmonary vein in which the blood is rich in oxygen.

WRONG CHOICES EXPLAINED:
(1) *Arteries* are blood vessels that transport blood away from the heart. Arteries are rather thick-walled and pump blood in rhythm with the heart. They have no valves.

(2) Small arteries are *arterioles.* This type of blood vessel functions similarly to arteries. Arterioles lead into capillaries.

(3) *Capillaries* are the smallest blood vessels. They are one cell thick and permit diffusion of water, nutrients, gases, and other substances into and out of the bloodstream. Capillaries have no valves; they are the connecting vessels between arterioles and venules.

53. **2** The *nephron* is the unit of structure of the kidney. Each nephron has a glomerulus, Bowman's capsule, and kidney tubules. The kidney tubules filter out excess water, salts, and the wastes from protein metabolism. Urea and salts dissolved in water form urine.

54. **1** The *alveolus* is an air sac in the lung. It not only permits the diffusion of oxygen from the lungs into the bloodstream but also aids in the diffusion of carbon dioxide and water vapor out of the blood into the lungs.

55. **4** The *liver* is the largest gland in the body. One of its functions is to destroy old red blood cells and change the waste products into bile. The liver also synthesizes the anticoagulant known as heparin.

56. **2** *Tendons* are tough connective tissues made strong by fibers. Tendons connect muscles to bones. The movable joints function when muscles pull on tendons.

WRONG CHOICES EXPLAINED:

(1) *Ligaments* are strong connective tissues that contain elastic muscle fibers. Ligaments connect bone to bone.

(3) *Cartilage* is a supporting tissue that provides strength to body structures without rigidity. Cartilage supports structures such as the ears and nose and covers the ends of bones that form joints. The ground substance, or matrix, of cartilage is made of protein.

(4) *Skin* is composed of epithelial tissue. Skin serves as a body covering and has no function in the movement of bones or muscles.

57. **3** *Diffusion* is the process that results in the movement of molecules from a region of higher concentration (area *A*) to a region of lower concentration (area *B*). This net movement occurs until the concentrations of molecules have reached equilibrium between area *A* and area *B*.

WRONG CHOICES EXPLAINED:

(1), (2) *Phagocytosis* and *pinocytosis* are processes by which certain protists engulf their food and enclose it within a vacuole for digestion.

(4) *Cyclosis* refers to the streaming of cytoplasm in the cell, a simple form of intracellular transport.

58. **3** Inorganic compounds are compounds that do not contain carbon atoms. *Water,* the universal solvent, is the principal inorganic compound of cytoplasm. Water is the medium through which all chemical reactions take place in the cell.

WRONG CHOICES EXPLAINED:
(1), (2), (4) *Lipids, carbohydrates,* and *nucleic acids* are organic compounds. Organic compounds are carbon-containing compounds.

59. 2 Glucose is the building block of *carbohydrate* molecules. The ratio of carbon to hydrogen to oxygen is 1:2:1 in glucose and all reducing sugars. By dehydration synthesis, many glucose molecules form complex carbohydrates. However, the 1:2:1 ratio holds.

WRONG CHOICES EXPLAINED:
(1) A *nucleic acid* is composed of a phosphate group, a protein base, and a five-carbon sugar. DNA and RNA are nucleic acids. The CHO 1:2:1 ratio is not applicable.

(3) *Proteins* are built from amino acids, which, in addition to carbon, hydrogen, and oxygen, contain nitrogen. Some protein molecules also contain sulfur. The 1:2:1 ratio of elements does not apply to proteins because proteins are tissue builders whereas carbohydrates are fuel molecules.

(4) *Lipids* are fats and are composed of three fatty acid molecules and one glycerol molecule. The 1:2:1 ratio of carbon to hydrogen to oxygen does not apply to fats.

60. 1 The end products of digestion are usually *smaller and more soluble* than the ingested food molecules. Digestion makes available nutrient molecules that can diffuse across cell membranes and enter the cytoplasm of cells. Carbohydrates are broken down into glucose molecules. Fats are hydrolyzed into fatty acids and glycerol. Proteins are digested into their component amino acid molecules. Each of these end products of digestion is able to diffuse across cell membranes and enter into the biochemical activities of cells.

WRONG CHOICES EXPLAINED:
(2) Synthesis produces *larger molecules.* Larger molecules are more complex and are usually less soluble than smaller, simpler ones. Digestion results in smaller nutrient molecules.

(3) *Smaller molecules* are usually more soluble than larger ones. Digestion produces molecules that are more soluble than the complex nutrient molecules that were ingested.

(4) *Molecules derived from digestion* of ingested food are not larger than the molecules from which they came. Molecules produced by the digestion of complex carbohydrates, proteins, and fats are more soluble and are able to dissolve in water. Thus, these molecules can cross cell membranes.

61. 1 The endoplasmic reticulum is a network of membranes that extends through-out the cell. The membranes form channels that *provide for the movement of materials through the cell.*

WRONG CHOICES EXPLAINED:

(2) *Urea* is a metabolic waste. It is a poisonous nitrogen compound. Urea must be removed from the cells if an organism is to survive.

(3) The *nucleus* is the organelle in the cell that regulates all cellular activities.

(4) The *chloroplasts* are organelles in plant cells that contain the green pigment chlorophyll. Chloroplasts are necessary for the process of photosynthesis.

62. 3 Proteins are polypeptide chains. Proteins are synthesized in the *ribosomes.*

WRONG CHOICES EXPLAINED:

(1) The *nuclei* contain the genetic material carried in the chromosomes.

(2) *Vacuoles* are saclike organelles in the cytoplasm. Food vacuoles and contrac-tile vacuoles are two common types of vacuoles.

(4) *Cilia* are microscopic hairs used for locomotion by some protozoans.

63. 1 The *nucleus* contains the hereditary factors. Nuclei of plant and animal cells house the chromosomes, which are composed of deoxyribonucleic acid. Molecules of DNA function as genes. Points on the chromosomes are genes. Genetic infor-mation is passed from parent to offspring by way of the genes. Chromosomes are part of the fine structure of the nucleus. DNA molecules contribute to the chemical structure. Genes are sites or points that dot the length of the chromosome. Genes, DNA molecules, and chromosomes function in passing along hereditary factors.

WRONG CHOICES EXPLAINED:

(2) The *cell membrane* encloses the contents of the cell and directs the flow of materials into and out of the cell. The cell membrane does not contribute to the passing of genetic material from one generation to the next. The function of the membrane is to control cellular transport.

(3) A *vacuole* is a fluid-filled space in the cytoplasm. Vacuoles help to regulate the internal pressure of the cell. The vacuoles in fat cells are filled with oil.

(4) The membranes that line the cytoplasmic canals within cells are known col-lectively as the *endoplasmic reticulum.* This cytoplasmic fine structure aids in the transport of molecules from the cell membrane to various sites within the cell. Neither the endoplasmic reticulum nor the vacuoles of the cell membrane contain hereditary structures.

64. 1 Centrioles are cell structures involved primarily in *cell division*. Centrioles are organelles that lie in the cytoplasm outside the nucleus; they are also found near the base of each flagellum and cilium. The centrioles of nonflagellated animal cells move to the spindle poles during cell division and seem to send out spindle fibers. The spindle fibers are attached to chromosomes and appear to pull the chromosomes from the center of the cell to the spindle poles.

WRONG CHOICES EXPLAINED:
(2) *Fats are stored* in cells. Fat in which the energy is channeled into heat production is stored in brown fat cells of hibernating mammals. At times, fat can be stored in arteries or accumulate around the heart. Fat cells are not involved in cell division.

(3) *Enzyme production* is controlled by the ribosomes that dot the membranes of the endoplasmic reticulum. Molecules of tRNA and mRNA regulate enzyme production.

(4) *Cellular respiration* is the process by which energy is released from glucose molecules. This process takes place in the mitochondria where oxygen is used as the final hydrogen carrier.

65. 3 Cellular respiration occurs in the mitochondria (plural of *mitochondrion*). Each step in the process of cellular respiration is regulated by enzymes. Respiratory enzymes are located in the mitochondria.

WRONG CHOICES EXPLAINED:
(1) The *cell wall* is composed of cellulose. Cell walls give shape and protection to plant cells.

(2) The *nucleolus* contains the materials needed for the synthesis of RNA. It is located in the nucleus of the cell.

(4) A *vacuole* is a rounded sac that serves as a storage place for food and waste products. Some vacuoles, such as contractile vacuoles, maintain a stable internal environment.

66. 2 *Synthesis* is the formation of complex molecules by combining simpler molecules. Water is removed from the simple molecules in this process.

WRONG CHOICES EXPLAINED:
(1) *Hydrolysis* is the addition of water to split complex molecules into simpler molecules. It is the opposite of synthesis.

(3) *Digestion* is another name for hydrolysis.

(4) *Respiration* is the process by which cells obtain energy. Glucose is converted to smaller molecules.

67. 3 Amino acids are the building blocks of proteins. The dehydration synthesis of amino acids *produces protein molecules.*

WRONG CHOICES EXPLAINED:
(1) *Cellulose* is a polysaccharide composed of hundreds of simple sugar molecules. The sugars were joined together by dehydration synthesis.

(2) Antibodies attack *invading bacteria.* Antibodies are protein molecules produced by special white blood cells.

(4) *Glycogen,* a polysaccharide, is a product of the dehydration synthesis of many glucose units.

68. 3 A *hydrolytic reaction* is a reaction in which a molecule is split. Enzymes are needed to speed up such a reaction. Any factor that affects the operation of the enzyme affects the speed at which the reaction takes place. The *concentration of carbon dioxide* has the least effect on enzyme activity.

WRONG CHOICES EXPLAINED:
(1) As the *temperature* is increased up to a point, the rate of the reaction increases. The increase in temperature increases the speed at which the enzyme and the substrate make contact with each other. The substrate is the molecule on which the enzyme acts. A very high temperature destroys the enzyme, and the reaction stops.

(2) Every enzyme works best at a particular *pH.* The enzymes in the stomach work in an acid environment, whereas the enzymes in the intestine work best in a basic medium.

(4) One molecule of an enzyme reacts with one molecule of a substrate. Increasing the *concentration of an enzyme* means that more substrate molecules will be acted on. The rate of the reaction will increase.

69. 1 A *neurotransmitter* is a chemical substance that is released by an impulse arriving at the terminal end of a neuron. The neurotransmitter diffuses across the synapse and stimulates the second nerve cell. Acetylcholine is an example of a neurotransmitter.

WRONG CHOICES EXPLAINED:
(2) A *synapse* is the space between the terminal end of one nerve cell and the dendrites of a second nerve cell. The area marked by an X in the diagram is a synapse.

(3) A *neuron* is a nerve cell that is specially adapted for the conduction of impulses.

(4) A *nerve* is made up of many neurons.

70. 1 A *neuron* is a single microscopic nerve cell. A *nerve* is composed of many nerve cells. A *ganglion* is a large mass of cell bodies of nerve cells; a ganglion functions as a coordinating center for impulses. A *receptor* is an organ specialized to receive environmental stimuli. The eye is an example of a receptor.

WRONG CHOICES EXPLAINED:
(2), (3), (4) In these three choices, either one or several structures are not arranged according to increasing size.

71. 2 *Gastric glands* are embedded in the walls of the stomach. They are duct glands that secrete gastric juice, a mixture of water, hydrochloric acid, rennin, and pepsin. Gastric juice begins the digestion of protein in the stomach. *Intestinal glands* are duct glands that line the walls of the small intestine. They secrete intestinal juice, a mixture of water, proteases, amylases, and lipases. Both types of glands lie within the digestive tube.

WRONG CHOICES EXPLAINED:
(1) Gastric glands are described above. *Thyroid glands* lie outside the digestive tract at the base of the neck, straddled across the larynx. The thyroid is a ductless gland that secretes the hormone known as thyroxin. Thyroid glands do not function in the biochemical process of digestion.

(3) *Thyroid glands* and intestinal glands are described above. Thyroxin controls the metabolism of cells. The explanation above shows why this choice is wrong.

(4) *Adrenal glands* are dual endocrine glands that lie on top of each kidney. They are not within the digestive tract. The adrenal medulla, the inner gland, secretes the hormone adrenaline, also known as epinephrine. This hormone enables the body to function in emergencies. The adrenal cortex secretes about six active hormones, including cortisone, the antiarthritis hormone.

72. 3 *Parathormone* is the hormone secreted by the parathyroid glands. The parathyroids are buried in the thyroids. Parathormone controls the level of calcium in the blood. Lack of blood calcium causes muscles to go into tetany. Tetany, or cramping, of the heart muscle causes death.

WRONG CHOICES EXPLAINED:

(1) *Adrenalin* is the hormone of the adrenal medulla, a ductless gland called the "gland of combat." Adrenaline stimulates the heart to beat faster, increases the rate of breathing, and controls the constriction and dilation of the arteriole walls.

(2) *Insulin* is secreted by the beta cells of the islets of Langerhans, which lie in the pancreas. Insulin controls sugar metabolism; specifically, it makes cell walls permeable to glucose and encourages the phosphorylation of fructose.

(4) *Thyroxin* is released by the thyroid gland. The rate of cellular metabolism is controlled by thyroxin. Iodine is used in the synthesis of thyroxin. People whose thyroid glands fail to develop become cretins; they are mentally retarded and physically undersized.

73. **2** Another name for anaerobic respiration is *fermentation.* In the process of fermentation, glucose is converted to energy, alcohol, and carbon dioxide.

WRONG CHOICES EXPLAINED:

(1) *Aerobic respiration* is another name for cellular respiration. This process requires oxygen. The following is an equation for aerobic respiration.

glucose + oxygen → pyruvic acid → carbon dioxide + water + energy

(3) *Chemosynthesis* is the synthesis of carbohydrates from inorganic compounds without the use of sunlight as a source of energy. Chemosynthesis is a form of autotrophic nutrition. It is carried out only by certain species of bacteria such as nitrifying bacteria.

(4) *Dehydration synthesis* is the method by which simple molecules are converted to complex molecules.

74. **1** *Cell membranes* and contractile vacuoles are excretory organelles of some unicellular organisms. The cell membrane is a selectively permeable membrane. It permits the diffusion of carbon dioxide and ammonia, two metabolic waste gases.

WRONG CHOICES EXPLAINED:

(2) *Cell walls* are composed of nonliving materials. Many canals penetrate through these walls, allowing the unrestricted passage of molecules.

(3) Proteins are synthesized in *ribosomes.*

(4) *Centrioles* are rodlike particles found in the centrosome. They function during the processes of mitosis and meiosis. Centrioles are found only in animal cells.

75. 3 A unicellular protist (e.g., an ameba) is composed of a single cell. When this cell divides by mitosis, the process is known as *binary fission.*

WRONG CHOICES EXPLAINED:

(1) *External fertilization* is an element of sexual reproduction in many aquatic multicellular species. Both the sexual nature of this process and the fact that it is carried out by multicellular animals eliminate this as a correct choice.

(2) *Tissue regeneration* implies a process that occurs in multicellular organisms.

(4) *Vegetative propagation* is a form of asexual reproduction common to certain species of multicellular plants; it cannot be carried out by unicellular protists.

Key Idea 2—Genetic Continuity

Organisms inherit genetic information in a variety of ways that result in continuity of structure and function between parents and offspring.

Performance Indicator	Description
2.1	The student should be able to explain how the structure and replication of genetic material result in offspring that resemble their parents.
2.2	The student should be able to describe and explain how the technology of genetic engineering allows humans to alter genetic makeup of organisms.

76. Corn plants grown in the dark will be white and usually much taller than genetically identical corn plants grown in light, which will be green and shorter. The most probable explanation for this is that the

    (1) corn plants grown in the dark were all mutants for color and height
    (2) expression of a gene may be dependent on the environment
    (3) plants grown in the dark will always be genetically albino
    (4) phenotype of a plant is independent of its genotype   76 _____

77. In order for a substance to act as a carrier of hereditary information, it must be

    (1) easily destroyed by enzyme action
    (2) exactly the same in all organisms
    (3) present only in the nuclei of cells
    (4) copied during the process of mitosis   77 _____

78. During synapsis in meiosis, portions of one chromosome may be exchanged for corresponding portions of its homologous chromosome. This process is known as

    (1) nondisjunction   (3) crossing-over
    (2) polyploidy   (4) hybridization   78 _____

79. A DNA nucleotide is composed of three parts. These three parts may be

   (1) phosphate, adenine, and thymine

   (2) phosphate, deoxyribose, and thymine

   (3) phosphate, glucose, and cytosine

   (4) adenine, thymine, and cytosine   79 _____

80. A double-stranded DNA molecule replicates as it unwinds and "unzips" along weak

   (1) hydrogen bonds   (3) phosphate groups

   (2) carbon bonds   (4) ribose groups   80 _____

    Base your answers to questions 81 through 84 on your knowledge of biology and the diagrams below. The diagram on the left represents a portion of a double-stranded DNA molecule. The diagrams at the right represent specific combinations of nitrogenous bases found in compounds transporting specific amino acids.

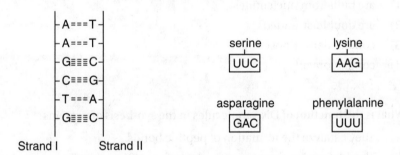

81. The amino acid whose genetic code is present in strand I is

   (1) lysine   (3) asparagine

   (2) serine   (4) phenylalanine   81 _____

82. The thymine (T) of strand I is accidentally replaced by adenine (A). This occurrence is called

   (1) segregation   (3) cytoplasmic inheritance

   (2) disjunction   (4) gene mutation   82 _____

83. The number of different amino acids coded by strand I is

(1) 1   (3) 8
(2) 2   (4) 12   83 _____

84. Which represents the sequence of nitrogenous bases in the molecule of messenger RNA synthesized by strand I?

(1) *-T-T-C-G-U-C-*   (3) *-U-U-C-G-A-C-*
(2) *-A-A-C-G-T-C-*   (4) *-A-A-G-C-U-G-*   84 _____

85. Molecules that transport amino acids to ribosomes are known as

(1) protein molecules   (3) mitochondria
(2) RNA molecules   (4) chromosomes   85 _____

86. A similarity between DNA molecules and RNA molecules is that they

(1) are built from nucleotides
(2) are double-stranded
(3) contain deoxyribose sugar
(4) contain uracil   86 _____

87. What is the function of DNA molecules in the synthesis of proteins?

(1) They catalyze the formation of peptide bonds.
(2) They determine the sequence of amino acids in a protein.
(3) They transfer amino acids from the cytoplasm to the nucleus.
(4) They supply energy for protein synthesis.   87 _____

88. In pea plants, the trait for smooth seeds is dominant over the trait for wrinkled seeds. When two hybrids are crossed, which results are most probable?

(1) 75% smooth and 25% wrinkled seeds
(2) 100% smooth seeds
(3) 50% smooth and 50% wrinkled seeds
(4) 100% wrinkled seeds   88 _____

89. A person who is homozygous for blood type A has a genotype that can be represented as

(1) $I^a I^b$ (3) $I^a i$

(2) $I^a I^a$ (4) ii 89 _____

90. Animal breeders often cross breed members of the same litter in order to maintain desirable traits. This procedure is known as

(1) hybridization

(2) inbreeding

(3) natural selection

(4) vegetative propagation   90 _____

Answers Explained

76. **2** Corn plants grown in the dark will be white. The most probable explanation for this is that expression of the gene for color may *depend on the environment*. The plants have the genetic information for chlorophyll production. This can be assumed because they are genetically identical to the plants grown in the light. Light is needed to activate the chlorophyll gene.

WRONG CHOICES EXPLAINED:

(1) *Mutations* are sudden changes in the genetic material. Mutations are inherited. Because the plants grown in the dark were genetically identical to those grown in the light, neither group lacked the genetic information for chlorophyll production.

(3) *Albinism* is a condition resulting from the absence of a normal gene for color. Both groups of plants had the normal gene for color.

(4) The *phenotype* is the physical appearance of the organism. The phenotype depends on the genotype.

77. **4** The hereditary information is contained in the chromosomes. During mitosis, the *chromosomes duplicate*. The duplication of chromosomes ensures the equal distribution of identical genetic material to the new cells.

WRONG CHOICES EXPLAINED:

(1) If the hereditary information is *destroyed*, the cells cannot function. The chromosomes contain the information necessary for carrying out all cellular activities.

(2) *No two organisms are exactly alike*. No two organisms have the same hereditary material. Identical twins are the only exception to these statements.

(3) Plasmagenes are *genes located outside the nucleus*. Drug resistance in some bacteria is transmitted through plasmagenes.

78. **3** *Crossing-over* is the exchange of chromosomal material between homologous pairs of chromosomes. This process occurs during synapsis in meiosis.

WRONG CHOICES EXPLAINED:

(1) *Nondisjunction* is the failure of homologous chromosomes to separate from each other during meiosis. Cells with extra chromosomes and cells with too few chromosomes result from nondisjunction.

(2) *Polyploidy* is a condition in which the cells have extra sets of chromosomes beyond the normal 2n number.

(4) *Hybridization* is the crossing of two organisms that are distinctly different from each other. The purpose is to bring together new combinations of genes. Usually the individual with the new gene combinations is more sturdy than either parent. A tangelo is a cross between a tangerine and a grapefruit.

79. **2** A DNA nucleotide is composed of a *deoxyribose sugar molecule, a phosphoric acid molecule, and a nitrogen base.*

    WRONG CHOICES EXPLAINED:
    (1) *Adenine and thymine* are bases.

    (3) *Glucose* is not the sugar molecule in DNA.

    (4) *Adenine, thymine, and cytosine* are bases.

80. **1** The two strands of DNA are held together by *hydrogen bonds.* The hydrogen bonds form weak links between the base pairs of each strand.

    WRONG CHOICES EXPLAINED:
    (2) The sugars and bases of the nucleotides of DNA are organic compounds. Each individual compound is made up of *carbon bonds.*

    (3) The nucleotides in each strand are joined together by *phosphate groups.*

    (4) There are no *ribose groups* in DNA.

81. **1** *Lysine* is the amino acid whose genetic code is present on DNA strand I. Strand I of the DNA molecule contains two triplet codons (a triplet codon is a three-base sequence): AAG-CTG. Each triplet codon represents a specific amino acid. The amino acids are carried by tRNA molecules. The tRNA codon matches the DNA codon (except that U replaces T). There are two possible tRNA codons that could match the DNA strand I sequence; they are AAG and CUG. Of these, only the AAG tRNA appears in the diagram. The AAG tRNA carries the amino acid known as lysine.

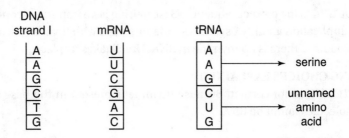

WRONG CHOICES EXPLAINED:

(2) The DNA triplet code for *serine* is TTC. This codon does not appear on strand I of the DNA molecule in the diagram.

(3) The DNA triplet code for *asparagine* is GAC. This codon does not appear on strand I of the DNA molecule in the diagram.

(4) The DNA triplet code for *phenylalenine* is TTT. This codon does not appear on strand I of the DNA molecule in the diagram.

82. 4 A gene controls the production of a protein. The substitution of one base for another changes the triplet code. One amino acid will be substituted for another. The result is a *mutation.* The replacement of glutamic acid by valine in a hemoglobin molecule causes sickle cell anemia.

WRONG CHOICES EXPLAINED:

(1) *Segregation* is the separation of alleles from each other during the formation of gametes.

(2) *Disjunction* is the separation of homologous chromosomes during the process of meiosis.

(3) *Cytoplasmic inheritance* is the inheritance of genes located in the cytoplasm, not in the nucleus. The cytoplasmic genes are called plasmagenes.

83. 2 *Two* different amino acids are coded by strand I. Strand I has two triplet codes, six bases.

WRONG CHOICES EXPLAINED:

(1) Only three bases would have to be shown in the diagram to code for *1* amino acid.

(3) 24 bases are needed for *8* amino acids.

(4) 48 bases are needed for *12* amino acids.

84. 3 *UUCGAC* is the correct sequence. Base pairing is an important concept in DNA duplication and RNA synthesis. Adenine pairs with thymine; cytosine pairs with guanine. There is no thymine in RNA. Uracil takes its place.

WRONG CHOICES EXPLAINED:

(1) *TTCGUC* is not correct. Because thymine is present in the base sequences, the molecule cannot be RNA.

(2) *AACGTC* is not correct. The base sequences are not complementary to either strand I or strand II.

(4) *AAGCUG* is not correct. The base sequences are complementary to strand II in the diagram not strand I.

85. **2** Amino acids are transported to the ribosomes by *RNA molecules* known as transfer RNA, tRNA.

WRONG CHOICES EXPLAINED:
(1) *Protein molecules* are synthesized in the ribosomes. The code for the synthesis is contained in mRNA.

(3) Cellular respiration takes place in the *mitochondria*.

(4) *Chromosomes* are structures found in the nucleus. They are composed of DNA and protein. The genes are located on the chromosomes.

86. **1** Both DNA and RNA *are built from nucleotides.*

WRONG CHOICES EXPLAINED:
(2) Only DNA is *double-stranded*. RNA is single-stranded.

(3) *Deoxyribose* is the sugar in the DNA nucleotides. Ribose is the sugar in the RNA nucleotides.

(4) The base thymine is replaced by *uracil* in RNA nucleotides.

87. **2** The *sequence of amino acids* in a protein is determined by the triplet codes in DNA. The codes are carried to the ribosomes when mRNA is synthesized. A DNA strand is the template in mRNA synthesis.

WRONG CHOICES EXPLAINED:
(1) Enzymes *catalyze the formation of peptide bonds*. A peptide bond is a C–N bond formed by the dehydration synthesis of amino acids.

(3) *Amino acids are transferred* from the cytoplasm to the ribosomes, not to the nucleus. Transfer RNA is the carrier molecule.

(4) ATP molecules *supply the energy for protein synthesis*.

88. **1** A *hybrid* is an individual that has two different alleles for a particular trait. The hybrid may be represented by the symbols *Ss*. The hybrid is smooth; the smooth trait is dominant over the wrinkled trait. When two hybrids are crossed, *75% of the offspring will have the smooth trait and 25% of the offspring will have the wrinkled trait.*

	S	s
S	SS	Ss
s	Ss	ss

WRONG CHOICES EXPLAINED:
(2), (3), (4) These choices are incorrect based on the information provided by the Punnett square.

89. **2** The term *homozygous* means *pure for the trait.* A homozygous individual has two identical alleles for a gene. There are three alleles for blood type: I^a, I^b, and i. The allele I^a produces a protein for blood type A; the allele I^b produces a protein for blood type B; the allele i does not produce either protein. The type of blood is determined by the combination of alleles. The genotype refers to the allelic combination. Because the person in the question is homozygous for type A blood, his genotype is I^aI^a.

WRONG CHOICES EXPLAINED:
(1) The I^a and I^b alleles are both dominant over the i allele. When both I^a and I^b alleles occur in the same person, the person has type AB blood.

(3) The type of blood represented by the genotype I^ai is type A.

(4) The genotype of a person with type O blood is *ii*.

90. **2** The mating of members of the same litter to maintain desirable traits is known as *inbreeding*. Because mating pairs come from the same litter, they are genetically similar to each other. Inbreeding is used to maintain pure breeds.

WRONG CHOICES EXPLAINED:
(1) *Hybridization,* or outbreeding, is the mating of organisms with contrasting traits. It is the opposite of inbreeding.

(3) Factors in the environment select the organisms that are best adapted to survive in the environment. This principle is known as *natural selection.* It is an essential feature in the theory of evolution.

(4) *Vegetative propagation* is asexual reproduction in plants.

Key Idea 3—Organic Evolution

Individual organisms and species change over time.

Performance Indicator	Description
3.1	The student should be able to explain the major patterns of evolution.

91. In modern classification, protozoa and algae are known as molds, and bacteria are known as

(1) bryophytes   (3) protists

(2) plants   (4) animals   91 _____

92. Most modern biologists agree that an ideal classification system should reflect

(1) nutritional similarities among organisms

(2) habitat requirements of like groups

(3) distinctions between organisms based on size

(4) evolutionary relationships among species   92 _____

93. Which term includes the other three?

(1) genus   (3) kingdom

(2) species   (4) phylum   93 _____

94. In one modern classification system, organisms are grouped into three

(1) kingdoms   (3) genera

(2) phyla   (4) species   94 _____

95. Which is one basic assumption of the heterotroph hypothesis?

   (1)   More complex organisms appeared before less complex organisms.

   (2)   Living organisms did not appear until there was oxygen in the atmosphere.

   (3)   Large autotrophic organisms appeared before small photosynthesizing organisms.

   (4)   Autotrophic activity added molecular oxygen to the environment.   95 _____

96. According to the heterotroph hypothesis, scientists believe that life arose in

   (1)   a desert environment

   (2)   a forest environment

   (3)   a vacuum

   (4)   an ocean environment   96 _____

97. From an evolutionary standpoint, the greatest advantage of sexual reproduction is the

   (1)   variety of organisms produced

   (2)   appearance of similar traits generation after generation

   (3)   continuity within a species

   (4)   small number of offspring produced   97 _____

98. According to modern theories of evolution, which of the following factors would be *least* effective in bringing about species changes?

   (1)   geographic isolation

   (2)   changing environments

   (3)   genetic recombination

   (4)   asexual reproduction   98 _____

99. A factor that tends to cause species to change is a

   (1)   stable environment

   (2)   lack of migration

   (3)   recombination of genes

   (4)   decrease in mutations   99 _____

100. If a fossil mammoth were discovered frozen in ice, its cells could be analyzed to determine whether its proteins were similar to those of the modern elephant. This type of investigation is known as comparative

(1) anatomy   (3) biochemistry

(2) embryology   (4) ecology   100 _____

101. If members of the same species have been geographically isolated from each other for an extended period of time, which will they most likely exhibit?

(1) mutations identical to each other

(2) random recombination occurring in the same manner

(3) evolution of traits of high adaptive value for their particular environments

(4) evolution into two new species that will have no problem interbreeding   101 _____

102. Skeletal similarities between two animals of different species are probably due to the fact that both species

(1) live in the same environment

(2) perform the same functions

(3) are genetically related to a common ancestor

(4) have survived until the present time   102 _____

103. The best means of discovering if there is a close evolutionary relationship between animals is to compare

(1) blood proteins   (3) foods consumed

(2) use of forelimbs   (4) habitats occupied   103 _____

104. In the process of evolution, the effect of the environment is to

(1) prevent the occurrence of mutations

(2) act as a selective force on variations in species

(3) provide conditions favorable for the formation of fossils

(4) provide stable conditions favorable to the survival of all species   104 _____

105. In a stable population in which the gene frequencies have been constant for a long time, the rate of evolution

    (1)   increases

    (2)   decreases

    (3)   remains the same

    (4)   increases, then decreases   105 _____

106. Certain strains of bacteria that were susceptible to penicillin in the past have now become resistant. The probable explanation for this is that

    (1)   the mutation rate must have increased naturally

    (2)   the strains have become resistant because they needed to do so for survival

    (3)   a mutation was retained and passed on to succeeding generations because it had high survival value

    (4)   the principal forces influencing the pattern of survival in a population are isolation and mating   106 _____

107. The frequency of traits that presently offer high adaptive value to a population may *decrease* markedly in future generations if

    (1)   conditions remain stable

    (2)   the environment changes

    (3)   all organisms with these traits survive

    (4)   mating remains random   107 _____

108. Since the publication of Darwin's theory, evolutionists have developed the concept that

    (1)   a species produces more offspring than can possibly survive

    (2)   the individuals that survive are those best fitted to the environment

    (3)   through time, favorable variations are retained in a species

    (4)   mutations are partially responsible for the variations within a species   108 _____

109. One factor that Darwin was unable to explain satisfactorily in his theory of evolution was

   (1)   natural selection

   (2)   overproduction

   (3)   survival of the fittest

   (4)   the source of variations   109 _____

Answers Explained

91. 3 The *protists* include all unicellular organisms and organisms that have both plant and animal features within one cell. Protozoa and algae are protists.

WRONG CHOICES EXPLAINED:
(1) *Bryophytes* are multicellular green plants that do not have vascular tissue. Mosses are examples of bryophytes.

(2) Multicellular photosynthetic organisms make up the *plant* kingdom, which includes both vascular and nonvascular plants.

(4) The *animal* kingdom is composed of multicellular organisms that cannot manufacture their own food. The organisms within this kingdom lack cell walls, and most are capable of some type of locomotion.

92. 4 A classification system should reflect *evolutionary relationships among species.* Evolutionary relationships are determined on the basis of the similarities in the anatomy, embryology, and biochemistry among organisms.

WRONG CHOICES EXPLAINED:
(1) All animals from protozoans to humans utilize the same nutrients in a similar manner. The process of photosynthesis is the same in tree cells and unicellular algae. *Nutritional similarities among organisms* are *not* useful in a system of classification.

(2) Both the whale and the fish live in an ocean environment. However, the whale is a mammal. Other than *sharing the same habitat*, the whale has no fish characteristics.

(3) Algae and protozoans are both microscopic organisms. However, algae have plant characteristics and protozoans have animal characteristics. *Size cannot be used* as the basis for a system of classification.

93. 3 According to the classification system, the largest grouping of organisms is the *kingdom.* Following this, the other groups, in order, are phylum, class, order, family, genus, species. Depending on its chief characteristics, an organism is placed in one of five kingdoms—Monera, Protist, Fungi, Animal, or Plant.

94. 1 In one modern classification system, organisms are grouped into three *kingdoms:* Animal, Plant, and Protist. Organisms that are not typical plants or animals are classified as protists. Examples are protozoa, slime molds, and bacteria.

WRONG CHOICES EXPLAINED:

(2) A *phylum* (plural phyla) is a large grouping that consists of classes, orders, families, genera, and species.

(3) A *genus* (plural genera) is a classification group composed of species. Members of a genus are more closely related than groups belonging to a given phylum.

(4) The *species* is the unit of classification. All members of a species are so closely related that they can mate and produce viable offspring.

95. **4** A heterotroph is an organism that must get its food from a source outside its own body cells; it cannot synthesize its food from inorganic materials. The heterotroph hypothesis proposes that the first living things on Earth were heterotrophs that obtained their food from the organic materials in the primitive seas. Autotrophs are organisms, such as green plants, that can synthesize their own food. At some stage in Earth's history, *autotrophic activity* used up the carbon dioxide in the air and, as a consequence of photosynthesis, *added molecular oxygen to the atmosphere.*

WRONG CHOICES EXPLAINED:

(1) Coacervates and then relatively *simple cells developed before more complex organisms.*

(2) Living heterotrophs *appeared before molecular oxygen was added to the atmosphere.*

(3) *Small cells that carried on photosynthesis appeared before the more complex vascular and seed plants.* The course of evolution is from the simple to the complex.

96. **4** According to the heterotroph hypothesis, life on earth evolved through a sequence of stages. The gases of the primitive atmosphere, such as methane, ammonia, and hydrogen, were washed down by heavy rains into the *early oceans.* They were acted on by ultraviolet radiation, cosmic rays, the earth's heat, and radioactivity. The bonding together of the molecules resulted in the formation of larger organic molecules.

WRONG CHOICES EXPLAINED:

(1) A *desert* environment could not support "first" life. The intense heat and the rapid evaporation of water are conditions that do not allow for the movement or maintenance of molecules in a fluid medium.

(2) A *forest* environment does not provide the pools of warm water on a continuous basis necessary for aggregate molecules to form.

(3) A *vacuum,* a place without air, cannot support life.

97. **1** Sexual reproduction helps to maximize the number of different allelic combinations that occur in offspring, leading to a greater *variety of organisms produced* within the species' population as a whole. Individuals displaying favorable traits in a changing environment are more likely to survive and to pass these traits on to their offspring, a fact that helps to promote evolutionary change.

WRONG CHOICES EXPLAINED:
(2), (3) *Appearance of similar traits generation after generation* and *continuity within a species* describe conditions that promote stability and uniformity within species, both of which are maximized during asexual reproduction.

(4) Evolutionarily speaking, the larger the number of offspring produced during reproduction, the more successful a particular species variety tends to be in competing with other varieties of the same species. *Small number of offspring produced,* therefore, is not an evolutionary advantage of sexual or any other type of reproduction.

98. **4** Variations among organisms are necessary for speciation. *Asexual reproduction* is least effective in bringing about changes in species. There are no variations among organisms that are reproduced asexually. These organisms are genetically like their parents.

WRONG CHOICES EXPLAINED:
(1) *Geographic isolation* increases the chance that a group of organisms will develop a new gene pool, which will give rise to a new species.

(2) *Changes in the environment* cause shifts in the gene pool. The genes that ensure the survival of organisms increase in the pool. Thus, the environment changes the characteristics of the original population, and a new species is formed.

(3) *Genetic recombination* occurs during meiosis and fertilization. The shuffling of genes results in the appearance of new characteristics in a population.

99. **3** The *recombination of genes* is one factor that tends to cause species to change. Mutations are changes in genes. When like mutations combine in the fertilized egg, the new characteristic will be expressed in the offspring. If this mutation adds to the survival value of the organism, the gene change will be passed on to progeny because individuals having this mutation will live to reproduce.

WRONG CHOICES EXPLAINED:

(1) A *stable environment* will probably not cause species to change. Beneficial mutations become effective in changing environments. For example, the mutations that produced white fur in the polar bear were beneficial. At one time, the polar regions were tropical. A change to a glacial environment was accompanied by a change or changes in the animal species that inhabited the region. Bears with a dark coat color became immediate targets for natural enemies.

(2) *Migration* aids the recombination of genes because organisms have greater opportunities for interbreeding.

(4) A *decrease in mutations* does not aid speciation but slows it.

100. 3 *Biochemistry* is the study of the chemistry of living organisms. Proteins are compounds found only in living organisms.

WRONG CHOICES EXPLAINED:

(1) *Anatomy* is the study of the structure of organisms.

(2) *Embryology* is the study of the development of embryos.

(4) *Ecology* is the study of the relationship of living organisms to each other and to their environment.

101. 3 Geographic isolation involves the separation of organisms by natural barriers. Each group of isolated individuals develops its own gene pool because each group lives under different environmental conditions. In the case of members of the same species who have been geographically isolated from each other, the *selection for individuals with special survival traits* is different in each environment.

WRONG CHOICES EXPLAINED:

(1) *Mutations* are the raw materials for evolution. Although mutations might have been the same in each group, they do not have the same adaptive value in each group.

(2) *Random recombination* occurs in the same manner in each group. However, the recombination process operates on two distinctly different gene pools.

(4) Usually, *different species cannot mate*. If their mating happens to be successful, their offspring will not be fertile.

102. 3 Morphology is the study of the structure and form of living things. When the arm of a human and the wing of a bird are studied, they are seen to have similar bone structure. This indicates that both organisms descended from a *common ancestor*. They have both undergone many changes since then and are now very different from each other. However, they still retain some of the same genes and therefore show a similarity in many parts of their bodies, including the arrangement of the bones in their forelimbs.

103. 1 A close evolutionary relationship between animals can be shown by a study of their *blood proteins*. The precipitin test is used to show such a relationship. A rabbit can be sensitized to human blood by being injected with human serum. When the sensitized rabbit serum is mixed with human serum, a white precipitate forms. If the sensitized rabbit serum is mixed with serum from a chicken, there is no reaction. However, a precipitate does form when the sensitized rabbit serum is mixed with the serum of a chimpanzee. In a like manner, the serum of a dog and a wolf show precipitation with serum sensitized to dog serum.

WRONG CHOICES EXPLAINED:

(2) The *forelimbs* of many unrelated animals may be used for the same purpose.

(3) Many unrelated animals *consume similar food*.

(4) Many unrelated animals *live in the same habitat*.

104. 2 There are many variations among the organisms of a species. Some variations allow an organism to survive best in a particular environment. Such factors as climate, food supply, and type of predators determine which organisms are *best adapted to that environment*.

WRONG CHOICES EXPLAINED:

(1) *Mutations* occur naturally and randomly. Mutations increase the variations among organisms.

(3) *Fossils* are the remains of organisms that lived in the past. They present evidence that evolution has occurred.

(4) There are many factors in an environment that influence the survival of different species. A *stable environment* preserves the species that have already adapted to that environment.

105. 3 Stable gene pools are a hallmark of nonevolving populations. The rate of evolution in such populations neither increases nor decreases, but *remains the same.*

WRONG CHOICES EXPLAINED:
(1), (2), and (4) all refer to changing rates of evolution that do not occur.

106. 3 Bacteria resistant to penicillin developed as a result of mutation. Organisms that did not receive the mutated gene were killed by the antibiotic. Those in which gene mutation occurred survived and *passed the mutation on to succeeding generations.*

WRONG CHOICES EXPLAINED:
(1) *The mutation rate did not increase.* The survivors had the mutated gene that allowed the bacteria to resist the effects of penicillin. These resistant strains reproduced, creating populations that replaced the nonresistant strains.

(2) *Need does not determine mutation.* Mutations are chance occurrences.

(4) Survival of a species depends on the ability of its members to obtain food, carry out respiration, and *reproduce successfully. Isolation* does not increase species survival.

107. 2 As long as environmental conditions remain stable, the alleles controlling traits that promote individual survival in that environment tend to be maintained at a high level in the population's gene pool. When *the environment changes,* however, the factors that promoted survival may no longer be present. Selection pressure may then operate to reduce the frequency of the once-prevailing alleles in favor of alleles controlling other, contrasting traits that increase individuals' chances for survival in the new environment.

WRONG CHOICES EXPLAINED:
(1) As long as *conditions remain stable,* selection pressures on individuals displaying favorable traits remain low, promoting the maintenance of a high frequency of alleles controlling these traits.

(3) If *all organisms with these traits survive,* their genes will be passed on to future generations at a high rate. This will help to maintain a high frequency of alleles controlling these traits in the population.

(4) If *mating remains random,* the probability that alleles will pair in unrestricted combinations will remain high, helping to ensure that the laws of probability will operate freely and that gene frequencies for existing traits will remain stable.

108. 4 Darwin proposed his theory of evolution in 1856. His theory did not explain how variation arose in organisms. In 1901, Hugo De Vries discovered the existence of mutations. *Mutations accounted for the rise of variations in organisms.*

WRONG CHOICES EXPLAINED:
(1) One of the principles of Darwin's theory of evolution stated that a species produced *more offspring than could possibly survive.*

(2) Another principle stated that the *individuals that survived were those best suited to the environment.*

(3) *The variations favored by the environment are retained within a species.* It is the environment that determines which variations are favorable.

109. 4 Darwin's theory of evolution was completed and published before Mendel completed his study of inheritance in the garden pea. Darwin could not explain *how variations occurred* or how they were passed on from parent to offspring.

WRONG CHOICES EXPLAINED:
(1) Darwin's theory of *natural selection* was divided into five distinct principles that formulated his concept of evolution. These ideas were set forth in his book *The Origin of the Species by Natural Selection.*

(2) *Overproduction* was one of the principles of Darwin's theory. He explained that for a species to continue in existence, it must overproduce in order to maintain the species number. For example, one female codfish lays about 9 million eggs. Not all of these eggs are fertilized and not all codfish lay about 9 million eggs. Not all of these eggs are fertilized and not all codfish fry reach adulthood. If the 9 million eggs per female were fertilized and if these zygotes developed into adult fish, the seas would be overrun with codfish. However, if the number of gametes produced by codfish were greatly reduced, the species would die out. This overproduction of gametes is necessary to maintain codfish survival.

(3) Another of Darwin's principles was *survival of the fittest.* No two organisms are alike; each has variations. These variations may either help or hinder the organism in its struggle for existence. An organism with variations that help it reach food faster is more fit and has a better potential for survival than a slower, less fit member of the species.

Key Idea 4—Reproductive Continuity

The continuity of life is sustained through reproduction and development.

Performance Indicator	Description
4.1	The student should be able to explain how organisms, including humans, reproduce their own kind.

110. Which occurs in a plant cell but *not* in an animal cell during mitotic cell division?

   (1)   formation of spindle fibers

   (2)   chromosome duplication

   (3)   formation of a cell plate

   (4)   cytoplasmic division   110 _____

111. A plant cell with 12 chromosomes undergoes normal mitosis. What is the total number of chromosomes in each of the resulting daughter cells?

   (1)   24   (3)   6

   (2)   12   (4)   4   111 _____

112. Asexual reproduction *differs* from sexual reproduction in that, in asexual reproduction,

   (1)   new organisms are usually genetically identical to the parent

   (2)   the reproductive cycle involves the production of gametes

   (3)   nuclei of sex cells fuse to form a zygote

   (4)   offspring show much genetic variation   112 _____

113. In most multicellular animals, meiotic cell division occurs in specialized organs known as

   (1)   gonads

   (2)   gametes

   (3)   kidneys

   (4)   cytoplasmic organelles   113 _____

114. Which is an important adaptation for reproduction among land animals?

   (1)   fertilization of gametes outside the body of the female

   (2)   fertilization of gametes within the body of the female

   (3)   production of sperm cells with thick cell walls

   (4)   production of sperm cells with thin cell walls   114 _____

115. In humans, a single primary sex cell may produce four gametes. These gametes are known as

   (1)   diploid egg cells   (3)   polar bodies

   (2)   monoploid egg cells   (4)   sperm cells   115 _____

116. In sexual reproduction, the $2n$ chromosome number is restored as a direct result of

   (1)   fertilization   (3)   cleavage

   (2)   gamete formation   (4)   meiosis   116 _____

117. In human females, the main function of the follicle-stimulating hormone (FSH) secreted by the pituitary gland is to

   (1)   stimulate the adrenal glands to produce cortisone

   (2)   stimulate activity in the ovaries

   (3)   control the metabolism of calcium

   (4)   regulate the rate of oxidation in the body   117 _____

118. If the first stage of an uninterrupted human menstrual cycle is the follicle stage, the last stage includes the

   (1)   formation of sperm cells in the testis

   (2)   release of a mature egg

   (3)   buildup of the uterine lining

   (4)   shedding of the uterine lining   118 _____

119. Which statement best describes internal fertilization?

(1) It does not require motile gametes.

(2) It helps to make terrestrial life possible.

(3) It requires the presence of many eggs.

(4) It normally occurs in the male. 119 _____

120. What are the normal chromosome numbers of a sperm, egg, and zygote, respectively?

(1) monoploid, monoploid, and monoploid

(2) monoploid, diploid, and diploid

(3) diploid, diploid, and diploid

(4) monoploid, monoploid, and diploid 120 _____

121. When compared with the number of gametes produced from a single primary sex cell during oogenesis, the number of gametes produced from a single human primary sex cell during spermatogenesis is usually

(1) four times as great   (3) half as great

(2) twice as great   (4) the same 121 _____

122. In human males, sperm cells are suspended in a fluid medium. The main advantage gained from this adaptation is that the fluid

(1) removes polar bodies from the surface of the sperm

(2) activates the egg nucleus so that it begins to divide

(3) acts as a transport medium for sperm

(4) provides currents that propel the egg down the oviduct 122 _____

Base your answers to questions 123 and 124 on the diagrams and the information below.

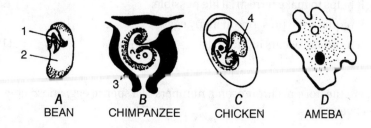

| A | B | C | D |
| BEAN | CHIMPANZEE | CHICKEN | AMEBA |

123. Which organisms were produced as a result of fertilization?

   (1)   *A, B,* and *C,* only   (3)   *C* and *D,* only

   (2)   *B* and *C,* only   (4)   *B, C,* and *D,* only   123 _____

124. Structures that function in the storage of food to be used by growing embryonic cells are indicated by

   (1)   1 and 3   (3)   2 and 4

   (2)   2 and 3   (4)   3 and 4   124 _____

Base your answers to questions 125 through 127 on your knowledge of biology and the information below.

A biologist cut a flap of ectoderm from the top of a developing embryo. He did not remove the piece of ectoderm but just folded it back. Then he cut out the mesoderm underneath and completely removed it. He folded the flap of ectoderm back in place. The ectoderm healed; however, a complete nervous system did not develop.

125. This experiment was most likely performed immediately after

   (1)   cleavage   (3)   fertilization

   (2)   gestation   (4)   gastrulation   125 _____

126. This experiment interfered with the process of

   (1)   differentiation   (3)   cleavage

   (2)   zygote formation   (4)   ovulation   126 _____

127. This experiment demonstrates that the

(1) ectoderm is solely responsible for development of the nervous system

(2) nervous system is destroyed during surgical operations

(3) mesoderm influences the development of the nervous system

(4) digestive enzymes have a major role in the development of embryonic layers 127 _____

128. In a developing embryo, the mesoderm layer normally gives rise to

(1) epidermal tissue

(2) skeletal tissue

(3) digestive tract lining

(4) respiratory tract lining 128 _____

129. What is the function of the placenta in a mammal?

(1) It surrounds the embryo and protects it from shock.

(2) It allows mixing of the maternal and fetal blood.

(3) It permits the passage of nutrients and oxygen from the mother to the fetus.

(4) It replaces the heart of the fetus until the fetus is born. 129 _____

Answers Explained

110. 3 Mitosis is the process by which two identical nuclei are formed. Mitotic cell division is usually followed by cytoplasmic division. A plant cell has a rigid cell wall. Division of the cytoplasm begins with the appearance of a *cell plate* between the two nuclei. The cell plate is composed of membrane fragments from the endoplasmic reticulum.

WRONG CHOICES EXPLAINED:

(1) *Spindle fibers* are elastic-like protein fibers. Chromosome movement is controlled by spindle fibers.

(2) In order for two nuclei to be identical, they must have the same number and kind of chromosomes. The *chromosomes duplicate* before the nucleus divides. The mitotic process is the same in both plant and animal cells.

(4) *Cytoplasmic division,* or cytokinesis, usually follows nuclear division in both plant and animal cells.

111. 2 Chromosomes are structures in the nucleus. During mitosis, two cells with identical chromosomes are formed. Because the cell had 12 chromosomes, the daughter cells must also have *12* chromosomes.

WRONG CHOICES EXPLAINED:

(1) A cell with *24* chromosomes has twice the diploid number. The condition in which there are extra sets of chromosomes is known as polyploidy.

(3) A cell with *6* chromosomes has one-half the diploid number. Monoploid cells arise through meiosis.

(4) A cell with *4* chromosomes can only arise through a complete breakdown of the mitotic or meiotic process.

112. 1 In asexual reproduction, new organisms are produced by a single parent. Asexual reproduction involves the mitotic process. The *genetic material of the offspring is identical to that of the parent.*

WRONG CHOICES EXPLAINED:

(2) *Gametes* are produced by sexually reproducing organisms.

(3) The fusion of sex cells (gametes) to *form a zygote* is characteristic of sexually reproducing organisms.

(4) *Genetic variation* among offspring is characteristic of sexually reproducing organisms. The process of meiosis through synapsis and segregation ensures new combinations of genetic material.

113. 1 *Gonads* are sex glands. In these glands, gametes, or sex cells, are produced from primary sex cells that undergo meiosis, also known as reduction division. Male gonads are called testes, and female gonads, ovaries.

WRONG CHOICES EXPLAINED:

(2) *Gametes* are sex cells and not organs. Sex cells are specialized for fertilization.

(3) *Kidneys* are organs of excretion and are specialized for filtering metabolic wastes out of the blood. The nephron is the unit of structure and function in the kidney. Meiosis does not take place in kidney cells.

(4) *Cytoplasmic organelles* such as mitochondria, ribosomes, lysosomes, and endoplasmic reticula are not organs.

114. 2 A gamete is a reproductive cell that must fuse with another gamete to produce a new individual. Sperm cells and egg cells are gametes. Fertilization is the fusion of an egg cell and a sperm cell. In land animals, *fertilization occurs within the body of the female* and is known as internal fertilization.

WRONG CHOICES EXPLAINED:

(1) External fertilization, the *union of gametes outside the female's body*, occurs in animals that live in a watery environment. Fish and amphibians reproduce by external fertilization.

(3), (4) The question refers to reproduction in animals. Animal cells, including gametes, do not have *cell walls*.

115. 4 Primary sex cells give rise to gametes. Gametes are formed by the process of meiosis. In meiosis, a diploid cell divides twice to form four monoploid cells. In humans the four gametes, which are identical in size, are known as *sperm cells*.

WRONG CHOICES EXPLAINED:

(1) Chromosomes occur in pairs. The diploid number of chromosomes is the full number of chromosomes of all the pairs. Meiosis is cell division in which the nucleus receives one member of each pair of chromosomes. The nucleus of an *egg cell* thus contains half the diploid chromosome number, or the monoploid number.

(2) In formation of the egg cell, the cytoplasm does not divide equally. One large monoploid cell, the egg cell, and three very small cells (polar bodies) are produced from one primary sex cell.

(3) The three small monoploid cells accompanying the egg cell are known as *polar bodies*. Polar bodies degenerate and do *not* function in fertilization.

116. 1 The diploid chromosome number is represented as $2n$, and the monoploid number as n. When two gametes in the n condition combine, a $2n$ cell is produced. *Fertilization* is the union of two gametes.

WRONG CHOICES EXPLAINED:
(2) *Gamete formation* reduces the chromosome number from $2n$ to n.

(3) *Cleavage* is mitotic cell division without growth. It is the process by which a fertilized egg cell becomes a multicellular embryo.

(4) *Meiosis,* or reduction division, reduces the chromosome number of diploid cells.

117. 2 Follicles contain immature egg cells. The follicles are found in the ovary. FSH *stimulates ovarian* follicle development.

WRONG CHOICES EXPLAINED:
(1) ACTH stimulates the *adrenal glands* to produce cortisone. ACTH is secreted by the pituitary gland.

(3) *Calcium metabolism* is controlled by parathormone. The parathyroid gland secretes parathormone.

(4) Thyroxin secreted by the thyroid gland is the major regulator of the rate of *oxidation*. The hormones from the adrenal glands and the pancreas also play a role in the oxidation of glucose.

118. 4 The menstrual cycle is a series of changes that occur within the female reproductive system. The events of the cycle prepare the uterus to receive an embryo. The lining of the uterus is built up. If the cycle is not interrupted, the egg is not fertilized and no embryo is formed. In the last stage of the cycle, the *lining of the uterus disintegrates and is shed*.

WRONG CHOICES EXPLAINED:
(1) *Sperm cells* are produced by the male.

(2) Ovulation (*release of a mature egg*) occurs midway through the menstrual cycle.

(3) Once a month the *uterus* is prepared to receive an embryo. What happens to the lining of the uterus depends upon presence or absence of an embryo. If an embryo is present, the uterus continues to develop and the menstrual cycle is interrupted.

119. 2 *It helps to make terrestrial life possible* is the correct response. Internal fertilization, as its name implies, occurs within the body of the parent (usually female). The conditions in the female reproductive tract provide an ideal environment for the survival and pairing of gametes, helping to ensure that fertilization occurs successfully. This method of reproduction is especially helpful in the survival of terrestrial animal species, who live where harsh conditions (such as drying, heat, and cold) can easily damage or kill gametes released into the environment for external fertilization.

WRONG CHOICES EXPLAINED:
(1) *Motile gametes* (such as human sperm cells) are common in species employing both external and internal fertilization. Motility (ability to move) enables the sperm cells to swim toward the egg cell in either environment.

(3) Because of the dangers posed to fragile gametes in any environment, the *presence of many eggs* is characteristic of species employing external fertilization. Species using internal fertilization produce relatively few eggs in the reproductive process.

(4) In most species internal *fertilization occurs within the body of the female*, not that of the *male*.

120. 4 *Monoploid, monoploid, and diploid* is the correct combination. Sperm cells and egg cells are monoploid (*n*) gametes formed during the process of meiotic cell division. A zygote results from the fusion of two monoploid nuclei in fertilization and so must be diploid (*2n*) in chromosome number.

WRONG CHOICES EXPLAINED:
(1), (2), (3) Each of these distracters contains an incorrect combination of choices (see above).

121. 1 In the process of oogenesis, a single primary sex cell gives rise to a single monoploid egg cell and three nonfunctional monoploid polar bodies. The process of spermatogenesis yields four functional monoploid sperm cells for each primary sex cell. Therefore, a comparison of these two processes leads to the conclusion that, per primary sex cell, spermatogenesis yields *four times* as many gametes as oogenesis does.

WRONG CHOICES EXPLAINED:
(2), (3), (4) Each of these distracters contains a mathematical comparison that is not consistent with the explanation above.

122. **3** The fluid surrounding human sperm cells *acts as a transport medium for sperm*. This fluid is known as semen. Its primary function is to provide a protective watery medium for sperm cells as they enter the female reproductive tract.

WRONG CHOICES EXPLAINED:
(1) *Removes polar bodies from the surface of the sperm* is a "nonsense" distracter. Polar bodies are not associated with sperm production.

(2) *Activates the egg nucleus so that it begins to divide* is not a function of semen. The egg is stimulated to divide by the act of fertilization. Semen is not directly involved in this process.

(4) *Provides currents that propel the egg down the oviduct* is not a function of semen. Cilia that line the oviduct are responsible for establishing fluid currents that both carry the egg downward toward the uterus and carry sperm upward toward the ovary. Semen is not directly involved in this process.

123. **1** Fertilization is one of the processes in sexual reproduction. Sexual reproduction is the method of reproduction in the bean plant, chimpanzee, and chicken (*A, B, and C, only*). The ameba reproduces asexually by binary fission.

WRONG CHOICES EXPLAINED:
(2) The bean plant (*B*) was omitted in this choice.

(3) The ameba (*D*) is an incorrect answer. Both the bean (*A*) and the chimpanzee (*B*) were omitted in this choice.

(4) The bean plant (*A*) was omitted and the ameba (*D*) is an incorrect answer.

124. **3** The structures that function in the storage of food for the embryonic cells are labeled *2* and *4*. Structure *2* is the cotyledon of the seed. The yolk sac is structure *4*.

WRONG CHOICES EXPLAINED:
(1) Structure *3* refers to the wall of the uterus. The embryo of a chimpanzee is nourished through the placenta, not the uterine wall. Structure *1* is the leaf of the embryo bean plant.

(2) Although structure *2* is a correct answer, structure *3* is incorrect.

(4) Although structure *4* is a correct answer, structure *3* is incorrect.

125. **4** The experiment was performed after *gastrulation.* Gastrulation is a stage in embryonic development that gives rise to three germ layers of cells. The three germ layers are the ectoderm, mesoderm, and endoderm.

WRONG CHOICES EXPLAINED:

(1) *Cleavage* is a stage of embryonic development in which the zygote undergoes rapid mitotic divisions. The final result is a ball of cells.

(2) *Gestation* is a prebirth period. It is the time a developing embryo spends in the uterus.

(3) The union of a sperm cell nucleus with an egg cell nucleus is called *fertilization.* The result of the process is a zygote.

126. **1** The experiment interfered with the development of a nervous system. The development of special tissues and organisms is known as *differentiation.*

WRONG CHOICES EXPLAINED:

(2) *Zygote formation* must occur before an embryo can develop.

(3) The process of *cleavage* provides the embryo with hundreds of undifferentiated cells.

(4) *Ovulation* is the release of an egg from the ovary.

127. **3** The experiment demonstrates that the development of the *nervous system is influenced by the presence of the mesoderm.* The nervous system does not develop when the mesoderm is removed.

WRONG CHOICES EXPLAINED:

(1) If the *ectoderm was solely responsible for the development of the nervous system*, the nervous system would have developed after the mesoderm was removed.

(2) There was no *nervous system* present when the surgery was performed.

(4) The experiment was not concerned with the reasons for the *development of the embryonic layers.*

128. **2** Each of the three germ layers of the embryo is responsible for the development of the systems of the body. The *skeletal system* develops from the mesoderm. The muscle system, circulatory system, and excretory system also evolve from the mesoderm.

WRONG CHOICES EXPLAINED:

(1) *Epidermal cells* form the outer covering or skin of the body. The skin and nervous system develop from the ectoderm.

(3), (4) The linings of the *digestive* and *respiratory tracts* develop from the endoderm.

120. **3** The placenta is an area of spongy tissue in the uterus. It is very rich in blood vessels. The placenta functions as a respiratory and excretory organ of the fetus. *Vital materials are exchanged between the capillaries of the fetus and the capillaries of the mother.*

WRONG CHOICES EXPLAINED:

(1) The amnion is a fluid-filled sac surrounding the embryo. The fluid bathes the cells of the fetus and *protects it against shock.*

(2) The circulatory systems of the mother and the fetus are separate from each other. *Blood does not flow from one system into the other.*

(4) The *embryo develops its own heart.* The placenta provides an area for the diffusion of materials into and out of the fetus.

Key Idea 5—Dynamic Equilibrium and Homeostasis

Organisms maintain a dynamic equilibrium that sustains life.

Performance Indicator	Description
5.1	The student should be able to explain the basic biochemical processes in living organisms and their importance in maintaining dynamic equilibrium.
5.2	The student should be able to explain disease as a failure of homeostasis.
5.3	The student should be able to relate processes at the system level to the cellular level in order to explain dynamic equilibrium.

130. Which energy conversion occurs in the process of photosynthesis?

(1) Light energy is converted to nuclear energy.

(2) Chemical bond energy is converted to nuclear energy.

(3) Light energy is converted to chemical bond energy.

(4) Mechanical energy is converted to light energy.   130 _____

131. An environmental change that would most likely increase the rate of photosynthesis in a bean plant would be an increase in the

(1) intensity of green light

(2) concentration of nitrogen in the air

(3) concentration of oxygen in the air

(4) concentration of carbon dioxide in the air   131 _____

132. During photosynthesis, molecules of oxygen are released as a result of the "splitting" of water molecules. This is a direct result of the

(1) dark reaction

(2) light reaction

(3) formation of PGAL

(4) formation of CO_2 132 _____

133. An organism that makes its own food without the direct need for any light energy is known as a

(1) chemosynthetic heterotroph

(2) chemosynthetic autotroph

(3) photosynthetic heterotroph

(4) photosynthetic autotroph   133 _____

134. While looking through a microscope at a section of a leaf from a freshwater plant, a student observed some cells in which chloroplasts were moving around with the cytoplasm. This type of movement is known as

(1) pinocytosis   (3) osmosis

(2) synapsis   (4) cyclosis   134 _____

135. By what process does carbon dioxide pass through the stomates into the leaf?

(1) diffusion   (3) respiration

(2) osmosis   (4) pinocytosis   135 _____

136. Two end products of aerobic respiration are

(1) oxygen and alcohol

(2) oxygen and water

(3) carbon dioxide and water

(4) carbon dioxide and oxygen   136 _____

137. Homeostatic regulation of the body is made possible through the coordination of all body systems. This coordination is achieved mainly by

(1) respiratory and reproductive systems

(2) skeletal and excretory systems

(3) nervous and endocrine systems

(4) circulatory and digestive systems   137 _____

138. Phenylketonuria (PKU) is an inherited condition characterized by mental retardation. The symptoms of the disorder result from an inability to synthesize a single type of

(1) enzyme

(2) nutrient

(3) blood cell

(4) brain cell   138 _____

139. All the children of a hemophiliac male and a normal female are normal with respect to blood clotting. However, some of their grandsons are hemophiliacs. This is an example of the pattern of hereditary known as

(1) sex determination

(2) sex linkage

(3) incomplete dominance

(4) multiple alleles   139 _____

140. What is the total number of chromosomes in a typical body cell of a person with Down's syndrome?

(1) 22

(2) 23

(3) 44

(4) 47   140 _____

Answers Explained

130. 3 In the process of photosynthesis, carbon dioxide and water molecules are converted to glucose. Light is the energy source for the reaction. *Light energy is transformed into the chemical bond energy* of the glucose molecules.

WRONG CHOICES EXPLAINED:
(1), (2), (4) The Law of Conservation of Energy states that energy cannot be created or destroyed, but it can be changed from one form to another. All the choices refer to this law, but only choice (3) occurs in living organisms.

131. 4 Experiments have shown that the rate of photosynthesis depends on the availability of carbon dioxide. The greater the *concentration of carbon dioxide*, the greater the rate of photosynthesis.

WRONG CHOICES EXPLAINED:
(1) Chlorophyll is a light-absorbing pigment found in chloroplasts. However, it does not absorb much of the green wavelength of light. An increase in the *intensity of green light* has no effect on the photosynthetic rate.

(2) Carbohydrates are the products of photosynthesis. They do not contain atoms of nitrogen. The nitrogen needed by plants comes from nitrates in the soil, not from *the concentration of nitrogen in the air*.

(3) Oxygen is released during photosynthesis. Therefore, *the concentration of oxygen in the air* has no direct effect on photosynthesis.

132. 2 Sunlight provides the energy needed to split water molecules into hydrogen and oxygen. Because light is required, this part of photosynthesis is known as the *light reaction*.

WRONG CHOICES EXPLAINED:
(1) The *dark reaction* does not use light energy. In this reaction, the hydrogen released from the light reaction is combined with carbon dioxide.

(3) *PGAL*, phosphoglyceric aldehyde, is the first stable compound formed during the dark reaction. This compound is later converted to glucose.

(4) *Carbon dioxide is not formed* but is used during photosynthesis.

133. 2 An autotroph manufactures its own food. A *chemosynthetic autotroph* produces its own food without the use of light energy. It obtains its energy from certain chemical reactions that take place in the cell.

WRONG CHOICES EXPLAINED:

(1) A *chemosynthetic organism* cannot be a heterotroph. Heterotrophs do not have the ability to manufacture their own food. All animals are heterotrophs.

(3) *Photosynthetic organisms* manufacture their own food. They cannot be heterotrophs.

(4) A *photosynthetic autotroph* utilizes light energy. All green plants are photosynthetic autotrophs.

134. 4 The movement of chloroplasts in the plant cell was due to the movement of cytoplasm in the cell. *Cyclosis* is the streaming of cytoplasm in a cell.

WRONG CHOICES EXPLAINED:

(1) *Pinocytosis* is the formation of a pocket by an infolding of the cell membrane. Large molecules are brought into the cell by this process.

(2) *Synapsis* is the pairing of homologous chromosomes during meiosis.

(3) *Osmosis* is the movement of water across a selectively permeable membrane. Osmosis is the diffusion of water.

135. 1 Carbon dioxide passes through the stomates of a leaf by *diffusion*. Diffusion is passive transport. Molecules move along a concentration gradient from an area of high density to an area of lower density.

WRONG CHOICES EXPLAINED:

(2) *Osmosis* is the diffusion of water.

(3) *Respiration* is an energy-releasing process.

(4) *Pinocytosis* is active transport. Cells use energy to draw in large molecules by the infolding of their cell membranes.

136. 3 Respiration is a process by which cells release energy from glucose molecules. Aerobic respiration requires the presence of oxygen. In the process of aerobic respiration, glucose is oxidized to *carbon dioxide and water*. Both compounds are end products, the results of a chemical reaction.

WRONG CHOICES EXPLAINED:

(1), (2), (4) All three choices are incorrect because oxygen is consumed in aerobic respiration. Oxygen is not the end product of the reaction.

137. 3 Homeostasis refers to the steady state of control of the cell and, in turn, the entire body. The biochemical processes that take place in body cells occur in even and regular sequences. The cells, tissues, and organs in all body systems must function cooperatively so that the organism can carry out its life functions effectively. The coordination of all these biochemical activities is made possible by the work of the *nervous and endocrine systems*. The nervous system carries impulses from sense organs to the brain or spinal cord and then to effector organs such as muscles or glands. The endocrine system secretes hormones that control the functions of certain glands, tissues, and organs. Together the nervous and endocrine systems maintain the homeostasis of the body.

WRONG CHOICES EXPLAINED:

(1) The *respiratory system* is specialized for the intake and distribution of oxygen. It also expels waste gases. The *reproductive system* is specialized for the developing of embryos. Both systems are controlled by the nervous and endocrine systems. In mammals, hormones control gestation and birth.

(2) The *skeletal system* gives support to the body. Hormones control the growth of long bones. Nerve cell fibers help the muscles to function. The *excretory system* coordinates waste removal from the body. Hormones control water loss and reabsorption by the kidney tubules.

(4) Blood and lymph circulate by way of the *circulatory system*. Hydrolysis of food takes place in the *digestive system*. Both systems depend on the nervous and endocrine systems for the coordination of mechanical and biochemical activities.

138. 1 According to the one gene–one enzyme theory, a single gene is responsible for the production of a single enzyme. Because PKU is an inherited defect, the gene for an *enzyme* is absent in the victim.

WRONG CHOICES EXPLAINED:

(2) *Nutrients* are *not* synthesized by animals. Nutrients must be taken in from organic sources.

(3) *Blood cells* are *not* affected nor involved in the PKU disorder.

(4) The development of *brain cells* is affected by the PKU condition. Because of the lack of an enzyme, phenylalanine is converted to phenylpyruvic acid, which accumulates in brain tissue. The result is mental retardation.

139. 2 A trait that appears more often in one sex than in the other sex is said to be *sex-linked*. The gene for the trait is located on a sex chromosome. Hemophilia and color blindness are examples of sex-linked traits.

WRONG CHOICES EXPLAINED:
(1) *Sex determination* is controlled by a pair of sex chromosomes. There are two kinds of sex chromosomes, an X and a Y chromosome. A female has two X chromosomes (XX). A male has one X chromosome and one Y chromosome (XY).

(3) *Incomplete dominance* is a type of inheritance in which neither allele in a hybrid is dominant. The hybrid shows a trait completely different from either parent. The inheritance of color in a Japanese four-o'clock flower is an example of incomplete dominance, or blending.

(4) Alleles are different forms of the same gene. The term *multiple alleles* implies that a gene has more than two forms. Inheritance of human blood type involves multiple alleles.

140. 4 The normal number of chromosomes in human body cells is 46. Down's syndrome results from the presence of an extra chromosome. The extra or 47th chromosome is due to meiotic nondisjunction.

WRONG CHOICES EXPLAINED:
(1), (3) The numbers 22 and 44 do not apply to any known normal human chromosome number.

(2) The number 23 is the monoploid number of chromosomes in human gametes.

Key Idea 6—Interdependence of Living Things

Plants and animals depend on each other and their physical environment.

Performance Indicator	Description
6.1	The student should be able to explain factors that limit growth of individuals and populations.
6.2	The student should be able to explain the importance of preserving diversity of species and habitats.
6.3	The student should be able to explain how the living and nonliving environments change over time and respond to disturbances.

141. Animals *cannot* synthesize nutrients from inorganic raw materials. Therefore, animals obtain their nutrients by

(1) combining carbon dioxide with water

(2) consuming preformed organic compounds

(3) hydrolyzing large quantities of simple sugars

(4) oxidizing inorganic molecules for energy   141 _____

142. Which organisms carry out heterotrophic nutrition?

(1) ferns   (3) fungi

(2) grasses   (4) mosses   142 _____

143. Which activity is an example of intracellular digestion?

(1) a grasshopper chewing blades of grass

(2) a maple tree converting starch to sugar in its roots

(3) an earthworm digesting proteins in its intestine

(4) a fungus digesting dead leaves   143 _____

144. A hydra ingests a daphnia, digests it, and later egests some materials. All these events are most closely associated with the life process known as

(1) transport   (3) growth

(2) synthesis   (4) nutrition   144 _____

145. Some bacteria are classified as saprophytes because they are organisms that

(1) feed on other living things

(2) feed on dead organic matter

(3) manufacture food by photosynthesis

(4) contain vascular bundles   145 _____

146. Aerobic organisms are dependent on autotrophs. One reason for this dependency is that most autotrophs provide the aerobic organisms with

(1) oxygen   (3) nitrogen gas

(2) carbon dioxide   (4) hydrogen   146 _____

147. Of the following, the greatest amount of the Earth's food production is thought to occur in

(1) coastal ocean waters   (3) taiga forests

(2) desert biomes   (4) tundra biomes   147 _____

148. Most of the minerals within an ecosystem are recycled and returned to the environment by the direct activities of organisms known as

(1) producers

(2) secondary consumers

(3) decomposers

(4) primary consumers   148 _____

149. Which type of organism is *not* shown in the following representation of a food chain?

grass → mouse → snake → hawk

(1) herbivore   (3) producer

(2) decomposer   (4) carnivore   149 _____

150. In the food chain shown below, which organism represents a primary consumer?

(1) grasshopper	(3) frog
(2) grass	(4) snake 150 _____

151. A lake contains minnows, mosquito larvae, sunfish, algae, and pike. Which of these organisms would probably be present in the largest number?

(1) minnows	(3) sunfish
(2) larvae	(4) algae 151 _____

152. An abiotic factor that affects the ability of pioneer organisms such as lichens to survive is the

(1) type of climax vegetation	(3) type of substratum
(2) species of algae	(4) species of bacteria 152 _____

153. In order to avoid predators, the clown fish hides unharmed in the stinging tentacles of the sea anemone. The clown fish attracts food to the sea anemone. This is an example of a type of relationship known as

(1) mutualism	(3) predator–prey
(2) commensalism	(4) parasitism 153 _____

154. Which world biome has the greatest number of organisms?

  (1) tundra   (3) temperate deciduous forest

  (2) tropical forest   (4) marine   154 _____

155. In a particular area, living organisms and the nonliving environment function together as

  (1) a population   (3) an ecosystem

  (2) a community   (4) a species   155 _____

Answers Explained

141. 2 The portions of food that are usable to an animal are known as nutrients. Carbohydrates, lipids, and proteins are the organic nutrients needed by all organisms. Animals cannot synthesize their own nutrients. They must eat other organisms that contain the *preformed organic nutrients.*

WRONG CHOICES EXPLAINED:
(1) *Carbon dioxide and water* are inorganic compounds. These compounds are converted to nutrients by plants only. The process is called photosynthesis.

(3) Hydrolysis is the breakdown of compounds to simpler molecules through the action of enzymes in the presence of water. Glucose is a *simple sugar.* The hydrolyzing of glucose results in the release of energy in a cell.

(4) The use of *inorganic molecules* for the production of energy occurs only in certain species of bacteria. Chemosynthesis does not occur in members of the animal kingdom.

142. 3 *Fungi* are nongreen plants. They are heterotrophs, which means that they cannot manufacture their own food. Organisms that carry out heterotrophic nutrition must take in preformed organic molecules.

WRONG CHOICES EXPLAINED:
(1), (2), (4) *Ferns, grasses,* and *mosses* are green plants or bryophytes. Green plants are *autotrophs,* which means that they manufacture their own food.

143. 2 Digestion that occurs within a cell is known as intracellular digestion. Plants do not have special digestive systems. Digestion, or the *conversion of starch to sugar,* occurs within the individual cells of a plant, including those of the root.

WRONG CHOICES EXPLAINED:
(1) *Grasshoppers* have a digestive system. Digestion is extracellular and takes place in a digestive tube outside the body cells. Chewing a blade of grass is an example of mechanical digestion taking place in the mouth.

(3) *Earthworms* also have a digestive system. Proteins are digested outside the body cells in a portion of the digestive system known as the intestine.

(4) *Fungi* demonstrate a special form of extracellular digestion. Digestive enzymes are secreted into the external environment. The nutrients from the digested food diffuse into the cells.

144. 4 *Nutrition* is the life process most closely associated with a hydra ingesting, digesting, and egesting a daphnia. Ingestion is the process by which food materials are taken into the body of an organism such as a hydra. Digestion is the process by which the complex food molecules within the daphnia are hydrolyzed to soluble end products. Egestion is the process by which the undigestible materials of the daphnia's body are expelled from the body of the hydra.

WRONG CHOICES EXPLAINED:

(1) *Transport* is the life process by which soluble foods and other materials are circulated through the body of an organism such that they reach all parts of the organism's body.

(2) *Synthesis* is a process by which complex materials are constructed from simpler chemical components. The processes described in the question represent hydrolysis, the opposite of synthesis.

(3) *Growth* is a process that involves an increase in cell number and cell size, leading to an increase in the size of the organism. The end products of digestion can be used to supply raw materials for such growth.

145. 2 *Saprophytes* are organisms that *feed on dead organic matter*. Fungi and the bacteria of decay are examples of saprophytes.

WRONG CHOICES EXPLAINED:

(1) Heterotrophs *live on or off other living organisms*. A dog flea is an example of a heterotroph also known as a parasite.

(3) Autotrophs *manufacture their own food by photosynthesis*. Algae, mosses, and grasses are examples of autotrophs.

(4) Higher plants *contain vascular bundles*. These plants, called *tracheophytes*, include ferns, conifers, and flowering plants.

146. 1 Aerobic organisms need *oxygen* for cellular respiration. Some autotrophs are photosynthetic organisms. Oxygen is released by photosynthesis. Aerobic organisms depend on the autotrophs to release oxygen into the environment.

WRONG CHOICES EXPLAINED:

(2) *Carbon dioxide* is a waste product from the cellular respiration of aerobic organisms.

(3) *Nitrogen gas* makes up 78% of the atmosphere. However, the nitrogen cannot be used in the gaseous form by aerobic organisms and most autotrophs.

(4) *Hydrogen* does *not* exist as a gas on our planet. It is combined with other elements. The hydrogen needed by organisms comes mostly from water and organic compounds.

147. 1 The area of greatest food production is in the region where the greatest rate of photosynthesis occurs. The area must be rich in minerals, water, gases, and light. The *coastal ocean waters* meet these requirements.

WRONG CHOICES EXPLAINED:
(2) There is very little precipitation in the *desert*. Water is the factor that limits plant growth.

(3) The water in the *taiga* is frozen part of the year.

(4) The *tundra* is a frozen plain. Water is frozen almost all year long on the tundra.

148. 3 Bacteria of decay are *decomposers* that release minerals from decaying plant and animal bodies and return them to the environment.

WRONG CHOICES EXPLAINED:
(1) *Producers* are autotrophs—that is, green plants that synthesize food by photosynthesis from carbon dioxide and water in the presence of sunlight.

(2) *Secondary consumers* are animals that eat other animals. For example, a frog feeds on flies. Thus, a frog is a secondary consumer.

(4) *Primary consumers* are organisms that feed on plants only. Herbivores are primary consumers.

149. 2 *Decomposers* are not shown in the food chain. They are organisms that live on dead things. Fungi and bacteria are decomposers.

WRONG CHOICES EXPLAINED:
(1) An *herbivore* is a primary consumer; that is, it eats vegetation. The mouse is the herbivore in the food chain.

(3) A *producer* is a green plant. It depends on sunlight to synthesize its own food. The grass is the producer in the food chain.

(4) A *carnivore* is an animal that eats the flesh of other animals. Both the snake and the hawk are carnivores.

150. 1 The *grasshopper* is a primary consumer because it feeds on vegetation.

WRONG CHOICES EXPLAINED:
(2) *Grass* is an autotroph—that is, a producer or self-feeder. Grass is a green plant that can make its own food.

(3) A *frog* is a secondary consumer; it eats insects that are plant eaters.

(4) A *snake* is a secondary consumer; it eats animals that are primary consumers.

151. 4 The organisms in the question make up a food chain. The number of organisms at each level of the food chain decreases as one moves down the chain. The pyramid of energy shown below represents this fact. The producers, which form the base of the pyramid, are the most numerous. *Algae* are the producers in this food chain.

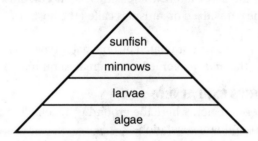

WRONG CHOICES EXPLAINED:
(1) The *minnows* are carnivores. They are the secondary consumers.

(2) The *larvae* are herbivores. They are primary consumers.

(3) The *sunfish* are tertiary consumers and occupy the top level of the pyramid. The organisms at the top of the pyramid are the least numerous.

152. 3 A *substratum* is the surface on which organisms grow. Lichens grow on rocks. Rocks are nonliving. The nonliving parts of the environment make up the abiotic factors.

WRONG CHOICES EXPLAINED:
(1) *Climax vegetation* is the type of vegetation that occupies an area in its final stage of succession. Plants make up the biotic, or living, environment.

(2) *Algae* are living organisms. They are part of the biotic (living) environment.

(4) *Bacteria* also make up the biotic environment.

153. 1 A relationship between two organisms in which both benefit from the association is known as *mutualism*. The clown fish is protected by the sea anemone. The sea anemone is sessile. The clown fish draws food to the sea anemone.

WRONG CHOICES EXPLAINED:

(2) *Commensalism* is a relationship between two organisms in which one organism is benefited by the association. The second organism is neither harmed nor benefited by the association. Barnacles attached to a whale are an example of a commensal relationship.

(3) A predator is a carnivore that hunts, kills, and eats its prey. The prey is the hunted organism. A *predator-prey* relationship is important in controlling the population of both.

(4) *Parasitism* is a relationship between two organisms in which one organism is benefited by the association whereas the second organism is harmed by the association. The parasitized organism is called the host.

154. 4 A biome is a large area dominated by one major type of vegetation and one type of climate. The *marine* biome has the greatest number of organisms.

WRONG CHOICES EXPLAINED:

(1) The *tundra* is a region where the ground is frozen all year long. Mosses and lichens are the dominant vegetation.

(2) The *tropical rain forest* is dominated by broadleaf plants. The region is always warm, and the rainfall is abundant and continuous.

(3) The *temperate deciduous forest* is made up of trees that shed their leaves once a year. The winters are cold, and the summers are warm. The rainfall is distributed throughout the year.

155. 3 *An ecosystem* is an area in which communities of living organisms interact with the nonliving environment.

WRONG CHOICES EXPLAINED:

(1) *A population* is all the organisms of a particular species living in a given area.

(2) *A community* is made up of populations of different species that interact with each other.

(4) *A species* is a group of organisms whose members are able to interbreed with each other. The offspring resulting from the matings are fertile and can reproduce.

Key Idea 7—Human Impact on the Environment

Human decisions and activities have a profound impact on the physical and living environment.

Performance Indicator	Description
7.1	The student should be able to describe the range of interrelationships of humans with the living and nonliving environment.
7.2	The student should be able to explain the impact of technological development and growth in the human population on the living and nonliving environment.
7.3	The student should be able to explain how individual choices and societal actions can contribute to improving the environment.

156. Human impact on the environment is most often more dramatic than the impact of most other living things because humans have a greater

(1) need for water

(2) need for food

(3) ability to adapt to change

(4) ability to alter the environment   156 _____

157. Which human activity would have the most direct impact on the oxygen–carbon dioxide cycle?

(1) reducing the rate of ecological succession

(2) decreasing the use of water

(3) destroying large forest areas

(4) enforcing laws that prevent the use of leaded gasoline   157 _____

158. Fertilizers used to improve lawns and gardens may interfere with the equilibrium of an ecosystem because they

(1) cause mutations in all plants

(2) cannot be absorbed by roots

(3) can be carried into local water supplies

(4) cause atmospheric pollution   158 _____

159. The tall wetland plant purple loosestrife was brought from Europe to the United States in the early 1800s as a garden plant. The plant's growth is now so widespread across the United States that it is crowding out a number of native plants. This situation is an example of

(1) the results of the use of pesticides

(2) the recycling of nutrients

(3) the flow of energy present in all ecosystems

(4) an unintended effect of adding a species to an ecosystem   159 _____

160. Choose *one* ecological problem from the list below.

Ecological Problems
Global warming
Destruction of the ozone shield
Loss of biodiversity

Discuss the ecological problem you chose. In your answer be sure to state:

■ the problem you selected and *one* human action that may have caused the problem [1]

■ *one* way in which the problem can negatively affect humans [1]

■ *one* positive action that could be taken to reduce the problem [1]

_____

_____

_____

Base your answers to questions 161 through 163 on the information below and on your knowledge of biology.

The planning board of a community held a public hearing in response to complaints by residents concerning a waste-recycling plant. The residents claimed that the waste-hauling trucks were polluting the air, land, and water and that the garbage has brought an increase in rats, mice, and pathogenic bacteria to the area. The residents insisted that the waste-recycling plant be closed permanently.

Other residents recognized the health risks but felt that the benefits of waste recycling outweighed the health issues.

161. Identify two specific health problems that could result from living near the waste-recycling plant. [2]

_____

162. Identify one specific contaminant that might be released into the environment from operation of the waste-recycling plant. [1]

_____

163. State one ecological benefit of recycling wastes. [1]

_____

164. Which organism is a near-extinct species?

    (1)   Japanese beetle   (3)   blue whale

    (2)   dodo bird   (4)   passenger pigeon   164 _____

165. Which human activity has probably contributed most to the acidification of lakes in the Adirondack region?

    (1)   passing environmental protection laws

    (2)   establishing reforestation projects in lumbered areas

    (3)   burning fossil fuels that produce air pollutants containing sulfur and nitrogen

    (4)   using pesticides for the control of insects that feed on trees   165 _____

166. Compared to a natural forest, the wheat field of a farmer lacks

    (1)   heterotrophs   (3)   autotrophs

    (2)   significant biodiversity   (4)   stored energy   166 _____

167. Which factor is not considered by ecologists when they evaluate the impact of human activities on an ecosystem?

    (1)   amount of energy released from the Sun

    (2)   quality of the atmosphere

    (3)   degree of biodiversity

    (4)   location of power plants   167 _____

168. A new type of fuel gives off excessive amounts of smoke. Before this type of fuel is widely used, an ecologist would most likely want to know

    (1)   what effect the smoke will have on the environment

    (2)   how much it will cost to produce the fuel

    (3)   how long it will take to produce the fuel

    (4)   if the fuel will be widely accepted by consumers   168 _____

169. Which of the following is the most ecologically promising method of insect control?

    (1)   interference with insect reproductive processes

    (2)   stronger insecticides designed to kill higher percentages of insects

    (3)   physical barriers to insect pests

    (4)   draining marshes and other insect habitats   169 _____

170. Which is an example of biological control of a pest species?

    (1)   DDT was used to destroy the red mite.

    (2)   Most of the predators of a deer population were destroyed by humans.

    (3)   Gypsy moth larvae (tree defoliators) are destroyed by beetle predators that were cultured and released.

    (4)   Drugs were used in the control of certain pathogenic bacteria.   170 _____

171. To ensure environmental quality for the future, each individual should

    (1)   acquire and apply knowledge of ecological principles

    (2)   continue to take part in deforestation

    (3)   use Earth's finite resources

    (4)   add and take away organisms from ecosystems   171 _____

172. Ladybugs were introduced as predators into an agricultural area of the United States to reduce the number of aphids (pests that feed on grain crops). Describe the positive and negative effects of this method of pest control. Your response must include at least:

- two advantages of this method of pest control [2]
- two possible dangers of using this method of pest control [2]

_____

_____

_____

_____

_____

173. Some people claim that certain carnivores should be destroyed because they kill beneficial animals. Explain why these carnivores should be protected. Your answer must include information concerning:

- prey population growth [1]
- extinction [1]
- the importance of carnivores in an ecosystem [1]

_____

_____

_____

_____

Answers Explained

156. 4 The fact that humans have a greater *ability to alter the environment* means that human impact on the environment is often more dramatic than that of most other living things. In addition to our ability to make physical changes in the environment, humans have the unique ability to alter the environment chemically, introducing many materials that are not found in nature and that cannot be converted to useful products by nature.

WRONG CHOICES EXPLAINED:
(1), (2) On an individual basis, humans' *need for water* and *need for food* are not significantly greater than those of other living things. However, the fact is that our large population places incredible demands on the environment to supply these basic resources. As a result, our tendency to destroy natural habitats to create additional water and agricultural resources is a significant factor affecting the natural world.

(3) On an individual basis, humans' *ability to adapt to change* is not significantly greater than that of other living things. However, as a species, we have created artificial environments to protect ourselves from harsh environmental conditions. To the extent that these artificial environments are dependent on energy and other natural resources, their construction and maintenance have resulted in significant alterations of the natural world.

157. 3 *Destroying large forest areas* is the human activity that would have the most direct impact on the oxygen-carbon cycle. Reducing the number of trees over a large area would decrease the forest's ability to absorb carbon dioxide and water and convert them to atmospheric oxygen and glucose. The millions of leaves in a forest are capable of releasing many tons of oxygen gas to the atmosphere. The massive bodies of forest trees can likewise store tons of carbon in the form of complex carbohydrates such as cellulose.

WRONG CHOICES EXPLAINED:
(1) *Reducing the rate of ecological succession* is not the human activity that would have the most direct impact on the oxygen-carbon cycle. Ecological succession is a process by which one plant-animal community is replaced over time by other plant-animal communities until a stable climax community is established. Reducing its rate would only have the effect of prolonging each successive community longer than might otherwise be expected but would not directly alter the cycling of carbon and oxygen.

(2) *Decreasing the use of water* is not the human activity that would have the most direct impact on the oxygen-carbon cycle. Water is a precious resource in many parts of the world. Reducing water use so as to conserve it would represent a positive impact of human activity on the environment but would not directly alter the cycling of carbon and oxygen.

(4) *Enforcing laws that prevent the use of leaded gasoline* is not the human activity that would have the most direct impact on the oxygen-carbon cycle. Lead is a dangerous heavy metal pollutant released when leaded gasoline is burned. Enforcing laws that limit its use would represent a positive impact of human activity on the environment, but would not directly alter the cycling of carbon and oxygen.

158. 3 Fertilizers used to improve lawns and gardens may interfere with the equilibrium of an ecosystem because they *can be carried into local water supplies*. Once dissolved fertilizers enter streams, ponds, wetlands, or lakes, they provide an abundant nutrient source for the growth of algae. As masses of algae die off in the water environment, their decomposition can rob the water of oxygen needed for the survival of fish and other water-dwelling populations, causing their elimination from the habitat. When these species disappear, other species that depend on them for food must migrate or starve. Because the changes caused by the entry of fertilizers into water environments are so significant, it can be said that ecosystem equilibrium is destroyed.

WRONG CHOICES EXPLAINED:
(1) It is not true that fertilizers *cause mutations in all plants*. Some compounds with chemical structures similar to that of fertilizers are known to stimulate rapid gene mutation in plant cells that may lead to the death of the plant. However, the class of chemical compounds known as fertilizers do not have this effect on all plants.

(2) It is not true that fertilizers *cannot be absorbed by all plants*. When dissolved in water, fertilizers can easily enter plants by being absorbed via simple diffusion into root hairs.

(4) It is not normally true that fertilizers *cause atmospheric pollution*. Most fertilizers are relatively stable chemical compounds that are solids at normal temperatures. For this reason fertilizers are not normally responsible for atmospheric pollution unless they are applied in a gaseous form (such as ammonia) or become airborne (when attached to dry soil particles).

159. 4 The situation described in the question is an example of *an unintended effect of adding a species to an ecosystem*. Although purple loosestrife has adapted well to North American habitats, its rapid growth in wetland environments has stressed or eliminated populations of cattail, pickerelweed, and other native plant species. The introduction of nonnative purple loosestrife to the North American continent has had an unintended negative effect on these native species and on the balance of nature established over many centuries.

WRONG CHOICES EXPLAINED:
(1), (2), (3) The situation described in the question is not an example of *the results of the use of pesticides, the recycling of nutrients*, or *the flow of energy present in all ecosystems*. The introduction of a nonnative plant (a living thing) is not the same as the introduction of a chemical pesticide, the recycling of nutrients, or the flow of energy (nonliving things).

160. A three-part response is required that must include the following points:

- One human activity that may have caused the ecological problem selected from the list [1]
- One way the problem may negatively affect humans [1]
- One positive action that could be taken to reduce the problem [1]

Note: No credit is awarded for discussing an ecological problem not on the list.

Acceptable responses include: [3]

- *Global warming is a worldwide ecological problem that may be caused by the release of carbon dioxide and other gases in automobile exhaust. [1] This problem may negatively affect humans if the warming conditions disrupt weather patterns and lead to droughts, floods, or other natural disasters. [1] One positive action that could be taken to help the problem would be to find an energy source for automobiles that would not release carbon dioxide into the atmosphere. [1]*

- *An ecological problem affecting humans is destruction of the ozone layer, which is caused by the use of chemicals known as CFCs as propellants in aerosol sprays. [1] This is a problem for humans because the ozone layer protects us from ultraviolet radiation from the sun; without this protection we would have an increased chance of getting skin cancer. [1] A way to help solve this problem would be to ban the use of CFCs in aerosols. [1]*

- *Loss of biodiversity is an ecological problem that negatively impacts humans. This problem is caused whenever humans destroy a natural habitat and convert it to other uses. [1] The overall health of our environment depends on the diversity of species that fill different roles in nature. When species diversity and environmental health are reduced, our health is threatened as well. [1] This problem can be reversed only if we use education to learn that protecting natural species is just as important as protecting our own. [1]*

161. Two responses are required. Acceptable responses include:

- *Asthma*
- *Respiratory infections*
- *Allergic reactions*
- *Cancer*
- *Bacterial infections*
- *Viral infections*
- *Disease linked to a pathogen*
- *Poisoning linked to toxic contamination of groundwater*

162. One response is required. Acceptable responses include:

- *Particles in the air*
- *Presence of viruses or bacteria on trucks*
- *Chemicals in air or water*
- *Carcinogens*
- *Mold and fungus spores*

163. One response is required. Acceptable responses include:

- *Conservation of natural resources*
- *Protection of finite resources*
- *Energy conservation*
- *Reduction in pollution*
- *Landfills last longer*
- *Preservation of open space resources*

164. 3 The blue whale is near extinction because of uncontrolled hunting by humans.

WRONG CHOICES EXPLAINED:
(1) The Japanese beetle is a plant pest that was accidentally introduced into the United States. Its population is kept in check by the praying mantis, its predator.

(2) The dodo bird became extinct because of hunting by humans.

(4) The passenger pigeon became extinct in the 1900s due to hunting by humans.

165. 3 *Burning fossil fuels that produce air pollutants containing sulfur and nitrogen* is the human activity that has probably contributed the most to the acidification of lakes in the Adirondack region. These pollutants combine with water in the atmosphere to form sulfuric and nitric acids. These acids then enter lakes in rainfall and runoff, adding to the acidic quality of the lake water and killing many susceptible species.

WRONG CHOICES EXPLAINED:
(1) *Passing environmental protection laws* is not an activity that results in the acidification of lakes. In fact, it is a positive human activity that can help to limit the production and release of such gases into the atmosphere.

(2) *Establishing reforestation projects in lumbered areas* is not an activity that results in the acidification of lakes. In fact, it is a positive human activity that can help to replace trees lost because of the acidification of soils by acid rain.

(4) *Using pesticides for the control of insects that feed on trees* is not an activity that results in the acidification of lakes. It is a negative human activity carried out to protect commercial crops from destruction and does not normally result in the production of sulfur and nitrogen gases.

166. 2 *Significant biodiversity* is the factor lacking in a wheat field as compared to a natural forest. *Biodiversity* is a term relating to the variety of life forms in an environment. Natural environments, including forests, are typically made up of thousands of species that interact to provide a balanced, ecologically responsive community. By contrast, farm fields are often limited to a small number of different species, and predominantly a single species. Communities lacking in biodiversity are unstable and prone to collapse when environmental conditions change.

WRONG CHOICES EXPLAINED:
(1) *Heterotrophs* are not lacking in a farm field compared to a forest. Heterotrophs are found within a wheat field, although their number and variety are normally limited to those that use wheat or its by-products as food.

(3) *Autotrophs* are not lacking in a farm field compared to a forest. Wheat is a type of autotroph, as are the various weed species that may be interspersed among the wheat plants in the field.

(4) *Stored energy* is not lacking in a farm field compared to a forest. As the wheat grows in the field, it absorbs the Sun's energy and stores it as the chemical bond energy of carbohydrates and other organic compounds.

167. 1 The *amount of energy released from the Sun* is not normally considered by an ecologist when evaluating the impact of human activities on an ecosystem. The amount of solar energy emitted by the Sun is generally constant and out of our direct control. Because it is not a variable that can be directly affected by human activities, it is usually not a consideration in decisions of this kind.

WRONG CHOICES EXPLAINED:
(2) The *quality of the atmosphere* is often a factor considered by ecologists in evaluating the impact of human activities on an ecosystem. Many human activities introduce chemical contaminants into the atmosphere. These chemicals may have a negative impact on the health and survival of humans and other species.

(3) The *degree of biodiversity* is often a factor considered by ecologists in evaluating the impact of human activities on an ecosystem. Human activities often put pressure on natural species, eliminating those unable to migrate or adapt. As biodiversity in an area declines, so does environmental stability. This situation threatens the health and survival of humans and other species.

(4) The *location of power plants* is often a factor considered by ecologists in evaluating the impact of human activities on an ecosystem. Fossil fuel plants can pollute the atmosphere and consume valuable petroleum products. Nuclear plants can release radiation and heat into the environment. Hydroelectric, solar, wind, and geothermal plants can destroy natural habitats because of space considerations. Each of these consequences can affect the health and survival of humans and other species.

168. 1 An ecologist would want to know *what effect the smoke will have on the environment* before a new type of fuel is widely used. By understanding this effect, the ecologist can make more informed judgments about whether the smoke will harm the environment and human health.

WRONG CHOICES EXPLAINED:
(2), (3), (4) An ecologist is less likely to want to know *how much it will cost to produce the fuel, how long it will take to produce the fuel,* and *if the fuel will be widely accepted by consumers*. Although these are important questions for the manufacturer, they do not provide critical information for the ecologist, whose main concern is the protection of environmental quality for humans and other organisms.

169. 1 Interference with insect reproductive processes is known as biological control. It is the most promising method of controlling insects because it is the least ecologically damaging.

WRONG CHOICES EXPLAINED:
(2) The use of insecticides is a chemical control of insects. Insecticides kill both harmful and helpful insects. The chemicals accumulate in the bodies of birds, fish, and mammals and interfere with their normal life activities.

(3) It is impossible to set up physical barriers for insects because they are motile and are also carried from place to place by animals and humans.

(4) Draining marshes and other insect habitats has helped to control many insects such as mosquitoes. However, this method interferes with the life cycles of useful organisms living in the area.

170. 3 Insecticides are chemical pest controls. Biological controls are other insect species that feed on or in some way prey on an insect pest species. The example given here is control of the gypsy moth larvae by a certain species of beetle.

WRONG CHOICES EXPLAINED:
(1) DDT is an insecticide and represents chemical control. DDT is no longer used because it destroyed the insect food of birds and other wildlife.

(2) Humans upset the balance of nature (the balance of natural communities) by killing off deer predators. The deer population then increased so dramatically that deer starved to death because there was not enough food to support them.

(4) The use of drugs to cure disease is an example of chemical control of pathogens.

171. 1 Each individual should *acquire and apply knowledge of ecological principles* in order to ensure environmental quality for the future. By understanding how environmental principles operate, we can make more informed judgments about activities that may harm the environment and human health.

WRONG CHOICES EXPLAINED:
(2) If each individual were to *continue to take part in deforestation*, environmental quality would be degraded. Because forests are a natural part of the environment, eliminating them disturbs the balance of nature and can have significant negative consequences for environmental quality.

(3) If each individual were to *use Earth's finite resources*, environmental quality would be threatened. As these resources are used up, fewer remain for future generations. In addition, processing these resources consumes energy, produces pollutants, and adds to the solid waste problem.

(4) If each individual were to *add and take away organisms from ecosystems*, environmental quality would be diminished. Each natural community has established itself based on the particular niches filled by each type of organism. Adding to or taking away from this community upsets the balance of nature and would likely cause negative consequences.

172. Write one or more paragraphs describing positive and negative effects of this method of pest control. Include the following points:

- Two advantages of this method of pest control [2]
  - Chemicals are not added to the environment.
  - Biological controls are more specific than chemical controls.
  - Ladybugs are less likely to kill beneficial organisms.
  - Desirable garden plants are protected from aphid attacks.
  - Birds and other unintended victims of pesticide use are spared.
  - Human health is protected against the toxic effects of pesticides.
- Two possible dangers of using this method of pest control [2]
  - The control insects may eat the food of other organisms.
  - The population of natural predators of the aphids may be eliminated or greatly reduced.
  - The control organism may become overpopulated.
  - The control organisms may themselves become pests.

Sample paragraph: The method of pest control described is known as "biological control." This method of insect control has some distinct advantages over chemical controls: First, biological controls don't release toxic chemicals into the air and water, a fact that helps to protect wildlife and humans from being unintended victims of chemical pesticides. Second, biological controls are usually specific, which means that beneficial insects such as ladybugs and preying mantises aren't harmed. [2] There are also some things we should be careful of in the use of biological controls: First, we should know a lot about the control organism to be sure that it doesn't crowd out our native beneficial organisms. Second, we should remember that the control organism could become a pest, too, if it gets too numerous in the environment. [2]

173. Write one or more paragraphs explaining why carnivores should be protected. Include the following points:

- Information concerning prey population growth [1]
  - If predators are destroyed, the prey population will increase.
  - If unchecked by predation or disease, a natural population will tend to increase in number geometrically.
- Information concerning extinction [1]
  - If too many carnivores of a particular species are killed, the species may become extinct.
  - Extinction is a definite possibility when any species has too few members alive to carry out effective breeding.
  - Complete elimination of any species from its natural range can destabilize the ecosystem.
- Information concerning the importance of carnivores in an environment [1]
  - By feeding on herbivores, carnivores help keep certain species of plants from being eliminated because of overgrazing in a particular area.
  - Without predators to limit its number, a prey population could exceed the capacity of its range, resulting in widespread starvation and death of the prey population.
  - Carnivorous animals are part of the natural scheme that promotes ecological equilibrium.

Sample paragraph: Carnivores are important in an ecosystem because by reducing the number of prey organisms, the food organisms of the prey are kept from being eliminated from the environment. [1] If the predators were destroyed, the prey population would increase [1], perhaps to the point of consuming so many of the plants that the prey feed on that these plants would become extinct. [1]

Standard/Key Idea	Question Numbers	Number of Correct Responses	Number of Incorrect Responses
1.1 Purpose of Scientific Inquiry	1–8		
1.2 Methods of Scientific Inquiry	9–28		
1.3 Analysis in Scientific Inquiry	29–47		
4.1 Application of Scientific Principles	48–75		
4.2 Genetic Continuity	76–90		
4.3 Organic Evolution	91–109		
4.4 Reproductive Continuity	110–129		
4.5 Dynamic Equilibrium and Homeostasis	130–140		
4.6 Interdependence of Living Things	141–155		
4.7 Human Impact on the Environment	156–173		

Glossary

Prominent Scientists

Crick, Francis A 20th-century British scientist who, with James Watson, developed the first workable model of DNA structure and function.

Darwin, Charles A 19th-century British naturalist whose theory of organic evolution by natural selection forms the basis for the modern scientific theory of evolution.

Fox, Sidney A 20th-century American scientist whose experiments showed that Stanley Miller's simple chemical precursors could be joined to form more complex biochemicals.

Hardy, G. H. A 20th-century British mathematician who, with W. Weinberg, developed the Hardy-Weinberg principle of gene frequencies.

Lamarck, Jean An 18th-century French scientist who devised an early theory of organic evolution based on the concept of "use and disuse."

Linnaeus, Carl An 18th-century Dutch scientist who developed the first scientific system of classification, based on similarity of structure.

Mendel, Gregor A 19th-century Austrian monk and teacher who was the first to describe many of the fundamental concepts of genetic inheritance through his work with garden peas.

Miller, Stanley A 20th-century American scientist whose experiments showed that the simple chemical precursors of life could be produced in the laboratory.

Morgan, Thomas Hunt A 20th-century American geneticist whose pioneering work with Drosophila led to the discovery of several genetic principles, including sex linkage.

Watson, James A 20th-century American scientist who, with Francis Crick, developed the first workable model of DNA structure and function.

Weinberg, W. A 20th-century German physician who, with G. H. Hardy, developed the Hardy-Weinberg principle of gene frequencies.

Weismann, August A 19th-century German biologist who tested Lamarck's theory of use and disuse and found it to be unsupportable by scientific methods.

Biological Terms

Abiotic factor Any of several nonliving, physical conditions that affect the survival of an organism in its environment.

Absorption The process by which water and dissolved solids, liquids, and gases are taken in by the cell through the cell membrane.

Accessory organ In human beings, any organ that has a digestive function but is not part of the food tube. (See **liver; gallbladder; pancreas.**)

Acid A chemical that releases hydrogen ion (H+) in solution with water.

Acid precipitation A phenomenon in which there is thought to be an interaction between atmospheric moisture and the oxides of sulfur and nitrogen that results in rainfall with low pH values.

Active immunity The immunity that develops when the body's immune system is stimulated by a disease organism or a vaccination.

Active site The specific area of an enzyme molecule that links to the substrate molecule and catalyzes its metabolism.

Active transport A process by which materials are absorbed or released by cells against the concentration gradient (from low to high concentration) with the expenditure of cell energy.

Adaptation Any structural, biochemical, or behavioral characteristic of an organism that helps it to survive potentially harsh environmental conditions.

Addition A type of chromosome mutation in which a section of a chromosome is transferred to a homologous chromosome.

Adenine A nitrogenous base found in DNA and RNA molecules.

Adenosine triphosphate (ATP) An organic compound that stores respiratory energy in the form of chemical-bond energy for transport from one part of the cell to another.

Adrenal cortex A portion of the adrenal gland that secretes steroid hormones which regulate various aspects of blood composition.

Adrenal gland An endocrine gland that produces several hormones, including adrenaline. (See **adrenal cortex; adrenal medulla.**)

Adrenal medulla A portion of the adrenal gland that secretes the hormone adrenaline, which regulates various aspects of the body's metabolic rate.

Adrenaline A hormone of the adrenal medulla that regulates general metabolic rate, the rates of heartbeat and breathing, and the conversion of glycogen to glucose.

Aerobic phase of respiration The reactions of aerobic respiration in which two pyruvic acid molecules are converted to six molecules of water and six molecules of carbon dioxide.

Aerobic respiration A type of respiration in which energy is released from organic molecules with the aid of oxygen.

Aging A stage of postnatal development that involves differentiation, maturation, and eventual deterioration of the body's tissues.

Air pollution The addition, due to technological oversight, of some unwanted factor (e.g., chemical oxides, hydrocarbons, particulates) to our air resources.

Albinism A condition, controlled by a single mutant gene, in which the skin lacks the ability to produce skin pigments.

Alcoholic fermentation A type of anaerobic respiration in which glucose is converted to ethyl alcohol and carbon dioxide.

Allantois A membrane that serves as a reservoir for wastes and as a respiratory surface for the embryos of many animal species.

Allele One of a pair of genes that exist at the same location on a pair of homologous chromosomes and exert parallel control over the same genetic trait.

Allergy A reaction of the body's immune system to the chemical composition of various substances.

Alveolus One of many "air sacs" within the lung that function to absorb atmospheric gases and pass them on to the bloodstream.

Amino acid An organic compound that is the component unit of proteins.

Amino group A chemical group having the formula $-NH_2$ that is found as a part of all amino acid molecules.

Ammonia A type of nitrogenous waste with high solubility and high toxicity.

Amniocentesis A technique for the detection of genetic disorders in human beings in which a small amount of amniotic fluid is removed and the chromosome content of its cells analyzed. (See **karyotyping**.)

Amnion A membrane that surrounds the embryo in many animal species and contains a fluid to protect the developing embryo from mechanical shock.

Amniotic fluid The fluid within the amnion membrane that bathes the developing embryo.

Amylase An enzyme specific for the hydrolysis of starch.

Anaerobic phase of respiration The reactions of aerobic respiration in which glucose is converted to two pyruvic acid molecules.

Anaerobic respiration A type of respiration in which energy is released from organic molecules without the aid of oxygen.

Anal pore The egestive organ of the paramecium.

Anemia A disorder of the human transport system in which the ability of the blood to carry oxygen is impaired, usually because of reduced numbers of red blood cells.

Angina pectoris A disorder of the human transport system in which chest pain signals potential damage to the heart muscle due to narrowing of the opening of the coronary artery.

Animal One of the five biological kingdoms; it includes multicellular organisms whose cells are not bounded by cell walls and which are incapable of photosynthesis (e.g., human being).

Annelida A phylum of the Animal Kingdom whose members (annelids) include the segmented worms (e.g., earthworm).

Antenna A receptor organ found in many arthropods (e.g., grasshopper), which is specialized for detecting chemical stimuli.

Anther The portion of the stamen that produces pollen.

Antibody A chemical substance, produced in response to the presence of a specific antigen, that neutralizes that antigen in the immune response.

Antigen A chemical substance, usually a protein, that is recognized by the immune system as a foreign "invader" and is neutralized by a specific antibody.

Anus The organ of egestion of the digestive tract.

Aorta The principal artery carrying blood from the heart to the body tissues.

Aortic arches A specialized part of the earthworm's transport system that serves as a pumping mechanism for the blood fluid.

Apical meristem A plant growth region located at the tip of the root or tip of the stem.

Appendicitis A disorder of the human digestive tract in which the appendix becomes inflamed as a result of bacterial infection.

Aquatic biome An ecological biome composed of many different water environments.

Artery A thick-walled blood vessel that carries blood away from the heart under pressure.

Arthritis A disorder of the human locomotor system in which skeletal joints become inflamed, swollen, and painful.

Arthropoda A phylum of the Animal Kingdom whose members (arthropods) have bodies with chitinous exoskeletons and jointed appendages (e.g., grasshopper).

Artificial selection A technique of plant/animal breeding in which individual organisms displaying desirable characteristics are chosen for breeding purposes.

Asexual reproduction A type of reproduction in which new organisms are formed from a single parent organism.

Asthma A disorder of the human respiratory system in which the respiratory tube becomes constricted by swelling brought on by some irritant.

Atrium In human beings, one of the two thin-walled upper chambers of the heart that receive blood.

Autonomic nervous system A subdivision of the peripheral nervous system consisting of nerves associated with automatic functions (e.g., heartbeat, breathing).

Autosome One of several chromosomes present in the cell that carry genes controlling "body" traits not associated with primary and secondary sex characteristics.

Autotroph An organism capable of carrying on autotrophic nutrition. Self-feeder.

Autotrophic nutrition A type of nutrition in which organisms manufacture their own organic foods from inorganic raw materials.

Auxin A biochemical substance, plant hormone, produced by plants that regulates growth patterns.

Axon An elongated portion of a neuron that conducts nerve impulses, usually away from the cell body of the neuron.

Base A chemical that releases hydroxyl ion (OH$^-$) in solution with water.

Bicarbonate ion The chemical formed in the blood plasma when carbon dioxide is absorbed from body tissues.

Bile In human beings, a secretion of the liver that is stored in the gallbladder and that emulsifies fats.

Binary fission A type of cell division in which mitosis is followed by equal cytoplasmic division.

Binomial nomenclature A system of naming, used in biological classification, that consists of the genus and species names (e.g., *Homo sapiens*).

Biocide use The use of pesticides that eliminate one undesirable organism but that have, due to technological oversight, unanticipated effects on beneficial species as well.

Biological controls The use of natural enemies of various agricultural pests for pest control, thereby eliminating the need for biocide use—a positive aspect of human involvement with the environment.

Biomass The total mass of living material present at the various trophic levels in a food chain.

Biome A major geographical grouping of similar ecosystems, usually named for the climax flora in the region (e.g., Northeast Deciduous Forest).

Biosphere The portion of the earth in which living things exist, including all land and water environments.

Biotic factor Any of several conditions associated with life and living things that affect the survival of living things in the environment.

Birth In placental mammals, a stage of embryonic development in which the baby passes through the vaginal canal to the outside of the mother's body.

Blastula In certain animals, a stage of embryonic development in which the embryo resembles a hollow ball of undifferentiated cells.

Blood The complex fluid tissue that functions to transport nutrients and respiratory gases to all parts of the body.

Blood typing An application of the study of immunity in which the blood of a person is characterized by its antigen composition.

Bone A tissue that provides mechanical support and protection for bodily organs, and levers for the body's locomotive activities.

Bowman's capsule A cup-shaped portion of the nephron responsible for the filtration of soluble blood components.

Brain An organ of the central nervous system that is responsible for regulating conscious and much unconscious activity in the body.

Breathing A mechanical process by which air is forced into the lungs by means of muscular contraction of the diaphragm and rib muscles.

Bronchiole One of several subdivisions of the bronchi that penetrate the lung interior and terminate in alveoli.

Bronchitis A disorder of the human respiratory system in which the bronchi become inflamed.

Bronchus One of the two major subdivisions of the breathing tube; the bronchi are ringed with cartilage and conduct air from the trachea to the lung interior.

Bryophyta A phylum of the Plant Kingdom that consists of organisms lacking vascular tissues (e.g., moss).

Budding A type of asexual reproduction in which mitosis is followed by unequal cytoplasmic division.

Bulb A type of vegetative propagation in which a plant bulb produces new bulbs that may be established as independent organisms with identical characteristics.

Cambium The lateral meristem tissue in woody plants responsible for annual growth in stem diameter.

Cancer Any of a number of conditions characterized by rapid, abnormal, and uncontrolled division of affected cells.

Capillary A very small, thin-walled blood vessel that connects an artery to a vein and through which all absorption into the blood fluid occurs.

Carbohydrate An organic compound composed of carbon, hydrogen, and oxygen in a 1:2:1 ratio (e.g., $C_6H_{12}O_6$).

Carbon-14 A radioactive isotope of carbon used to trace the movement of carbon in various biochemical reactions, and also used in the "carbon dating" of fossils.

Carbon-fixation reactions A set of biochemical reactions in photosynthesis in which hydrogen atoms are combined with carbon and oxygen atoms to form PGAL and glucose.

Carbon-hydrogen-oxygen cycle A process by which these three elements are made available for use by other organisms through the chemical reactions of respiration and photosynthesis.

Carboxyl group A chemical group having the formula —COOH and found as part of all amino acid and fatty acid molecules.

Cardiac muscle A type of muscle tissue in the heart and arteries that is associated with the rhythmic nature of the pulse and heartbeat.

Cardiovascular disease In human beings, any disease of the circulatory organs.

Carnivore A heterotrophic organism that consumes animal tissue as its primary source of nutrition. (See **secondary consumer**.)

Carrier An individual who, though not expressing a particular recessive trait, carries this gene as part of his/her heterozygous genotype.

Carrier protein A specialized molecule embedded in the cell membrane that aids the movement of materials across the membrane.

Cartilage A flexible connective tissue found in many flexible parts of the body (e.g., knee); common in the embryonic stages of development.

Catalyst Any substance that speeds up or slows down the rate of a chemical reaction. (See **enzyme.**)

Cell plate A structure that forms during cytoplasmic division in plant cells and serves to separate the cytoplasm into two roughly equal parts.

Cell theory A scientific theory that states, "All cells arise from previously existing cells" and "Cells are the unit of structure and function of living things."

Cell wall A cell organelle that surrounds and gives structural support to plant cells; cell walls are composed of cellulose.

Central nervous system The portion of the vertebrate nervous system that consists of the brain and the spinal cord.

Centriole A cell organelle found in animal cells that functions in the process of cell division.

Centromere The area of attachment of two chromatids in a double-stranded chromosome.

Cerebellum The portion of the human brain responsible for the coordination of muscular activity.

Cerebral hemorrhage A disorder of the human regulatory system in which a broken blood vessel in the brain may result in severe dysfunction or death.

Cerebral palsy A disorder of the human regulatory system in which the motor and speech centers of the brain are impaired.

Cerebrum The portion of the human brain responsible for thought, reasoning, sense interpretation, learning, and other conscious activities.

Cervix A structure that bounds the lower end of the uterus and through which sperm must pass in order to fertilize the egg.

Chemical digestion The process by which nutrient molecules are converted by chemical means into a form usable by the cells.

Chemosynthesis A type of autotrophic nutrition in which certain bacteria use the energy of chemical oxidation to convert inorganic raw materials to organic food molecules.

Chitin A polysaccharide substance that forms the exoskeleton of the grasshopper and other arthropods.

Chlorophyll A green pigment in plant cells that absorbs sunlight and makes possible certain aspects of the photosynthetic process.

Chloroplast A cell organelle found in plant cells that contains chlorophyll and functions in photosynthesis.

Chordata A phylum of the Animal Kingdom whose members (chordates) have internal skeletons made of cartilage and/or bone (e.g., human being).

Chorion A membrane that surrounds all other embryonic membranes in many animal species, protecting them from mechanical damage.

Chromatid One strand of a double-stranded chromosome.

Chromosome mutation An alteration in the structure of a chromosome involving many genes. (See **nondisjunction; translocation; addition; deletion**.)

Cilia Small, hairlike structures in paramecia and other unicellular organisms that aid in nutrition and locomotion.

Classification A technique by which scientists sort, group, and name organisms for easier study.

Cleavage A series of rapid mitotic divisions that increase cell number in a developing embryo without corresponding increase in cell size.

Climax community A stable, self-perpetuating community that results from an ecological succession.

Cloning A technique of genetic investigation in which undifferentiated cells of an organism are used to produce new organisms with the same set of traits as the original cells.

Closed transport system A type of circulatory system in which the transport fluid is always enclosed within blood vessels (e.g., earthworm, human).

Clot A structure that forms as a result of enzyme-controlled reactions following the rupturing of a blood vessel and serves as a plug to prevent blood loss.

Codominance A type of intermediate inheritance that results from the simultaneous expression of two dominant alleles with contrasting effects.

Codon See **triplet codon**.

Coelenterata A phylum of the Animal Kingdom whose members (coelenterates) have bodies that resemble a sack (e.g., hydra, jellyfish).

Coenzyme A chemical substance or chemical subunit that functions to aid the action of a particular enzyme. (See **vitamin.**)

Cohesion A force binding water molecules together that aids in the upward conduction of materials in the xylem.

Commensalism A type of symbiosis in which one organism in the relationship benefits and the other is neither helped nor harmed.

Common ancestry A concept central to the science of evolution that postulates that all organisms share a common ancestry whose closeness varies with the degree of shared similarity.

Community A level of biological organization that includes all of the species populations inhabiting a particular geographic area.

Comparative anatomy The study of similarities in the anatomical structures of organisms, and their use as an indicator of common ancestry and as evidence of organic evolution.

Comparative biochemistry The study of similarities in the biochemical makeups of organisms, and their use as an indicator of common ancestry and as evidence of organic evolution.

Comparative cytology The study of similarities in the cell structures of organisms, and their use as an indicator of common ancestry and as evidence of organic evolution.

Comparative embryology The study of similarities in the patterns of embryological development of organisms, and their use as an indicator of common ancestry and as evidence of organic evolution.

Competition A condition that arises when different species in the same habitat attempt to use the same limited resources.

Complete protein A protein that contains all eight essential amino acids.

Compound A substance composed of two or more different kinds of atoms (e.g., water: H_2O).

Compound light microscope A tool of biological study capable of producing a magnified image of a biological specimen by using a focused beam of light.

Conditioned behavior A type of response that is learned, but that becomes automatic with repetition.

Conservation of resources The development and application of practices to protect valuable and irreplaceable soil and mineral resources—a positive aspect of human involvement with the environment.

Constipation A disorder of the human digestive tract in which fecal matter solidifies and becomes difficult to egest.

Consumer Any heterotrophic animal organism (e.g., human being).

Coronary artery An artery that branches off the aorta to feed the heart muscle.

Coronary thrombosis A disorder of the human transport system in which the heart muscle becomes damaged as a result of blockage of the coronary artery.

Corpus luteum A structure resulting from the hormone-controlled transformation of the ovarian follicle that produces the hormone progesterone.

Corpus luteum stage A stage of the menstrual cycle in which the cells of the follicle are transformed into the corpus luteum under the influence of the hormone LH.

Cotyledon A portion of the plant embryo that serves as a source of nutrition for the young plant before photosynthesis begins.

Cover-cropping A proper agricultural practice in which a temporary planting (cover crop) is used to limit soil erosion between seasonal plantings of main crops.

Crop A portion of the digestive tract of certain animals that stores food temporarily before digestion.

Cross-pollination A type of pollination in which pollen from one flower pollinates flowers of a different plant of the same species.

Crossing-over A pattern of inheritance in which linked genes may be separated during synapsis in the first meiotic division, when sections of homologous chromosomes may be exchanged.

Cuticle A waxy coating that covers the upper epidermis of most leaves and acts to help the leaf retain water.

Cutting A technique of plant propagation in which vegetative parts of the parent plant are cut and rooted to establish new plant organisms with identical characteristics.

Cyclosis The circulation of the cell fluid (cytoplasm) within the cell interior.

Cyton The "cell body" of the neuron, which generates the nerve impulse.

Cytoplasm The watery fluid that provides a medium for the suspension of organelles within the cell.

Cytoplasmic division The separation of daughter nuclei into two new daughter cells.

Cytosine A nitrogenous base found in both DNA and RNA molecules.

Daughter cell A cell that results from mitotic cell division.

Daughter nucleus One of two nuclei that form as a result of mitosis.

Deamination A process by which amino acids are broken down into their component parts for conversion into urea.

Death The irreversible cessation of bodily functions and cellular activities.

Deciduous A term relating to broadleaf trees, which shed their leaves in the fall.

Decomposer Any saprophytic organism that derives its energy from the decay of plant and animal tissues (e.g., bacteria of decay, fungus); the final stage of a food chain.

Decomposition bacteria In the nitrogen cycle, bacteria that break down plant and animal protein and produce ammonia as a by-product.

Dehydration synthesis A chemical process in which two organic molecules may be joined after removing the atoms needed to form a molecule of water as a by-product.

Deletion A type of chromosome mutation in which a section of a chromosome is separated and lost.

Dendrite A cytoplasmic extension of a neuron that serves to detect an environmental stimulus and carry an impulse to the cell body of the neuron.

Denitrifying bacteria In the nitrogen cycle, bacteria that convert excess nitrate salts into gaseous nitrogen.

Deoxygenated blood Blood that has released its transported oxygen to the body tissues.

Deoxyribonucleic acid (DNA) A nucleic acid molecule known to be the chemically active agent of the gene; the fundamental hereditary material of living organisms.

Deoxyribose A five-carbon sugar that is a component part of the nucleotide unit in DNA only.

Desert A terrestrial biome characterized by sparse rainfall, extreme temperature variation, and a climax flora that includes cactus.

Diabetes A disorder of the human regulatory system in which insufficient insulin production leads to elevated blood sugar concentrations.

Diarrhea A disorder of the human digestive tract in which the large intestine fails to absorb water from the waste matter, resulting in watery feces.

Diastole The lower pressure registered during blood pressure testing. (See **systole**.)

Differentiation The process by which embryonic cells become specialized to perform the various tasks of particular tissues throughout the body.

Diffusion A form of passive transport by which soluble substances are absorbed or released by cells.

Digestion The process by which complex foods are broken down by mechanical or chemical means for use by the body.

Dipeptide A chemical unit composed of two amino acid units linked by a peptide bond.

Diploid chromosome number The number of chromosomes found characteristically in the cells (except gametes) of sexually reproducing species.

Disaccharidase Any disaccharide-hydrolyzing enzyme.

Disaccharide A type of carbohydrate known also as a "double sugar"; all disaccharides have the molecular formula $C_{12}H_{22}O_{11}$.

Disjunction The separation of homologous chromosome pairs at the end of the first meiotic division.

Disposal problems Problems, due to technological oversight, that result when commercial and technological activities produce solid and/or chemical wastes that must be disposed of.

Dissecting microscope A tool of biological study that magnifies the image of a biological specimen up to 20 times normal size for purposes of gross dissection.

Dominance A pattern of genetic inheritance in which the effects of a dominant allele mask those of a recessive allele.

Dominant allele (gene) An allele (gene) whose effect masks that of its recessive allele.

Double-stranded chromosome The two-stranded structure that results from chromosomal replication.

Down's syndrome In human beings, a condition, characterized by mental and physical retardation, that may be caused by the nondisjunction of chromosome number 21.

Drosophila The common fruit fly, an organism that has served as an object of genetic research in the development of the gene-chromosome theory.

Ductless gland See **endocrine gland**.

Ecology The science that studies the interactions of living things with each other and with the nonliving environment.

Ecosystem The basic unit of study in ecology, including the plant and animal community in interaction with the nonliving environment.

Ectoderm An embryonic tissue that differentiates into skin and nerve tissue in the adult animal.

Effector An organ specialized to produce a response to an environmental stimulus; effectors may be muscles or glands.

Egestion The process by which undigested food materials are eliminated from the body.

Electron microscope A tool of biological study that uses a focused beam of electrons to produce an image of a biological specimen magnified up to 25,000 times normal size.

Element The simplest form of matter; an element is a substance (e.g., nitrogen) made up of a single type of atom.

Embryo An organism in the early stages of development following fertilization.

Embryonic development A series of complex processes by which animal and plant embryos develop into adult organisms.

Emphysema A disorder of the human respiratory system in which lung tissue deteriorates, leaving the lung with diminished capacity and efficiency.

Emulsification A process by which fat globules are surrounded by bile to form fat droplets.

Endocrine ("ductless") gland A gland (e.g., thyroid, pituitary) specialized for the production of hormones and their secretion directly into the bloodstream; such glands lack ducts.

Endoderm An embryonic tissue that differentiates into the digestive and respiratory tract lining in the adult animal.

Endoplasmic reticulum (ER) A cell organelle known to function in the transport of cell products from place to place within the cell.

Environmental laws Federal, state, and local legislation enacted in an attempt to protect environmental resources—a positive aspect of human involvement with the environment.

Enzymatic hydrolysis An enzyme-controlled reaction by which complex food molecules are broken down chemically into simpler subunits.

Enzyme An organic catalyst that controls the rate of metabolism of a single type of substrate; enzymes are protein in nature.

Enzyme-substrate complex A physical association between an enzyme molecule and its substrate within which the substrate is metabolized.

Epicotyl A portion of the plant embryo that specializes to become the upper stem, leaves, and flowers of the adult plant.

Epidermis The outermost cell layer in a plant or an animal.

Epiglottis In a human being, a flap of tissue that covers the upper end of the trachea during swallowing and prevents inhalation of food.

Esophagus A structure in the upper portion of the digestive tract that conducts the food from the pharynx to the midgut.

Essential amino acid An amino acid that cannot be synthesized by the human body, but must be obtained by means of the diet.

Estrogen A hormone, secreted by the ovary, that regulates the production of female secondary sex characteristics.

Evolution Any process of gradual change through time.

Excretion The life function by which living things eliminate metabolic wastes from their cells.

Exoskeleton A chitinous material that covers the outside of the bodies of most arthropods and provides protection for internal organs and anchorage for muscles.

Exploitation of organisms Systematic removal of animals and plants with commercial value from their environments, for sale—a negative aspect of human involvement with the environment.

Extensor A skeletal muscle that extends (opens) a joint.

External development Embryonic development that occurs outside the body of the female parent (e.g., birds).

External fertilization Fertilization that occurs outside the body of the female parent (e.g., fish).

Extracellular digestion Digestion that occurs outside the cell.

Fallopian tube See **oviduct.**

Fatty acid An organic molecule that is a component of certain lipids.

Fauna The animal species comprising an ecological community.

Feces The semisolid material that results from the solidification of undigested foods in the large intestine.

Fertilization The fusion of gametic nuclei in the process of sexual reproduction.

Filament The portion of the stamen that supports the anther.

Flagella Microscopic, whiplike structures found on certain cells that aid in locomotion and circulation.

Flexor A skeletal muscle that flexes (closes) a joint.

Flora The plant species comprising an ecological community.

Flower The portion of a flowering plant that is specialized for sexual reproduction.

Fluid-mosaic model A model of the structure of the cell membrane in which large protein molecules are thought to be embedded in a bilipid layer.

Follicle One of many areas within the ovary that serve as sites for the periodic maturation of ova.

Follicle stage The stage of the menstrual cycle in which an ovum reaches its final maturity under the influence of the hormone FSH.

Follicle-stimulating hormone (FSH) A pituitary hormone that regulates the maturation of, and the secretion of estrogen by, the ovarian follicle.

Food chain A series of nutritional relationships in which food energy is passed from producer to herbivore to carnivore to decomposer; a segment of a food web.

Food web A construct showing a series of interrelated food chains and illustrating the complex nutritional interrelationships that exist in an ecosystem.

Fossil The preserved direct or indirect remains of an organism that lived in the past, as found in the geologic record.

Fraternal twins In human beings, twin offspring that result from the simultaneous fertilization of two ova by two sperm; such twins are not genetically identical.

Freshwater biome An aquatic biome made up of many separate freshwater systems that vary in size and stability and may be closely associated with terrestrial biomes.

Fruit Any plant structure that contains seeds; a mechanism of seed dispersal.

Fungi One of the five biological kingdoms; it includes organisms unable to manufacture their own organic foods (e.g., mushroom).

Gallbladder An accessory organ that stores bile.

Gallstones A disorder of the human digestive tract in which deposits of hardened cholesterol lodge in the gallbladder.

Gamete A specialized reproductive cell produced by organisms of sexually reproducing species. (See **sperm; ovum; pollen; ovule**.)

Gametogenesis The process of cell division by which gametes are produced. (See **meiosis; spermatogenesis; oogenesis**.)

Ganglion An area of bunched nerve cells that acts as a switching point for nerve impulses traveling from receptors and to effectors.

Garden pea The research organism used by Mendel in his early scientific work in genetic inheritance.

Gastric cecum A gland in the grasshopper that secretes digestive enzymes.

Gastrula A stage of embryonic development in animals in which the embryo assumes a tube-within-a-tube structure and distinct embryonic tissues (ectoderm, mesoderm, endoderm) begin to differentiate.

Gastrulation The process by which a blastula becomes progressively more indented, forming a gastrula.

Gene A unit of heredity; a discrete portion of a chromosome thought to be responsible for the production of a single type of polypeptide; the "factor" responsible for the inheritance of a genetic trait.

Gene frequency The proportion (percentage) of each allele for a particular trait that is present in the gene pool of a population.

Gene linkage A pattern of inheritance in which genes located along the same chromosome are prevented from assorting independently, but are linked together in their inheritance.

Gene mutation An alteration of the chemical nature of a gene that changes its ability to control the production of a polypeptide chain.

Gene pool The sum total of all the inheritable genes for the traits in a given sexually reproducing population.

Gene-chromosome theory A theory of genetic inheritance that is based on current understanding of the relationships between the biochemical control of traits and the process of cell division.

Genetic counseling Clinical discussions concerning inheritance patterns that are designed to inform prospective parents of the potential for expression of a genetic disorder in their offspring.

Genetic engineering The use of various techniques to move genes from one organism to another.

Genetic screening A technique for the detection of human genetic disorders in which bodily fluids are analyzed for the presence of certain marker chemicals.

Genotype The particular combination of genes in an allele pair.

Genus A level of biological classification that represents a subdivision of the phylum level; having fewer organisms with great similarity (e.g., *Drosophila,* paramecium).

Geographic isolation The separation of species populations by geographical barriers, facilitating the evolutionary process.

Geologic record A supporting item of evidence of organic evolution, supplied within the earth's rock and other geological deposits.

Germination The growth of the pollen tube from a pollen grain; the growth of the embryonic root and stem from a seed.

Gestation The period of prenatal development of a placental mammal; human gestation requires approximately 9 months.

Gizzard A portion of the digestive tract of certain organisms, including the earthworm and the grasshopper, in which food is ground into smaller fragments.

Glomerulus A capillary network lying within Bowman's capsule of the nephron.

Glucagon A hormone, secreted by the islets of Langerhans, that regulates the release of blood sugar from stored glycogen.

Glucose A monosaccharide produced commonly in photosynthesis and used by both plants and animals as a "fuel" in the process of respiration.

Glycerol An organic compound that is a component of certain lipids.

Glycogen A polysaccharide synthesized in animals as a means of storing glucose; glycogen is stored in the liver and in the muscles.

Goiter A disorder of the human regulatory system in which the thyroid gland enlarges because of a deficiency of dietary iodine.

Golgi complex Cell organelles that package cell products and move them to the plasma membrane for secretion.

Gonad An endocrine gland that produces the hormones responsible for the production of various secondary sex characteristics. (See **ovary; testis.**)

Gout A disorder of the human excretory system in which uric acid accumulates in the joints, causing severe pain.

Gradualism A theory of the time frame required for organic evolution which assumes that evolutionary change is slow, gradual, and continuous.

Grafting A technique of plant propagation in which the stems of desirable plants are attached (grafted) to rootstocks of related varieties to produce new plants for commercial purposes.

Grana The portion of the chloroplast within which chlorophyll molecules are concentrated.

Grassland A terrestrial biome characterized by wide variation in temperature and a climax flora that includes grasses.

Growth A process by which cells increase in number and size, resulting in an increase in size of the organism.

Growth-stimulating hormone (GSH) A pituitary hormone regulating the elongation of the long bones of the body.

Guanine A nitrogenous base found in both DNA and RNA molecules.

Guard cell One of a pair of cells that surround the leaf stomate and regulate its size.

Habitat The environment or set of ecological conditions within which an organism lives.

Hardy-Weinberg principle A hypothesis, advanced by G. H. Hardy and W. Weinberg, which states that the gene pool of a population should remain stable as long as a set of "ideal" conditions is met.

Heart In human beings, a four-chambered muscular pump that facilitates the movement of blood throughout the body.

Helix Literally a spiral; a term used to describe the "twisted ladder" shape of the DNA molecule.

Hemoglobin A type of protein specialized for the transport of respiratory oxygen in certain organisms, including earthworms and human beings.

Herbivore A heterotrophic organism that consumes plant matter as its primary source of nutrition. (See **primary consumer.**)

Hermaphrodite An animal organism that produces both male and female gametes.

Heterotroph An organism that typically carries on heterotrophic nutrition.

Heterotroph hypothesis A scientific hypothesis devised to explain the probable origin and early evolution of life on earth.

Heterotrophic nutrition A type of nutrition in which organisms must obtain their foods from outside sources of organic nutrients.

Heterozygous A term used to refer to an allele pair in which the alleles have different contrasting effects (e.g., *Aa, RW*).

High blood pressure A disorder of the human transport system in which systolic and diastolic pressures register higher than normal because of narrowing of the artery opening.

Histamine A chemical product of the body that causes irritation and swelling of the mucous membranes.

Homeostasis The condition of balance and dynamic stability that characterizes living systems under normal conditions.

Homologous chromosomes A pair of chromosomes that carry corresponding genes for the same traits.

Homologous structures Structures present within different species that can be shown to have had a common origin, but that may or may not share a common function.

Homozygous A term used to refer to an allele pair in which the alleles are identical in terms of effect (e.g., *AA, aa*).

Hormone A chemical product of an endocrine gland which has a regulatory effect on the cell's metabolism.

Host The organism that is harmed in a parasitic relationship.

Hybrid A term used to describe a heterozygous genotype. (See **heterozygous**.)

Hybridization A technique of plant/animal breeding in which two varieties of the same species are crossbred in the hope of producing offspring with the favorable traits of both varieties.

Hydrogen bond A weak electrostatic bond that holds together the twisted strands of DNA and RNA molecules.

Hydrolysis The chemical process by which a complex food molecule is split into simpler components through the addition of a molecule of water to the bonds holding it together.

Hypocotyl A portion of the plant embryo that specializes to become the root and lower stem of the adult plant.

Hypothalamus An endocrine gland whose secretions affect the pituitary gland.

Identical twins In human beings, twin offspring resulting from the separation of the embryonic cell mass of a single fertilization into two separate masses; such twins are genetically identical.

Importation of organisms The introduction of nonactive plants and animals into new areas where they compete strongly with native species—a negative aspect of human involvement with the environment.

In vitro fertilization A laboratory technique in which fertilization is accomplished outside the mother's body using mature ova and sperm extracted from the parents' bodies.

Inbreeding A technique of plant/animal breeding in which a "purebred" variety is bred only with its own members, so as to maintain a set of desired characteristics.

Independent assortment A pattern of inheritance in which genes on different, nonhomologous chromosomes are free to be inherited randomly and regardless of the inheritance of the others.

Ingestion The mechanism by which an organism takes in food from its environment.

Inorganic compound A chemical compound that lacks the element carbon or hydrogen (e.g., table salt: NaCl).

Insulin A hormone, secreted by the islets of Langerhans, that regulates the storage of blood sugar as glycogen.

Intercellular fluid (ICF) The fluid that bathes cells and fills intercellular spaces.

Interferon A substance, important in the fight against human cancer, that may now be produced in large quantities through techniques of genetic engineering.

Intermediate inheritance Any pattern of inheritance in which the offspring expresses a phenotype different from the phenotypes of its parents and usually representing a form intermediate between them.

Internal development Embryonic development that occurs within the body of the female parent.

Internal fertilization Fertilization that occurs inside the body of the female parent.

Interneuron A type of neuron, located in the central nervous system, that is responsible for the interpretation of impulses received from sensory neurons.

Intestine A portion of the digestive tract in which chemical digestion and absorption of digestive end-products occur.

Intracellular digestion A type of chemical digestion carried out within the cell.

Iodine A chemical stain used in cell study; an indicator used to detect the presence of starch. (See **staining**.)

Islets of Langerhans An endocrine gland, located within the pancreas, that produces the hormones insulin and glucagon.

Karyotype An enlarged photograph of the paired homologous chromosomes of an individual cell that is used in the detection of certain genetic disorders involving chromosome mutation.

Karyotyping A technique for the detection of human genetic disorders in which a karyotype is analyzed for abnormalities in chromosome structure or number.

Kidney The excretory organ responsible for maintaining the chemical composition of the blood. (See **nephron**.)

Kidney failure A disorder of the human excretory system in which there is a general breakdown of the kidney's ability to filter blood components.

Kingdom A level of biological classification that includes a broad grouping of organisms displaying general structural similarity; five kingdoms have been named by scientists.

Lacteal A small extension of the lymphatic system, found inside the villus, that absorbs fatty acids and glycerol resulting from lipid hydrolysis.

Lactic acid fermentation A type of anaerobic respiration in which glucose is converted to two lactic acid molecules.

Large intestine A portion of the digestive tract in which undigested foods are solidified by means of water absorption to form feces.

Lateral meristem A plant growth region located under the epidermis or bark of a stem. (See **cambium**.)

Latin The language used in biological classification for naming organisms by means of binomial nomenclature.

Lenticel A small pore in the stem surface that permits the absorption and release of respiratory gases within stem tissues.

Leukemia A disorder of the human transport system in which the bone marrow produces large numbers of abnormal white blood cells. (See **cancer**.)

Lichen A symbiosis of alga and fungus that frequently acts as a pioneer species on bare rock.

Limiting factor Any abiotic or biotic condition that places limits on the survival of organisms and on the growth of species populations in the environment.

Lipase Any lipid-hydrolyzing enzyme.

Lipid An organic compound composed of carbon, hydrogen, and oxygen in which hydrogen and oxygen are *not* in a 2:1 ratio (e.g., a wax, plant oil); many lipids are constructed of a glycerol and three fatty acids.

Liver An accessory organ that stores glycogen, produces bile, destroys old red blood cells, deaminates amino acids, and produces urea.

Lock-and-key model A theoretical model of enzyme action that attempts to explain the concept of enzyme specificity.

Lung The major organ of respiratory gas exchange.

Luteinizing hormone (LH) A pituitary hormone that regulates the conversion of the ovarian follicle into the corpus luteum.

Lymph Intercellular fluid (ICF) that has passed into the lymph vessels.

Lymph node One of a series of structures in the body that act as reservoirs of lymph and also contain white blood cells as part of the body's immune system.

Lymph vessel One of a branching series of tubes that collect ICF from the tissues and redistribute it as lymph.

Lymphatic circulation The movement of lymph throughout the body.

Lymphocyte A type of white blood cell that produces antibodies.

Lysosome A cell organelle that houses hydrolytic enzymes used by the cell in the process of chemical digestion.

Malpighian tubules In arthropods (e.g., grasshopper), an organ specialized for the removal of metabolic wastes.

Maltase A specific enzyme that catalyzes the hydrolysis (and dehydration synthesis) of maltose.

Maltose A type of disaccharide; a maltose molecule is composed of two units of glucose joined together by dehydration synthesis.

Marine biome An aquatic biome characterized by relatively stable conditions of moisture, salinity, and temperature.

Marsupial mammal See **nonplacental mammal**.

Mechanical digestion Any of the processes by which foods are broken apart physically into smaller particles.

Medulla The portion of the human brain responsible for regulating the automatic processes of the body.

Meiosis The process by which four monoploid nuclei are formed from a single diploid nucleus.

Meningitis A disorder of the human regulatory system in which the membranes of the brain or spinal cord become inflamed.

Menstrual cycle A hormone-controlled process responsible for the monthly release of mature ova.

Menstruation The stage of the menstrual cycle in which the lining of the uterus breaks down and is expelled from the body via the vaginal canal.

Meristem A plant tissue specialized for embryonic development. (See **apical meristem; lateral meristem; cambium**.)

Mesoderm An embryonic tissue that differentiates into muscle, bone, the excretory system, and most of the reproductive system in the adult animal.

Messenger RNA (mRNA) A type of RNA that carries the genetic code from the nuclear DNA to the ribosome for transcription.

Metabolism All of the chemical processes of life considered together; the sum total of all the cell's chemical activity.

Methylene blue A chemical stain used in cell study. (See **staining**.)

Microdissection instruments Tools of biological study that are used to remove certain cell organelles from within cells for examination.

Micrometer (μm) A unit of linear measurement equal in length to 0.001 millimeter (0.000001 meter), used for expressing the dimensions of cells and cell organelles.

Mitochondrion A cell organelle that contains the enzymes necessary for aerobic respiration.

Mitosis A precise duplication of the contents of a parent cell nucleus, followed by an orderly separation of these contents into two new, identical daughter nuclei.

Mitotic cell division A type of cell division that results in the production of two daughter cells identical to each other and to the parent cell.

Monera One of the five biological kingdoms; it includes simple unicellular forms lacking nuclear membranes (e.g., bacteria).

Monohybrid cross A genetic cross between two organisms both heterozygous for a trait controlled by a single allele pair. The phenotypic ratio resulting is 3:1; the genotypic ratio is 1:2:1.

Monoploid chromosome number The number of chromosomes commonly found in the gametes of sexually reproducing species.

Monosaccharide A type of carbohydrate known also as a "simple sugar"; all monosaccharides have the molecular formula $C_6H_{22}O_6$.

Motor neuron A type of neuron that carries "command" impulses from the central nervous system to an effector organ.

Mucus A protein-rich mixture that bathes and moistens the respiratory surfaces.

Multicellular Having a body that consists of large groupings of specialized cells (e.g., human being).

Multiple alleles A pattern of inheritance in which the existence of more than two alleles is hypothesized, only two of which are present in the genotype of any one individual.

Muscle A type of tissue specialized to produce movement of body parts.

Mutagenic agent Any environmental condition that initiates or accelerates genetic mutation.

Mutation Any alteration of the genetic material, either a chromosome or a gene, in an organism.

Mutualism A type of symbiosis beneficial to both organisms in the relationship.

Nasal cavity A series of channels through which outside air is admitted to the body interior and is warmed and moistened before entering the lung.

Natural selection A concept, central to Darwin's theory of evolution, to the effect that the individuals best adapted to their environment tend to survive and to pass their favorable traits on to the next generation.

Negative feedback A type of endocrine regulation in which the effects of one gland may inhibit its own secretory activity, while stimulating the secretory activity of another gland.

Nephridium An organ found in certain organisms, including the earthworm, specialized for the removal of metabolic wastes.

Nephron The functional unit of the kidney. (See **glomerulus; Bowman's capsule.**)

Nerve A structure formed from the bundling of neurons carrying sensory or motor impulses.

Nerve impulse An electrochemical change in the surface of the nerve cell.

Nerve net A network of "nerve" cells in coelenterates such as the hydra.

Neuron A cell specialized for the transmission of nerve impulses.

Neurotransmitter A chemical substance secreted by a neuron that aids in the transmission of the nerve impulse to an adjacent neuron.

Niche The role that an organism plays in its environment.

Nitrifying bacteria In the nitrogen cycle, bacteria that absorb ammonia and convert it into nitrate salts.

Nitrogen cycle The process by which nitrogen is recycled and made available for use by other organisms.

Nitrogen-fixing bacteria A type of bacteria responsible for absorbing atmospheric nitrogen and converting it to nitrate salts in the soil.

Nitrogenous base A chemical unit composed of carbon, hydrogen, and nitrogen that is a component part of the nucleotide unit.

Nitrogenous waste Any of a number of nitrogen-rich compounds that result from the metabolism of proteins and amino acids in the cell. (See **ammonia; urea; uric acid.**)

Nondisjunction A type of chromosome mutation in which the members of one or more pairs of homologous chromosomes fail to separate during the disjunction phase of the first meiotic division.

Nonplacental mammal A species of mammal in which internal development is accomplished without the aid of a placental connection (marsupial mammals).

Nucleic acid An organic compound composed of repeating units of nucleotide.

Nucleolus A cell organelle located within the nucleus that is known to function in protein synthesis.

Nucleotide The repeating unit making up the nucleic acid polymer (e.g., DNA, RNA).

Nucleus A cell organelle that contains the cell's genetic information in the form of chromosomes.

Nutrition The life function by which living things obtain food and process it for their use.

Omnivore A heterotrophic organism that consumes both plant and animal matter as sources of nutrition.

One gene-one polypeptide A scientific hypothesis concerning the role of the individual gene in protein synthesis.

Oogenesis A type of meiotic cell division in which one ovum and three polar bodies are produced from each primary sex cell.

Open transport system A type of circulatory system in which the transport fluid is *not* always enclosed within blood vessels (e.g., grasshopper).

Oral cavity In human beings, the organ used for the ingestion of foods.

Oral groove The ingestive organ of the paramecium.

Organ transplant An application of the study of immunity in which an organ or tissue of a donor is transplanted into a compatible recipient.

Organelle A small, functional part of a cell specialized to perform a specific life function (e.g., nucleus, mitochondrion).

Organic compound A chemical compound that contains the elements carbon and hydrogen (e.g., carbohydrate, protein).

Organic evolution The mechanism thought to govern the changes in living species over geologic time.

Osmosis A form of passive transport by which water is absorbed or released by cells.

Ovary A female gonad that secretes the hormone estrogen, which regulates female secondary sex characteristics; the ovary also produces ova, which are used in reproduction.

Overcropping A negative aspect of human involvement with the environment in which soil is overused for the production of crops, leading to exhaustion of soil nutrients.

Overgrazing The exposure of soil to erosion due to the loss of stabilizing grasses when it is overused by domestic animals—a negative aspect of human involvement with the environment.

Overhunting A negative aspect of human involvement with the environment in which certain species have been greatly reduced or made extinct by uncontrolled hunting practices.

Oviduct A tube that serves as a channel for conducting mature ova from the ovary to the uterus; the site of fertilization and the earliest stages of embryonic development.

Ovulation The stage of the menstrual cycle in which the mature ovum is released from the follicle into the oviduct.

Ovule A structure located within the flower ovary that contains a monoploid egg nucleus and serves as the site of fertilization.

Ovum A type of gamete produced as a result of oogenesis in female animals; the egg, the female sex cell.

Oxygen-18 A radioactive isotope of oxygen that is used to trace the movement of this element in biochemical reaction sequences.

Oxygenated blood Blood that contains a high percentage of oxyhemoglobin.

Oxyhemoglobin Hemoglobin that is loosely bound to oxygen for purposes of oxygen transport.

Palisade layer A cell layer found in most leaves that contains high concentrations of chloroplasts.

Pancreas An accessory organ which produces enzymes that complete the hydrolysis of foods to soluble end-products; also the site of insulin and glucagon production.

Parasitism A type of symbiosis from which one organism in the relationship benefits, while the other (the "host") is harmed, but not ordinarily killed.

Parathormone A hormone of the parathyroid gland that regulates the metabolism of calcium in the body.

Parathyroid gland An endocrine gland whose secretion, parathormone, regulates the metabolism of calcium in the body.

Passive immunity A temporary immunity produced as a result of the injection of preformed antibodies.

Passive transport Any process by which materials are absorbed into the cell interior from an area of high concentration to an area of low concentration, without the expenditure of cell energy (e.g., osmosis, diffusion).

Penis A structure that permits internal fertilization through direct implantation of sperm into the female reproductive tract.

Peptide bond A type of chemical bond that links the nitrogen atom of one amino acid with the terminal carbon atom of a second amino acid in the formation of a dipeptide.

Peripheral nerves Nerves in the earthworm and grasshopper that branch from the ventral nerve cord to other parts of the body.

Peripheral nervous system A major subdivision of the nervous system that consists of all the nerves of all types branching through the body. (See **autonomic nervous system; somatic nervous system**.)

Peristalsis A wave of contraction of the smooth muscle lining; the digestive tract that causes ingested food to pass along the food tube.

Petal An accessory part of the flower that is thought to attract pollinating insects.

pH A chemical unit used to express the concentration of hydrogen ion (H^+), or the acidity, of a solution.

Phagocyte A type of white blood cell that engulfs and destroys bacteria.

Phagocytosis The process by which the ameba surrounds and ingests large food particles for intracellular digestion.

Pharynx The upper part of the digestive tube that temporarily stores food before digestion.

Phenotype The observable trait that results from the action of an allele pair.

Phenylketonuria (PKU) A genetically related human disorder in which the homozygous combination of a particular mutant gene prevents the normal metabolism of the amino acid phenylalanine.

Phloem A type of vascular tissue through which water and dissolved sugars are transported in plants from the leaf downward to the roots for storage.

Phosphate group A chemical group made up of phosphorus and oxygen that is a component part of the nucleotide unit.

Phosphoglyceraldehyde (PGAL) An intermediate product formed during photosynthesis that acts as the precursor of glucose formation.

Photochemical reactions A set of biochemical reactions in photosynthesis in which light is absorbed and water molecules are split. (See **photolysis**.)

Photolysis The portion of the photochemical reactions in which water molecules are split into hydrogen atoms and made available to the carbon fixation reactions.

Photosynthesis A type of autotrophic nutrition in which green plants use the energy of sunlight to convert carbon dioxide and water into glucose.

Phylum A level of biological classification that is a major subdivision of the kingdom level, containing fewer organisms with greater similarity (e.g., Chordata).

Pinocytosis A special type of absorption by which liquids and particles too large to diffuse through the cell membrane may be taken in by vacuoles formed at the cell surface.

Pioneer autotrophs The organisms supposed by the heterotroph hypothesis to have been the first to evolve the ability to carry on autotrophic nutrition.

Pioneer species In an ecological succession, the first organisms to inhabit a barren environment.

Pistil The female sex organ of the flower. (See **stigma; style; ovary**.)

Pituitary gland An endocrine gland that produces hormones regulating the secretions of other endocrine glands; the "master gland."

Placenta In placental mammals, a structure composed of both embryonic and maternal tissues that permits the diffusion of soluble substances to and from the fetus for nourishment and the elimination of fetal waste.

Placental mammal A mammal species in which embryonic development occurs internally with the aid of a placental connection to the female parent's body.

Plant One of the five biological kingdoms; it includes multicellular organisms whose cells are bounded by cell walls and which are capable of photosynthesis (e.g., maple tree).

Plasma The liquid fraction of blood, containing water and dissolved proteins.

Plasma membrane A cell organelle that encloses the cytoplasm and other cell organelles and regulates the passage of materials into and out of the cell.

Platelet A cell-like component of the blood that is important in clot formation.

Polar body One of three nonfunctional cells produced during oogenesis that contain monoploid nuclei and disintegrate soon after completion of the process.

Polio A disorder of the human regulatory system in which viral infection of the central nervous system may result in severe paralysis.

Pollen The male gamete of the flowering plant.

Pollen tube A structure produced by the germinating pollen grain that grows through the style to the ovary and carries the sperm nucleus to the ovule for fertilization.

Pollination The transfer of pollen grains from anther to stigma.

Pollution control The development of new procedures to reduce the incidence of air, water, and soil pollution—a positive aspect of human involvement with the environment.

Polyploidy A type of chromosome mutation in which an entire set of homologous chromosomes fail to separate during the disjunction phase of the first meiotic division.

Polysaccharide A type of carbohydrate composed of repeating units of monosaccharide that form a polymeric chain.

Polyunsaturated fat A type of fat in which many bonding sites are unavailable for the addition of hydrogen atoms.

Population All the members of a particular species in a given geographical location at a given time.

Population control The use of various practices to slow the rapid growth in the human population—a positive aspect of human interaction with the environment.

Population genetics A science that studies the genetic characteristics of a sexually reproducing species and the factors that affect its gene frequencies.

Postnatal development The growth and maturation of an individual from birth, through aging, to death.

Prenatal development The embryonic development that occurs within the uterus before birth. (See **gestation.**)

Primary consumer Any herbivorous organism that receives food energy from the producer level (e.g., mouse); the second stage of a food chain.

Primary sex cell The diploid cell that undergoes meiotic cell division to produce monoploid gametes.

Producer Any autotrophic organism capable of trapping light energy and converting it to the chemical bond energy of food (e.g., green plants); the organisms forming the basis of the food chain.

Progesterone A hormone produced by the corpus luteum and/or placenta that has the effect of maintaining the uterine lining and suppressing ovulation during gestation.

Protease Any protein-hydrolyzing enzyme.

Protein A complex organic compound composed of repeating units of amino acid.

Protista One of the five biological kingdoms; it includes simple unicellular forms whose nuclei are surrounded by nuclear membranes (e.g., ameba, paramecium).

Pseudopod A temporary, flowing extension of the cytoplasm of an ameba that is used in nutrition and locomotion.

Pulmonary artery One of two arteries that carry blood from the heart to the lungs for reoxygenation.

Pulmonary circulation Circulation of blood from the heart through the lungs and back to the heart.

Pulmonary vein One of four veins that carry oxygenated blood from the lungs to the heart.

Pulse Rhythmic contractions of the artery walls that help to push the blood fluid through the capillary networks of the body.

Punctuated equilibrium A theory of the time frame required for evolution which assumes that evolutionary change occurs in "bursts" with long periods of relative stability intervening.

Pyramid of biomass A construct used to illustrate the fact that the total biomass available in each stage of a food chain diminishes from producer level to consumer level.

Pyramid of energy A construct used to illustrate the fact that energy is lost at each trophic level in a food chain, being most abundant at the producer level.

Pyruvic acid An intermediate product in the aerobic or anaerobic respiration of glucose.

Receptor An organ specialized to receive a particular type of environmental stimulus.

Recessive allele (gene) An allele (gene) whose effect is masked by that of its dominant allele.

Recombinant DNA DNA molecules that have been moved from one cell to another in order to give the recipient cell a genetic characteristic of the donor cell.

Recombination The process by which the members of segregated allele pairs are randomly recombined in the zygote as a result of fertilization.

Rectum The portion of the digestive tract in which digestive wastes are stored until they can be released to the environment.

Red blood cell Small, nonnucleated cells in the blood that contain hemoglobin and carry oxygen to bodily tissues.

Reduction division See **meiosis**.

Reflex A simple, inborn, involuntary response to an environmental stimulus.

Reflex arc The complete path, involving a series of three neurons (sensory, interneuron, and motor), working together, in a reflex action.

Regeneration A type of asexual reproduction in which new organisms are produced from the severed parts of a single parent organism; the replacement of lost or damaged tissues.

Regulation The life process by which living things respond to changes within and around them, and by which all life processes are coordinated.

Replication An exact self-duplication of the chromosome during the early stages of cell division; the exact self-duplication of a molecule of DNA.

Reproduction The life process by which new cells arise from preexisting cells by cell division.

Reproductive isolation The inability of species varieties to interbreed and produce fertile offspring, because of variations in behavior or chromosome structure.

Respiration The life function by which living things convert the energy of organic foods into a form more easily used by the cell.

Response The reaction of an organism to an environmental stimulus.

Rhizoid A rootlike fiber produced by fungi that secrete hydrolytic enzymes and absorb digested nutrients.

Ribonucleic acid (RNA) A type of nucleic acid that operates in various ways to facilitate protein synthesis.

Ribose A five-carbon sugar found as a component part of the nucleotides of RNA molecules only.

Ribosomal RNA (rRNA) The type of RNA that makes up the ribosome.

Ribosome A cell organelle that serves as the site of protein synthesis in the cell.

Root A plant organ specialized to absorb water and dissolved substances from the soil, as well as to anchor the plant to the soil.

Root hair A small projection of the growing root that serves to increase the surface area of the root for absorption.

Roughage A variety of undigestible carbohydrates that add bulk to the diet and facilitate the movement of foods through the intestine.

Runner A type of vegetative propagation in which an above-ground stem (runner) produces roots and leaves and establishes new organisms with identical characteristics.

Saliva A fluid secreted by salivary glands that contains hydrolytic enzymes specific to the digestion of starches.

Salivary gland The gland that secretes saliva, important in the chemical digestion of certain foods.

Salt A chemical composed of a metal and a nonmetal joined by means of an ionic bond (e.g., sodium chloride).

Saprophyte A heterotrophic organism that obtains its nutrition from the decomposing remains of dead plant and animal tissues (e.g., fungus, bacteria).

Saturated fat A type of fat molecule in which all available bonding sites on the hydrocarbon chains are taken up with hydrogen atoms.

Scrotum A pouch extending from the wall of the lower abdomen that houses the testes at a temperature optimum for sperm production.

Secondary consumer Any carnivorous animal that derives its food energy from the primary consumer level (e.g., a snake); the third level of a food chain.

Secondary sex characteristics The physical features, different in males and females, that appear with the onset of sexual maturity.

Seed A structure that develops from the fertilized ovule of the flower and germinates to produce a new plant.

Seed dispersal Any mechanism by which seeds are distributed in the environment so as to widen the range of a plant species. (See **fruit**.)

Segregation The random separation of the members of allele pairs that occurs during meiotic cell division.

Self-pollination A type of pollination in which the pollen of a flower pollinates another flower located on the same plant organism.

Sensory neuron A type of neuron specialized for receiving environmental stimuli, which are detected by receptor organs.

Sepal An accessory part of the flower that functions to protect the bud during development.

Sessile A term that relates to the "unmoving" state of certain organisms, including the hydra.

Seta One of several small, chitinous structures (setae) that aid the earthworm in its locomotor function.

Sex chromosomes A pair of homologous chromosomes carrying genes that determine the sex of an individual; these chromosomes are designated as X and Y.

Sex determination A pattern of inheritance in which the conditions of maleness and femaleness are determined by the inheritance of a pair of sex chromosomes (XX = female; XY = male).

Sex linkage A pattern of inheritance in which certain nonsex genes are located on the X sex chromosome, but have no corresponding alleles on the Y sex chromosome.

Sex-linked trait A genetic trait whose inheritance is controlled by the genetic pattern of sex linkage (e.g., color blindness).

Sexual reproduction A type of reproduction in which new organisms are formed as a result of the fusion of gametes from two parent organisms.

Shell An adaptation for embryonic development in many terrestrial, externally developing species that protects the developing embryo from drying and physical damage (e.g., birds).

Sickle cell anemia A genetically related human disorder in which the homozygous combination of a mutant gene leads to the production of abnormal hemoglobin and crescent-shaped red blood cells.

Skeletal muscle A type of muscle tissue associated with the voluntary movements of skeletal levers in locomotion.

Small intestine In human beings, the longest portion of the food tube, in which final digestion and absorption of soluble end-products occur.

Smooth muscle See **visceral muscle**.

Somatic nervous system A subdivision of the peripheral nervous system that is made up of nerves associated with voluntary actions.

Speciation The process by which new species are thought to arise from previously existing species.

Species A biological grouping of organisms so closely related that they are capable of interbreeding and producing fertile offspring (e.g., human being).

Species presentation The establishment of game lands and wildlife refuges that have permitted the recovery of certain endangered species—a positive aspect of human involvement with the environment.

Sperm A type of gamete produced as a result of spermatogenesis in male animals; the male reproductive cell.

Spermatogenesis A type of meiotic cell division in which four sperm cells are produced for each primary sex cell.

Spinal cord The part of the central nervous system responsible for reflex action, as well as impulse conduction between the peripheral nervous system and the brain.

Spindle apparatus A network of fibers that form during cell division and to which centromeres attach during the separation of chromosomes.

Spiracle One of several small pores in arthropods, including the grasshopper, that serve as points of entry of respiratory gases from the atmosphere to the tracheal tubes.

Spongy layer A cell layer found in most leaves that is loosely packed and contains many air spaces to aid in gas exchange.

Spore A specialized asexual reproductive cell produced by certain plants.

Sporulation A type of asexual reproduction in which spores released from special spore cases on the parent plant germinate and grow into new adult organisms of the species.

Staining A technique of cell study in which chemical stains are used to make cell parts more visible for microscopic study.

Stamen The male reproductive structure in a flower. (See **anther; filament**.)

Starch A type of polysaccharide produced and stored by plants.

Stem A plant organ specialized to support the leaves and flowers of a plant, as well as to conduct materials between the roots and the leaves.

Stigma The sticky upper portion of the pistil, which serves to receive pollen.

Stimulus Any change in the environment to which an organism responds.

Stomach A muscular organ that acts to liquefy food and that produces gastric protease for the hydrolysis of protein.

Stomate A small opening that penetrates the lower epidermis of a leaf and through which respiratory and photosynthetic gases diffuse.

Strata The layers of sedimentary rock that contain fossils, whose ages may be determined by studying the patterns of sedimentation.

Stroke A disorder of the human regulatory system in which brain function is impaired because of oxygen starvation of brain centers.

Stroma An area of the chloroplast within which the carbon-fixation reactions occur; stroma lie between pairs of grana.

Style The portion of the pistil that connects the stigma to the ovary.

Substrate A chemical that is metabolized by the action of a specific enzyme.

Succession A situation in which an established ecological community is gradually replaced by another until a climax community is established.

Survival of the fittest The concept, frequently associated with Darwin's theory of evolution, that in the intraspecies competition among naturally occurring species the organisms best adapted to the particular environment will survive.

Sweat glands In human beings, the glands responsible for the production of perspiration.

Symbiosis A term which refers to a variety of biotic relationships in which organisms of different species live together in close physical association.

Synapse The gap that separates the terminal branches of one neuron from the dendrites of an adjacent neuron.

Synapsis The intimate, highly specific pairing of homologous chromosomes that occurs in the first meiotic division, forming tetrads.

Synthesis The life function by which living things manufacture the complex compounds required to sustain life.

Systemic circulation The circulation of blood from the heart through the body tissues (except the lungs) and back to the heart.

Systole The higher pressure registered during blood pressure testing. (See **diastole**.)

Taiga A terrestrial biome characterized by long, severe winters and climax flora that includes coniferous trees.

Tay-Sachs A genetically related human disorder in which fatty deposits in the cells, particularly of the brain, inhibit proper functioning of the nervous system.

Technological oversight A term relating to human activities that adversely affect environmental quality due to failure to adequately assess the environmental impact of a technological development.

Teeth Structures located in the mouth that are specialized to aid in the mechanical digestion of foods.

Temperate deciduous forest A terrestrial biome characterized by moderate climatic conditions and climax flora that includes deciduous trees.

Template A pattern or design provided by the DNA molecule for the synthesis of protein molecules.

Tendon A type of connective tissue that attaches a skeletal muscle to a bone.

Tendonitis A disorder of the human locomotor system in which the junction between a tendon and a bone becomes irritated and inflamed.

Tentacle A grasping structure in certain organisms, including the hydra, that contains stinging cells and is used for capturing prey.

Terminal branch A cytoplasmic extension of the neuron that transmits a nerve impulse to adjacent neurons via the secretion of neurotransmitters.

Terrestrial biome A biome that comprises primarily land ecosystems, the characteristics of which are determined by the major climate zone of the earth.

Test cross A genetic cross accomplished for the purpose of determining the genotype of an organism expressing a dominant phenotype; the unknown is crossed with a homozygous recessive.

Testis A gonad in human males that secretes the hormone testosterone, which regulates male secondary sex characteristics; the testis also produces sperm cells for reproduction.

Testosterone A hormone secreted by the testis that regulates the production of male secondary sex characteristics.

Tetrad A grouping of four chromatids that results from synapsis.

Thymine A nitrogenous base found only in DNA.

Thyroid gland An endocrine gland that regulates the body's general rate of metabolism through secretion of the hormone thyroxin.

Thyroid-stimulating hormone (TSH) A pituitary hormone that regulates the secretions of the thyroid gland.

Thyroxin A thyroid hormone that regulates the body's general metabolic rate.

Tongue A structure that aids in the mechanical digestion of foods.

Trachea A cartilage-ringed tube that conducts air from the mouth to the bronchi.

Tracheal tube An adaptation in arthropods (e.g., grasshopper) which functions to conduct respiratory gases from the environment to the moist internal tissues.

Tracheophyta A phylum of the Plant Kingdom whose members (tracheophytes) contain vascular tissues and true roots, stems, and leaves (e.g., geranium, fern, bean, maple tree, corn).

Transfer RNA (tRNA) A type of RNA that functions to transport specific amino acids from the cytoplasm to the ribosome for protein synthesis.

Translocation A type of chromosome mutation in which a section of a chromosome is transferred to a nonhomologous chromosome.

Transpiration The evaporation of water from leaf stomates.

Transpiration pull A force that aids the upward conduction of materials in the xylem by means of the evaporation of water (transpiration) from leaf surfaces.

Transport The life function by which substances are absorbed, circulated, and released by living things.

Triplet codon A group of three nitrogenous bases that provide information for the placement of amino acids in the synthesis of proteins.

Tropical forest A terrestrial biome characterized by a warm, moist climate and a climax flora that includes many species of broadleaf trees.

Tropism A plant growth response to an environmental stimulus.

Tuber A type of vegetative propagation in which an underground stem (tuber) produces new tubers, each of which is capable of producing new organisms with identical characteristics.

Tundra A terrestrial biome characterized by permanently frozen soil and climax flora that includes lichens and mosses.

Tympanum A receptor organ in arthropods (e.g., grasshopper) that is specialized to detect vibrational stimuli.

Ulcer A disorder of the human digestive tract in which a portion of its lining erodes and becomes irritated.

Ultracentrifuge A tool of biological study that uses very high speeds of centrifugation to separate cell parts for examination.

Umbilical cord In placental mammals, a structure containing blood vessels that connects the placenta to the embryo.

Unicellular Having a body that consists of a single cell (e.g., paramecium).

Uracil A nitrogenous base that is a component part of the nucleotides of RNA molecules only.

Urea A type of nitrogenous waste with moderate solubility and moderate toxicity.

Ureter In human beings, a tube that conducts urine from the kidney to the urinary bladder.

Urethra In human beings, a tube that conducts urine from the urinary bladder to the exterior of the body.

Uric acid A type of nitrogenous waste with low solubility and low toxicity.

Urinary bladder An organ responsible for the temporary storage of urine.

Urine A mixture of water, salts, and urea excreted from the kidney.

Use and disuse A term associated with the evolutionary theory of Lamarck, since proved incorrect.

Uterus In female placental mammals, the organ within which embryonic development occurs.

Vaccination An inoculation of dead or weakened disease organisms that stimulates the body's immune system to produce active immunity.

Vacuole A cell organelle that contains storage materials (e.g., starch, water) housed inside the cell.

Vagina In female placental mammals, the portion of the reproductive tract into which sperm are implanted during sexual intercourse and through which the baby passes during birth.

Variation A concept, central to Darwin's theory of evolution, that refers to the range of adaptation which can be observed in all species.

Vascular tissues Tubelike plant tissues specialized for the conduction of water and dissolved materials within the plant. (See **xylem; phloem.**)

Vegetative propagation A type of asexual reproduction in which new plant organisms are produced from the vegetative (nonfloral) parts of the parent plant.

Vein (human) A relatively thin-walled blood vessel that carries blood from capillary networks back toward the heart.

Vein (plant) An area of vascular tissues located in the leaf that aid the upward transport of water and minerals through the leaf and the transport of dissolved sugars to the stem and roots.

Vena cava One of two major arteries that return blood to the heart from the body tissues.

Ventral nerve cord The main pathway for nerve impulses between the brain and peripheral nerves of the grasshopper and earthworm.

Ventricle One of two thick-walled, muscular chambers of the heart that pump blood out to the lungs and body.

Villi Microscopic projections of the lining of the small intestine that absorb the soluble end-products of digestion. (See **lacteal.**)

Visceral muscle A type of muscle tissue associated with the involuntary movements of internal organs (e.g., peristalsis in the small intestine).

Vitamin a type of nutrient that acts as a coenzyme in various enzyme-controlled reactions.

Water cycle The mechanism by which water is made available to living things in the environment through the processes of precipitation, evaporation, runoff, and percolation.

Water pollution A type of technological oversight that involves the addition of some unwanted factor (e.g., sewage, heavy metals, heat, toxic chemicals) to our water resources.

Watson-Crick model A model of DNA structure devised by J. Watson and F. Crick that hypothesizes a "twisted ladder" arrangement for the DNA molecule.

White blood cell A type of blood cell that functions in disease control. (See **phagocyte; lymphocyte.**)

Xylem A type of vascular tissue through which water and dissolved minerals are transported upward through a plant from the root to the stems and leaves.

Yolk A food substance, rich in protein and lipid, found in the eggs of many animal species.

Yolk sac The membrane that surrounds the yolk food supply of the embryos of many animal species.

Zygote The single diploid cell that results from the fusion of gametes in sexual reproduction; a fertilized egg.

Regents Examinations, Answers, and Student Self-Appraisal Guides

Examination August 2019
Living Environment

PART A

Answer all questions in this part. [30]

Directions (1–30): For *each* statement or question, record in the space provided the *number* of the word or expression that, of those given, best completes the statement or answers the question.

1. The diagram below represents an energy pyramid.

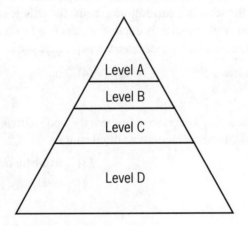

In this pyramid, the greatest amount of stored energy is found at level

(1) *A*

(2) *B*

(3) *C*

(4) *D*   1 _____

2. A certain species of plant serves as the only food for the young larvae of a particular species of butterfly. In a large field, a disease kills all the members of this plant species. As a result of the plant disease, the butterfly population will most likely

    (1) quickly adapt to eat other plants

    (2) disappear from the area

    (3) evolve to form a new species

    (4) enter the adult stage more quickly   2 _____

3. When handling cat litter, humans can potentially be exposed to a harmful single-celled protozoan. Its primary host is the common domestic cat, but it can also live in humans. This protozoan is an example of a

    (1) predator   (3) parasite

    (2) producer   (4) scavenger   3 _____

4. Certain seaweeds contain a greater concentration of iodine inside their cells than there is in the seawater surrounding them. The energy required to maintain this concentration difference is most closely associated with the action of

    (1) ribosomes   (3) vacuoles

    (2) mitochondria   (4) nuclei   4 _____

5. Doctors sometimes use a vaccine to prepare the body to defend itself against future infections. These vaccines most often contain

    (1) antibodies   (3) white blood cells

    (2) antibiotics   (4) weakened pathogens   5 _____

6. Building large manufacturing facilities can affect ecosystems by increasing the

    (1) atmospheric quality

    (2) biodiversity in the area

    (3) demand for resources such as fossil fuels

    (4) availability of space and resources for organisms   6 _____

7. An ameba is a single-celled organism. It uses its cell membrane to obtain food from its environment, digests the food with the help of organelles called lysosomes, and uses other organelles to process the digested food. From this, we can best infer that

  (1)   all single-celled organisms have lysosomes to digest food

  (2)   amebas are capable of digesting any type of food molecule

  (3)   single-celled organisms are as complex as multicellular organisms

  (4)   structures in amebas have functions similar to organs in multicellular organisms   7 _____

8. White blood cells are most closely associated with which two body systems?

  (1)   circulatory and digestive

  (2)   immune and circulatory

  (3)   digestive and excretory

  (4)   excretory and immune   8 _____

9. Carnivorous plants, such as pitcher plants and sundews, live in bogs where many other organisms cannot. Due to the high rate of decomposition occurring in bogs, the environment is acidic and contains very little oxygen and nutrients. The bogs only support certain types of organisms because

  (1)   organisms in an environment are not limited by available energy and resources

  (2)   the growth and survival of organisms depends upon specific physical conditions

  (3)   favorable gene mutations only occur when organisms live in harsh environments

  (4)   photosynthetic organisms can only inhabit environments that have a low acidity   9 _____

10. Anhidrosis is the inability to sweat normally. If the human body cannot sweat properly, it cannot cool itself, which is potentially harmful. Anhidrosis most directly interferes with

  (1)   a feedback mechanism that maintains homeostasis

  (2)   an immune system response to harmless antigens

  (3)   the synthesis of hormones in the circulatory system

  (4)   the enzymatic breakdown of water in cells   10 _____

11. The hair colors of the members of a family are listed below.

> mother – brown hair
> father – blond hair
> older son – brown hair
> younger son – blond hair

The hair colors of the sons are most likely a direct result of

(1) natural selection in males
(2) heredity
(3) evolution
(4) environmental influences   11 _____

12. A sample of DNA from a human skin cell contains 32% cytosine (C) bases. Approximately what percentage of the bases in this sample will be thymine (T)?

(1) 18   (3) 32
(2) 24   (4) 36   12 _____

13. Carmine, a compound that comes from the cochineal beetle, shown below, is used as a food coloring.

Source: https://alibi.com/events/256770/
Cochineal-Empire-making-Insect.html

The food coloring is not harmful to most people, but in a small number of individuals, it causes a reaction and affects their ability to breathe. This response to carmine is known as

(1) a stimulus   (3) natural selection
(2) an allergy   (4) an adaptation   13 _____

14. As a way to reduce the number of cases of malaria, a human tropical disease, a specific DNA sequence is inserted into the reproductive cells of *Anopheles* mosquitoes. Which process was most likely used to alter these mosquitoes?

(1) cloning studies   (3) natural selection

(2) genetic engineering   (4) random mutations   14 _____

15. Which row in the chart below accurately identifies two causes of mutations and the cells that must be affected in order for the mutations to be passed on to offspring?

Row	Cause of Mutations	Cells Affected
(1)	infections and antigens	body cells
(2)	meiosis and mitosis	body cells
(3)	disease and differentiation	sex cells
(4)	chemicals and radiation	sex cells

15 _____

16. Many tiny plants can be seen developing asexually along the edge of the mother-of-thousands plant leaf, as shown in the photo below. The tiny plants eventually drop to the ground and grow into new plants of the same species.

Source: http://www.plantamundo.com/produto_completo.asp?IDProduto=255

One way this form of reproduction differs from sexual reproduction is

(1)   more genetic variations are seen in the offspring

(2)   there is a greater chance for mutations to occur

(3)   the offspring and the parents are genetically identical

(4)   the new plants possess the combined genes of both parents   16 _____

17. A food web is represented in the diagram below.

Ocean Food Web

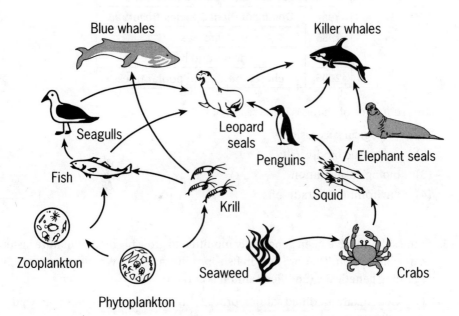

Adapted from: www.siyavula.com/gr7-9-websites/natural-sciences/gr8/gr8-11-02.html

If the fish population *decreases*, what is the most direct effect this will have on the aquatic ecosystem?

(1) The leopard seals will all die from lack of food.

(2) The krill population will only be consumed by seagulls.

(3) The zooplankton population will increase in size.

(4) The phytoplankton population will increase in size.   17 _____

18. The chart below shows a sequence of events that was observed at an abandoned ski center over a period of years.

Changes in Plant Species Over Time	
Year	**Dominant Plant Species Observed**
1985	grasses
1995	shrubs and bushes
2005	cherry, birch, and poplar trees

This sequence of changes is the result of

(1) ecological succession

(2) decreased biodiversity

(3) biological evolution

(4) environmental trade-offs   18 _____

19. Some salmon have been genetically modified to grow bigger and mature faster than wild salmon. They are kept in fish-farming facilities. Which statement regarding genetically modified salmon is correct?

(1) Genetically modified salmon produce more of some proteins than wild salmon.

(2) Genetically modified salmon and wild salmon would have identical DNA.

(3) Wild salmon reproduce asexually while genetically modified salmon reproduce sexually.

(4) Wild salmon have an altered protein sequence, but genetically modified salmon do not.   19 _____

20. Which group of organisms in an ecosystem fills the niche of recycling organic matter back to the environment?

(1) carnivores   (3) producers

(2) decomposers   (4) predators   20 _____

21. The use of solar panels has increased in the last ten years. A benefit of using solar energy would include

(1) adding more carbon dioxide to the atmosphere

(2) using less fossil fuel to meet energy needs

(3) using a nonrenewable source of energy

(4) releasing more gases for photosynthesis 21 _____

22. In a sewage treatment facility, an optimal environment is maintained for the survival of naturally occurring species of microorganisms. These organisms can then break the sewage down into relatively harmless wastewater. For these microorganisms, the wastewater facility serves as

(1) its carrying capacity (3) an ecosystem

(2) a food chain (4) an energy pyramid 22 _____

23. The diagram below represents a process taking place in a cell.

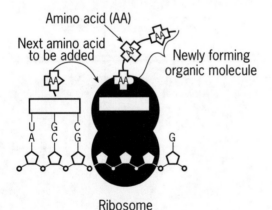

The type of organic molecule that is being synthesized is

(1) DNA (3) protein

(2) starch (4) fat 23 _____

24. The governments of many countries have regulations that are designed to prevent the accidental introduction of nonnative insects into their countries. This is because, in these new habitats, the nonnative insects might

(1) become food for birds

(2) not survive a cold winter

(3) not have natural predators

(4) add to the biodiversity 24 _____

25. The process of transferring energy during respiration occurs in a series of steps. This prevents too much heat from being released at one time. Maintaining an appropriate temperature is beneficial to an organism because

(1) enzymes need a proper range of temperatures to catalyze vital reactions

(2) cellular waste products can only be excreted in cooler temperatures

(3) hormones can only produce antibodies if temperatures are not excessive

(4) nutrients diffuse faster into cells when temperatures are lower 25 _____

26. The diagram below represents some structures in the human female reproductive system.

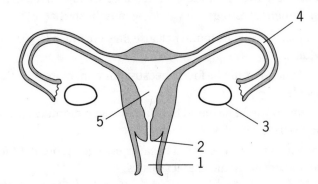

The processes of meiosis and fertilization are essential in human reproduction. Which row in the chart correctly identifies where in the female reproductive system these two processes occur?

Row	Meiosis	Fertilization
(1)	1	3
(2)	2	5
(3)	3	4
(4)	4	5

26 _____

27. Fruits and vegetables exposed to air begin to brown because of a chemical reaction in their cells. This may result in these foods being thrown out. Some people have found that adding lemon juice (citric acid) to apple slices keeps them from turning brown. The prevention of browning is likely the result of

(1) increasing the concentration of enzymes
(2) increasing the temperature
(3) slowing the rate of enzyme action
(4) maintaining the pH

27 _____

28. Scientists monitoring frog populations have noticed that the ratio of male frogs to female frogs varies when certain chemicals are present in the environment. The influence of estrogen, for example, has a noticeable effect. In the presence of a higher amount of estrogen, it would be most likely that

(1) fewer males would be found because they are much larger and fewer are produced

(2) fewer females would be found because they are more sensitive to pesticides

(3) more males would be found because estrogen promotes the development of male characteristics

(4) more females would be found because estrogen promotes the development of female characteristics   28 _____

29. Which action could humans take to slow the rate of global warming?

(1) Cut down trees for more efficient land use.

(2) Increase the consumption of petroleum products.

(3) Use alternate sources of energy such as wind.

(4) Reduce the use of fuel-efficient automobiles.   29 _____

30. The role of antibodies in the human body is to

(1) stimulate pathogen reproduction to produce additional white blood cells

(2) increase the production of guard cells to defend against pathogens

(3) promote the production of antigens to stimulate an immune response

(4) recognize foreign antigens and mark them for destruction   30 _____

PART B-1

Answer all questions in this part.   **[13]**

Directions **(31–43): For *each* statement or question, record in the space provided the *number* of the word or expression that, of those given, best completes the statement or answers the question.**

Base your answers to questions 31 and 32 on the information below and on your knowledge of biology.

Fossil Footprints

Scientists examined a trail of fossil footprints left by early humans in soft volcanic ash in Eastern Africa. A drawing of the trail of footprints is shown below. Each footprint is represented as a series of lines indicating the depth that different parts of the foot sank into the volcanic ash.

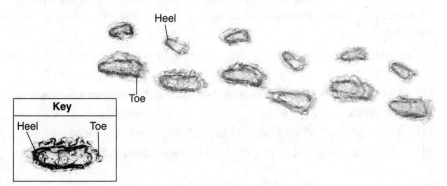

Source: http://www.indiana.edu/~eniweb/lessons/foot-topo-10inch.pdf

31. Which statement is an accurate observation that can be made based on this trail of footprints?

(1)   The individuals were running from a predator.

(2)   The volcano was about to erupt again.

(3)   One individual was much taller than the other.

(4)   One individual had larger feet than the other.   31 _____

32. The type of information directly provided by these fossil footprints is useful because it

(1) offers details about how these individuals changed during their lifetime

(2) offers data regarding their exposure to ultraviolet (UV) radiation

(3) is a record of information about what these individuals ate during their lifetime

(4) is a record of some similarities and differences they share with present-day species

32 _____

33. Since the early 1990s, proton pump inhibitors (PPIs) have been widely used to treat acid reflux disease. Although clinical tests in the 1980s deemed PPIs to be safe for humans, in 2012 the FDA announced warnings that long-term use of PPIs could increase the risk of bone fractures, kidney disease, and some intestinal infections.

Which statement best explains why the safety of PPIs is now in question when clinical experiments in the 1980s provided evidence that they were safe?

(1) Researchers have been able to collect more data than were available in the 1980s.

(2) Fewer people had acid reflux in the 1980s compared to today.

(3) The medication containing PPIs has changed since the 1980s when tests were done.

(4) The original experiments in the 1980s used only test animals and did not use human subjects.

33 _____

34. The process of embryonic development is represented in the diagram below.

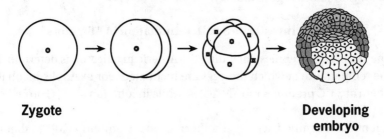

       Zygote   **Developing**
                                                      embryo

The three arrows in the diagram each represent a process known as

(1)   mitotic cell division

(2)   meiotic cell division

(3)   fertilization of gamete cells

(4)   differentiation of tissues   34 _____

35. A cell with receptors for two different hormones is represented below.

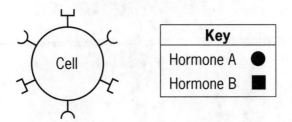

Key	
Hormone A	●
Hormone B	■

Which chemical would most likely interfere with the activity of hormone *A*, but *not* hormone *B*?

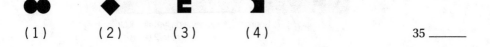

(1)   (2)   (3)   (4)   35 _____

Base your answers to questions 36 through 39 on the information and photograph below and on your knowledge of biology.

Scientists Investigate Sex Determination in Alligators

The sex of some reptiles, including the American alligator, is determined by the temperature at which the eggs are incubated. For example, incubating them at 33°C produces mostly males, while incubation at 30°C produces mostly females.

Scientists recently discovered a thermosensor protein, TRPV4, that is associated with this process in American alligators. TRPV4 is activated by temperatures near the mid-30s, and increases the movement of calcium ions into certain cells involved with sex determination.

A baby alligator emerges from its egg shell during hatching

Source: http://www.dailymail.co.uk/
news/article-2190839/

36. The results of this scientific investigation will most likely lead other scientists to hypothesize that

   (1) human sex cells also contain the TRPV4 protein
   (2) other reptiles may have the TRPV4 protein in their eggs
   (3) the TRPV4 protein affects the growth of plants
   (4) the TRPV4 protein is present in all of the foods eaten by alligators

36 _____

37. Which information was most essential in preparing to carry out this scientific investigation?

(1) a knowledge of the variety of mutations found in American alligator populations

(2) the arrangement of the DNA bases found in the TRPV4 protein

(3) the effects of temperature on the incubation of alligator eggs

(4) a knowledge of previous cloning experiments conducted on alligators and other reptiles   37 _____

38. The movement of the calcium ions into certain cells is most likely due to

(1) the destruction of the TRPV4 when it contacts the cell membrane

(2) the action of TRPV4 proteins on the cells involved with sex determination

(3) the sex of the alligator embryo present in that particular egg

(4) the action of receptor proteins attached to the mitochondria in alligator sex cells   38 _____

39. Environmental changes, such as global warming, could affect species such as the American alligator because even slight increases in environmental temperature could

(1) lead to an overabundance of females and few, if any, males

(2) lead to an overabundance of males and few, if any, females

(3) cause the breakdown of the TRPV4 protein in female alligators

(4) increase the rate at which calcium ions exit the sex cells   39 _____

_____

40. Which human activity can have a *negative* impact on the stability of a mature ecosystem?

(1) replanting trees in areas where forests have been cut down for lumber

(2) building dams to control the flow of water in rivers, in order to produce electricity

(3) preserving natural wetlands, such as swamps, to reduce flooding after heavy rainfalls

(4) passing laws that limit the dumping of pollutants in forests   40 _____

Base your answer to question 41 on the diagram below and on your knowledge of biology. The diagram represents a pond ecosystem.

Source: freshwaterecosystemswebquest.wikispaces.com/ ponds,+lakes,+and+inland+seas

41. Energy in this ecosystem passes directly from the Sun to

(1) herbivores

(3) heterotrophs

(2) consumers

(4) autotrophs

41 _____

Base your answers to questions 42 and 43 on the information and photograph below and on your knowledge of biology.

Wild Horse Roundup

Source: http://tuesdayshorse.wordpress.com/2012/10/
31outrageover-advisory-board-proposal-to-sterilize-wild-mustangs/

Wild horses called mustangs roam acres of federally owned land in the western United States. These horses have overgrazed the local vegetation to the extent that plants and soils are being lost entirely.

When the number of mustangs that roam the land exceeds the number of horses that the land can sustain, the government organizes helicopter-driven roundups. The horses are guided into a roped-off area and then are sold to the public or brought to pastures in the Midwest. About one percent of the horses captured die from injuries or accidents that occur during roundups.

42. The risk to the horses during the roundups compared to the entire loss of plants and soils is considered

(1) selective breeding   (3) direct harvesting

(2) a technological fix   (4) a trade-off   42 _____

43. The number of organisms that an area of land can sustain over a long period of time is known as

(1) ecological succession

(2) its finite resources

(3) its carrying capacity

(4) evolutionary change   43 _____

PART B-2

Answer all questions in this part.   [12]

Directions (44–55): **For those questions that are multiple choice, record your answers in the spaces provided. For all other questions in this part, record your answers in accordance with the directions.**

Base your answers to questions 44 through 48 on the information and data table below and on your knowledge of biology.

White Nose Syndrome Found in Bats

White nose syndrome (WNS) is a disease found in bats. The disease, first detected in bats during the winter of 2006, is characterized by the appearance of a white fungus on the nose, skin, and wings of some bats, which live in and around caves and mines. It affects the cycle of hibernation and is responsible for the deaths of large numbers of bats of certain species. In some areas, 80–90% of bats have died. Not all bats in an area are affected, and certain bats that are susceptible in one area are not affected in other areas.

The roles of temperature and humidity in the environment of the bats are two of the many factors being investigated to help control the disease. Over the past few years, the Conserve Wildlife Foundation of New Jersey conducted summer bat counts of two bat species at 22 different sites, totaled the number, and reported the results. The approximate numbers of bats counted (to the nearest hundred) are listed in the table below.

Summer Bat Count (Total Number of Bats)		
Year	Big Brown Bats (*Eptesicus fuscus*)	Little Brown Bats (*Myotis lucifugus*)
2009	900	6100
2010	1000	1700
2011	1000	500
2012	1000	400
2013	1300	300

Directions (44–46): **Using the information in the data table, construct a line graph on the grid below, following the directions below.**

44. Mark an appropriate scale, without any breaks in the data, on the axis labeled "Number of Bats."   [1]

45. Plot the data for big brown bats on the grid, connect the points, and surround each point with a small circle.   [1]

Example:

46. Plot the data for little brown bats on the grid, connect the points, and surround each point with a small triangle.   [1]

Example:

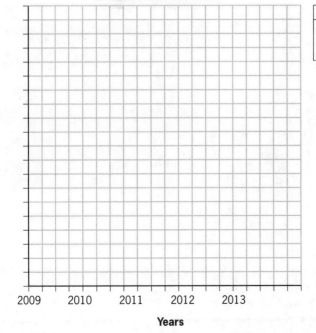

Summer Brown Bat Count

Key
⊙ =Big brown bats
△ =Little brown bats

Note: The answer to question 47 should be recorded in the space provided.

47. Biologists in New York and Vermont have noted that, in recent years, a higher percentage of the little brown bats are now surviving. Which statement best explains this increased survival rate?

   (1)   A few of the bats possessed an immunity to the WNS disease and produced offspring that were immune.

   (2)   The bats needed to reproduce in greater numbers, otherwise they would have died out completely.

   (3)   The people that performed the recent counts did not identify the bats correctly and were counting bats of a different species.

   (4)   The original decline in the bat population due to WNS was a natural occurrence and is part of a natural cycle.   47 _____

48. Conservation groups have promoted the building and placing of bat houses in areas thought to be most suitable for bat populations. Explain why this might have a positive effect on the control of WNS in bats.   [1]

   _____

   _____

Note: The answer to question 49 should be recorded in the space provided.

49. In the coastal waters off western North America, there is a starfish species that feeds primarily on mussels, another marine organism. In an experimental area, the starfish were removed from the waters. The effect of this removal is shown in the graph below.

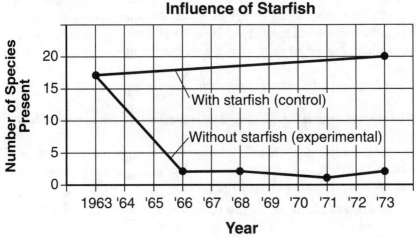

Influence of Starfish

Source: *Biology, 8th Ed.*, Campbell, Reese, et al. Pearson, San Francisco, CA, 2009, p. 1208.

What conclusion can be made regarding the role of the starfish in this ecosystem?

(1) The biodiversity of this ecosystem increased within ten years as organisms adjusted to the loss of the starfish.

(2) The starfish is important in maintaining the biodiversity of this ecosystem.

(3) When the starfish were removed, the ecosystem decreased in stability and increased in biodiversity.

(4) Biodiversity in this ecosystem is not dependent on the presence of starfish.   49 _____

Base your answers to questions 50 through 52 on the information below and on your knowledge of biology.

Biomass Energy

Biomass is the term for all living, or recently living, materials coming from plants and animals that can be used as a source of energy. Biomass can be burned to produce heat and used to make electricity. The most common materials used for biomass energy are wood, plants, decaying materials, and wastes, including garbage and food waste. Burning the wood and plant matter does produce some air pollutants. Biomass contains energy that originally came from the Sun. Some biomass can be converted into liquid biofuels. These biofuels can be used to power cars and machinery.

Note: The answer to question 50 should be recorded in the space provided.

50. In a community, before biomass is widely used as an energy source, several experts, including an ecologist, are hired to provide specific information. The ecologist would most likely be asked about

    (1)   the cost of producing the fuel compared with the profit when the fuel is sold
    (2)   whether the fuel will be widely accepted by consumers
    (3)   what effect the production of the fuel will have on the environment
    (4)   the time it will take to produce large amounts of the fuel   50 _____

51. Explain why biomass is considered a renewable energy source.   [1]

    _____

    _____

    _____

52. State *one* specific advantage and *one* specific disadvantage of the use of biofuels as an energy source.   [1]

Advantage: _____

_____

Disadvantage: _____

_____

Base your answers to questions 53 and 54 on the information below and on your knowledge of biology.

Photosynthesis is a process that is important to the survival of many organisms on Earth.

53. Identify *two* raw materials necessary for photosynthesis.   [1]

_____ and _____

54. State *one* reason why photosynthesis is necessary for animals to survive.   [1]

_____

_____

55. The diagram below represents a pond ecosystem.

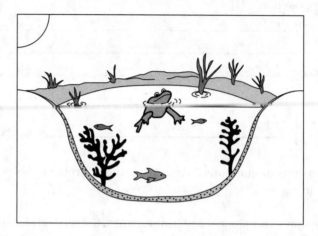

Identify *one* abiotic factor present in the pond ecosystem and explain how this abiotic factor would affect the frogs in the pond.   [1]

Abiotic factor: _____

_____

_____

PART C

Answer all questions in this part.   [17]

Directions (56–72): **Record your answers in the spaces provided.**

Base your answers to questions 56 through 58 on the information below and on your knowledge of biology.

Bye–Bye Bananas?

The world's most popular type of banana is facing a major health crisis. According to a new study, a disease caused by a powerful fungus is killing the Cavendish banana, which accounts for 99% of the banana market around the globe. The disease, called tropical race 4 (TR4), has affected banana crops in southeast Asia for decades. In recent years, it has spread to the Middle East and the African nation of Mozambique. Now experts fear the disease will show up in Latin America, where the majority of the world's bananas are grown. ...

...Once a banana plant is infected with TR4, it cannot get nourishment from water and nutrients, and basically dies of thirst. TR4 lives in soil, and can easily end up on a person's boots. If the contaminated boots are then worn on a field where Cavendish bananas are grown, the disease could be transferred. "Once a field has been contaminated with the disease, you can't grow Cavendish bananas there anymore," Randy Ploetz [scientist] says. "The disease lasts a long time in the soil."...

...But Cavendish [banana] is also particularly vulnerable to TR4. The banana is grown in what is called monoculture. "You see a big field of bananas and each one is genetically identical to its neighbor" Ploetz says. "And they are all uniformly susceptible to this disease. So once one plant gets infected, it just runs like wildfire throughout that entire plantation."...

Source: http://www.timeforkids.com/new/bye-bye-bananas/3311666

56. State how the TR4 fungus threatens homeostasis within the banana plant.   [1]

_____

_____

57. Explain why the entire Cavendish banana crop worldwide is particularly vulnerable to the TR4 fungus.   [1]

_____

_____

58. If the fungus cannot be stopped by chemical treatment of the soil, describe *one* other possible way that the growers may be able to combat the disease. [1]

_____

_____

Base your answers to questions 59 through 61 on the passage below and on your knowledge of biology.

Lead Poisoning

Two pathways by which lead can enter the human body are ingestion and inhalation. Once in the bloodstream, lead is distributed to parts of the body including the brain, bones, and teeth.

One reason that lead is toxic is that it interferes with the functioning of a variety of enzymes. It acts like metals such as calcium and iron and replaces them, changing the molecular structure of these enzymes. In the case of calcium, lead is absorbed through the same cell membrane channels that take in calcium.

Lead affects children and adults in different ways. Even low lead levels in children can cause many different problems, including nervous system damage, learning disabilities, decreased intelligence, poor bone growth, and death. In adults, high levels of lead can cause hearing problems, memory and concentration problems, muscle and joint pain, brain damage, and death. It wasn't until 1971 that steps were taken against the use of lead with the passage of the Lead Poisoning Prevention Act. However, lead is still a public health risk today.

59. It is recommended that children eat foods high in calcium and iron as a way to reduce the accumulation of lead in their cells and enzymes. Explain why this is a scientifically valid recommendation. [1]

_____

_____

---

60. Describe how the presence of lead in body cells could interfere with the ability of enzymes to function.   [1]

_____

_____

_____

61. Based on the parts of the body that are most affected by lead intake, identify *one* type of cell that would be expected to have numerous calcium channels. Support your answer.   [1]

Type of cell: _____

Support: _____

_____

_____

Base your answers to questions 62 and 63 on the illustration and passage below and on your knowledge of biology.

The Telltale DNA of Manx Cats

Source: http://commons.wikimedia.org/wiki/File:
Manx cat (stylizes) 1885.jpg

A few breeds of cat have no tails. Manx cats have extremely short tails and may even appear to have no tail at all. Manx cats were first discovered several hundred years ago.

Scientists have determined that a certain mutation in a group of genes (called T-box genes) interferes with the development of the spine in the cat embryo. Mutations in these T-box genes can cause abnormalities in the number, shape, and/or size of bones in the spines of Manx cats, which results in smaller spines and shorter tails.

If a Manx cat inherits one copy of the mutated T-box gene and one copy of the normal gene, it will have a very short tail or no tail at all.

If the cat embryo inherits two copies of these mutated genes, it will stop developing and die. Therefore, all surviving Manx cats have only one copy of the mutated gene.

62. State *one* reason why the mutation in Manx cat embryos causes them to have such very short tails.   [1]

_____

_____

63. Two Manx cats have several litters of offspring. Explain how the genes that the kittens inherit determine whether they will have a normal tail or a short tail.   [1]

_____

_____

_____

Base your answer to questions 64–66 on the passage below and on your knowledge of biology.

Hummingbirds Are Sugar Junkies

Source: http://bug-bird.com/hummingbirds-large-images/

Most humans enjoy candy, cake, and ice cream. As a result of evolutionary history, we have a wide variety of tastes. This is not true of all animals. Cats do not seek sweets. Over the course of their evolutionary history, the cat family tree lost a gene to detect sweet flavors. Most birds also lack this gene, with a few exceptions. Hummingbirds are sugar junkies.

Hummingbirds evolved from an insect-eating ancestor. The genes that detect the savory flavor of insects underwent changes, making hummingbirds more sensitive to sugars. These new sweet-sensing genes give hummingbirds a preference for high-calorie flower nectar. Hummingbirds actually reject certain flowers whose nectar is not sweet enough!

64–66. Discuss how sweet sensitivity in hummingbirds has developed. In your answer, be sure to:

- identify the initial event responsible for the new sweet-sensing gene   [1]
- explain how the presence of the sweet-sensing gene increased in the hummingbird population over time   [1]
- describe how the fossil record of hummingbird ancestors might be used to learn more about the evolution of food preferences in hummingbirds   [1]

_____

_____

_____

_____

_____

_____

_____

_____

_____

_____

_____

_____

_____

Base your answers to questions 67 through 70 on the information below and on your knowledge of biology.

Folic acid is a type of vitamin that is essential for the normal growth and development of cells in the body. If a woman consumes folic acid in her diet before and during the earliest stages of pregnancy, it can help to reduce her baby's risk for developing a type of birth defect called a neural tube defect. Early in pregnancy, the neural tube forms the brain and spinal cord. If the neural tube does not form properly, serious birth defects may result.

67. Explain why taking folic acid early in pregnancy is important to the prevention of neural tube defects.   [1]

_____

_____

_____

68. Describe how a fetus receives folic acid and other essential materials directly from its mother for its development.   [1]

_____

_____

69. Identify *one* factor, other than a lack of folic acid, that may interfere with the proper development of essential organs during pregnancy.   [1]

_____

_____

70. Many foods, such as breads, cereals, pastas, and rice, are fortified or enriched with folic acid. Explain why adding folic acid to foods is an advantage to people other than pregnant women.   [1]

_____

_____

_____

Base your answers to questions 71 and 72 on the diagram below and on your knowledge of biology. The diagram represents an evolutionary tree.

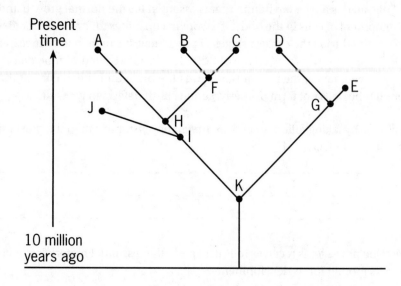

71. Are species *A* and *B* more closely related than *A* and *D*? Circle yes *or* no and support your answer with information from the diagram.   [1]

Circle one: Yes   *or*   No

_____

_____

72. State *one* possible cause for the extinction of species *E*.   [1]

_____

PART D

Answer all questions in this part.   [13]

Directions **(73–85): For those questions that are multiple choice, record your answers in the spaces provided. For all other questions in this part, record your answers in accordance with the directions.**

Note: The answer to question 73 should be recorded in the space provided.

Base your answer to question 73 on the information and diagram below and on your knowledge of biology.

During the *Relationships and Biodiversity* lab, simulated pigments from three plant species were compared to those in *Botana curus*. The results were similar to those represented below.

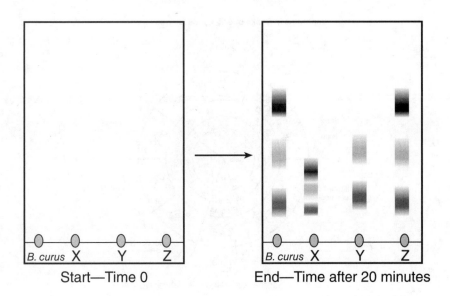

Start—Time 0   End—Time after 20 minutes

73. Based on the results of this comparison alone, is there enough information to conclude which of the other three species is most closely related to *Botana curus*?

(1) Yes. Only species *X* has the same bands as *Botana curus*.

(2) Yes. Species *Z* has only two of the bands that *Botana curus* has.

(3) No. Additional tests should be done to test for other chemical similarities.

(4) No. Other rainforest plant species should be tested.   73 _____

Note: The answer to question 74 should be recorded in the space provided.

Base your answer to question 74 on the diagram below and on your knowledge of biology.

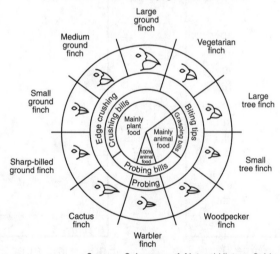

Variations in Beaks of Galapagos Islands Finches

Source: *Galapagos: A Natural History Guide*

74. Insects can get diseases just like other organisms. A deadly bacteria infected the insects on one Galapagos Island. Among the birds living there, the finches most likely to experience a drastic *decrease* in population size would be the

(1) warbler finches

(2) cactus finches

(3) large ground finches

(4) medium ground finches   74 _____

Note: The answer to question 75 should be recorded in the space provided.

75. A variety of species of Galapagos finches evolved from one original species long
    ago through the process of

    (1)   asexual reproduction   (3)   natural selection
    (2)   ecological succession   (4)   selective breeding   75 _____

Note: The answer to question 76 should be recorded in the space provided.

76. If scientists want to determine the similarities in the DNA fragments in several
    plant species, they should

    (1)   add salt water to cells from each plant
    (2)   analyze electrophoresis results
    (3)   compare seed structures of the plants
    (4)   examine their chromosomes with a microscope   76 _____

77. One coach of an Olympic rowing team makes his athletes warm up by doing
    30 minutes of stretching and jogging in place before practicing each day.
    Another coach suggests that resting before practicing will result in better perfor-
    mance by her team. They decide to conduct an experiment to see which practice
    is correct. One team rests before practice, the other team warms up for thirty
    minutes, and they then record the time that it takes each team to row a specific
    distance. Identify the dependent variable in this experiment.   [1]

    _____

78. A student squeezes a clothespin 82 times in a minute. Then, using the same
    hand and the same clothespin, he squeezes the clothespin 68 times in a minute.
    State *one* biological reason for the decrease in the number of squeezes during
    the second trial.   [1]

    _____

    _____

79. During rest, an adult's heart rate averages 60–100 beats per minute. When exercising, an adult's heart rate may increase to 100–170 beats per minute. State *one* reason why the heart rate increased during exercise.   [1]

_____

_____

80. In some single-celled protozoans living in fresh water, such as the paramecium, contractile vacuoles are organelles used to pump excess water out of the cell. Explain why a paramecium would require contractile vacuoles while a similar protozoan living in salt water would *not*.   [1]

_____

_____

_____

Base your answers to questions 81 through 83 on the information and diagram below and on your knowledge of biology.

A cube cut from a potato is placed in a beaker of distilled water. The potato cells have a relatively high concentration of starch and a relatively low concentration of water. The diagram represents the water and starch molecules in and around one of the potato cells in contact with the water in the beaker.

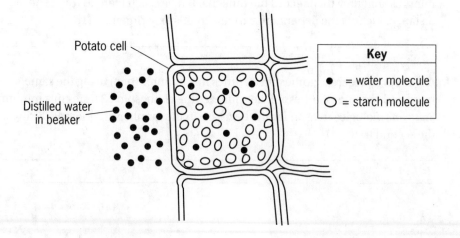

Note: The answer to question 81 should be recorded in the space provided.

81. Which row in the chart correctly describes what would be expected to occur in the potato cells, with regard to both the starch and water molecules?

Row	Water	Starch
(1)	More water will move into the cell than will leave the cell.	Starch will remain inside the potato cell.
(2)	More water will leave the cell than will enter.	Starch will move out of the cell.
(3)	Water content of the cell will not change.	Starch will move out of the cell.
(4)	More water will leave the cell than will enter.	Starch will remain inside the potato cell.

81 _____

Note: The answer to question 82 should be recorded in the space provided.

82. Which statement correctly describes a possible result if starch indicator is added to the water in the beaker one hour after the potato cube was added?

(1)   The indicator solution would turn to an amber color in the water if starch molecules were present in the water in the beaker.

(2)   The indicator would remain amber in color if starch molecules were not present in the water in the beaker.

(3)   The indicator would change to a black color if starch molecules were not present in the water in the beaker.

(4)   The indicator solution would remain black in color if starch molecules were present in the water in the beaker.

82 _____

83. Before placing the potato in the beaker, the student used an electronic balance to determine the mass of the potato cube. The mass of the cube was determined again after it was in the beaker for an hour. Describe how this information could specifically be used to determine if water moved during the investigation.   [1]

_____

_____

_____

84. Select *one* row in the chart below and explain how the systems in that row work together during exercise.   [1]

	System	System	System
Row 1	Respiratory	Circulatory	Muscular
Row 2	Muscular	Circulatory	Excretory
Row 3	Digestive	Circulatory	Muscular

Row: _____

_____

_____

_____

Base your answer to question 85 on the information below and on your knowledge of biology.

In an experiment, a membrane bag containing 95% water and 5% salt was placed in a beaker containing 80% water and 20% salt, as shown below. The setup was put aside until the next day.

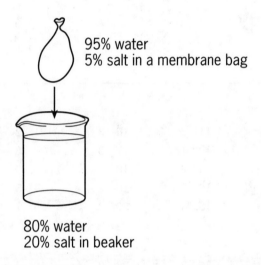

95% water
5% salt in a membrane bag

80% water
20% salt in beaker

85. Based on the information given, state *one* way the bag or its contents will have changed by the next day. Support your answer with an explanation.   [1]

_____

_____

_____

Answers August 2019
Living Environment

Answer Key

PART A

1. 4	6. 3	11. 2	16. 3	21. 2	26. 3
2. 2	7. 4	12. 1	17. 3	22. 3	27. 3
3. 3	8. 2	13. 2	18. 1	23. 3	28. 4
4. 2	9. 2	14. 2	19. 1	24. 3	29. 3
5. 4	10. 1	15. 4	20. 2	25. 1	30. 4

PART B-1

31. 4	34. 1	36. 2	38. 2	40. 2	42. 4
32. 4	35. 1	37. 3	39. 2	41. 4	43. 3
33. 1					

PART B-2

44. *See* Answers Explained.

45. *See* Answers Explained.

46. *See* Answers Explained.

47. 1

48. *See* Answers Explained.

49. 2

50. 3

51. *See* Answers Explained.

52. *See* Answers Explained.

53. *See* Answers Explained.

54. *See* Answers Explained.

55. *See* Answers Explained.

PART C. *See* **Answers Explained**.

PART D

73. 3

74. 1

75. 3

76. 2

77. *See* Answers Explained.

78. *See* Answers Explained.

79. *See* Answers Explained.

80. *See* Answers Explained.

81. 1

82. 2

83. *See* Answers Explained.

84. *See* Answers Explained.

85. *See* Answers Explained.

Answers Explained

PART A

1. **4** In this pyramid, the greatest amount of stored energy is found at level *D*. In any traditional energy pyramid, the base of the pyramid (producer level) is presumed to contain the greatest quantity of energy. This level represents green plants that absorb solar energy and convert it to the chemical bond energy of glucose.

   WRONG CHOICES EXPLAINED:
   (1) In this pyramid, the greatest amount of stored energy is *not* found at level *A*. Level *A*, which represents consumers such as top predators, contains the smallest amount of energy of the levels shown in the diagram.

   (2) In this pyramid, the greatest amount of stored energy is *not* found at level *B*. Level *B*, which represents carnivores (meat eaters), contains less energy than the levels below it in the diagram.

   (3) In this pyramid, the greatest amount of stored energy is *not* found at level *C*. Level *C*, which represents herbivores (plant eaters), contains less energy than the level below it in the diagram.

2. **2** As a result of the plant disease, the butterfly population will most likely *disappear from the area*. Absent their main food source, the butterfly larvae will likely die of starvation before they can pupate. Adult butterflies of this particular species will lay their eggs on members of this certain plant species that may be surviving in nearby fields.

   WRONG CHOICES EXPLAINED:
   (1), (3), (4) As a result of the plant disease, the butterfly population will *not* most likely *quickly adapt to eat other plants, evolve to form a new species,* or *enter the adult stage more quickly*. These responses all imply evolutionary modifications based on need rather than on natural selection. These changes are unlikely to occur as a result of natural selection, which requires random genetic mutations coupled with environmental selection pressures over substantial time.

3. **3** This protozoan is an example of a *parasite*. Parasites are other feeders that feed on the bodies of living plants and animals in their ecosystems. This protozoan may enter the bodies of humans and feed on their living tissues.

WRONG CHOICES EXPLAINED:

(1) This protozoan is *not* an example of a *predator*. Predators are other feeders that hunt and feed on the killed bodies of herbivores and carnivores in their ecosystems. The protozoan described in the question does not fit the definition of a predator.

(2) This protozoan is *not* an example of a *producer*. Producers are photosynthetic self-feeders that generally do not feed on any other organisms in their ecosystems. The protozoan described in the question does not fit the definition of a producer.

(4) This protozoan is *not* an example of a *scavenger*. Scavengers are other feeders that feed on the dead bodies of animals in their ecosystems. The protozoan described in the question does not fit the definition of a scavenger.

4. 2 The energy required to maintain this concentration difference is most likely associated with the action of *mitochondria*. Mitochondria are cell organelles responsible for the conversion of chemical bond energy in glucose to the chemical bond energy of ATP. ATP is used in the cell as a source of energy to operate cell functions such as active transport. Active transport is a process by which substances are transported across cell membranes from an area of low relative concentration to an area of high relative concentration, requiring the expenditure of cell energy. The seaweed's ability to concentrate iodine in its cells is an example of the energy-consuming action of active transport in nature.

WRONG CHOICES EXPLAINED:

(1) The energy required to maintain this concentration difference is *not* most likely associated with the action of *ribosomes*. Ribosomes function in protein synthesis by capturing mRNA molecules, attracting tRNA molecules, and linking amino acids in the arrangement determined by the genetic code. Ribosomes are not directly associated with the process of active transport of iodine in cells.

(3) The energy required to maintain this concentration difference is *not* most likely associated with the action of *vacuoles*. Vacuoles are cell organelles that store food or wastes in the cell. Such vacuoles are not directly associated with the process of active transport of iodine in cells.

[NOTE: The action of the contractile vacuole, found in certain protozoa, pumps water from inside the cell (an area of low relative concentration of water) to the cell's aquatic environment (an area of high relative concentration of water) with the expenditure of ATP energy. This process is similar to but distinct from the process of active transport that allows seaweed to concentrate iodine in its cells.]

(4) The energy required to maintain this concentration difference is *not* most likely associated with the action of *nuclei*. Nuclei are cell organelles that store the genetic codes of the cell. Nuclei are not directly associated with the process of active transport of iodine in cells.

5. **4** These vaccines most often contain *weakened pathogens*. Proteins known as antigens are embedded in the cell membrane of pathogenic organisms such as bacteria and viruses. By adding weakened or dead pathogens to a vaccine and injecting the vaccine into the tissues of a human, the body's immune system is stimulated to produce specific antibodies to neutralize the antigen. This immune response protects the body against future attacks by live pathogens of the same type as are found in the vaccine.

WRONG CHOICES EXPLAINED:
(1) These vaccines do *not* most often contain *antibodies*. Antibodies are substances produced by the immune system in response to the presence of a foreign antigen. Antibodies are not found in vaccines.

(2) These vaccines do *not* most often contain *antibiotics*. Antibiotics are biochemical compounds produced naturally by certain molds that are lethal to specific bacterial species. Antibiotics are not found in vaccines.

(3) These vaccines do *not* most often contain *white blood cells*. White blood cells are structures produced in the circulatory system that engulf and devour pathogens as a means of protecting the body against infectious disease. White blood cells are not found in vaccines.

6. **3** Building large manufacturing facilities can affect ecosystems by increasing the *demand for resources such as fossil fuels*. Human manufacturing activities require the expenditure of large quantities of electrical energy. Much of the electrical energy produced in the United States comes from power plants that burn fossil fuels such as coal, oil, and natural gas. Producing electrical energy in this manner uses the limited carbon-based resources, pollutes the air and water, and produces greenhouse gases that are known to contribute to climate change.

WRONG CHOICES EXPLAINED:
(1), (2), (4) It is *not* true that building large manufacturing facilities can affect ecosystems by increasing the *atmospheric quality*, the *biodiversity in the area*, or the *availability of space and resources for organisms*. Manufacturing activities typically decrease, not increase, these environmental factors in the area in which factories are constructed.

7. 4 From this, we can best infer that *structures in amebas have functions similar to organs in multicellular organisms.* In order to survive in their environment, all organisms must perform the same basic life functions. In multicellular organisms such as the human, these life functions are performed by specialized tissues, organs, and organ systems. In single-celled organisms such as the ameba, these life functions are performed by specialized organelles.

WRONG CHOICES EXPLAINED:
(1) It is *not* true that from this, we can best infer that *all single-celled organisms have lysosomes to digest food.* The information presented in the question provides information about only amebas, not all single-celled organisms.

(2) It is *not* true that from this, we can best infer that *amebas are capable of digesting any type of food molecule.* No information presented is in the question concerning the types of food molecules the ameba is capable of digesting.

(3) It is *not* true that from this, we can best infer that *single-celled organisms are as complex as multicellular organisms.* Single-celled organisms such as the ameba are among the simplest life forms that inhabit Earth. By definition, multicellular organisms are more complex than single-celled organisms.

8. 2 White blood cells are most closely associated with the *immune and circulatory* systems. White blood cells are immune cells that travel through the circulatory system to engulf and devour pathogens as a means of protecting the body against infectious disease. In this way, white blood cells function in both the immune and circulatory systems of the human body.

WRONG CHOICES EXPLAINED:
(1), (3), (4) White blood cells are *not* most closely associated with the *circulatory and digestive* systems, the *digestive and excretory* systems, or the *excretory and immune* systems. The digestive system processes food for the body's metabolic activities, while the excretory system eliminates waste materials resulting from those activities. White blood cells do not directly function in either of these systems.

9. 2 The bogs only support certain types of organisms because *the growth and survival of organisms depends on specific physical conditions.* Each of Earth's species functions best under a specific set of abiotic (nonliving or physical) conditions unique to that species. Carnivorous bog plant species have evolved in a manner that allows them to thrive under conditions of low oxygen, low mineral nutrients, and high acidity.

WRONG CHOICES EXPLAINED:
(1) It is *not* true that the bogs support only certain types of organisms because *organisms in an environment are not limited by available energy and resources.* All organisms in all environments are subject to limitations imposed by the amounts of energy and other resources available in those environments.

(3) It is *not* true that the bogs support only certain types of organisms because *favorable gene mutations only occur when organisms live in harsh environments.* Mutations are random events in which genetic material (DNA) is altered, sometimes in favorable ways and sometimes in unfavorable ways. Mutations occur in the cells of organisms irrespective of the environments those organisms inhabit.

(4) It is *not* true that the bogs support only certain types of organisms because *photosynthetic organisms can only inhabit environments that have low acidity.* Bogs and other acidic environments support photosynthetic organisms adapted to survive and thrive in those environments.

10. **1** Anhidrosis most directly interferes with *a feedback mechanism that maintains homeostasis.* Under normal circumstances, the human body maintains its internal temperature within a narrow range around 98.68°F (37°C). The release and evaporation of perspiration (sweat) from tiny sweat glands embedded in the skin is a key factor in lowering body temperature when it gets too high from exercise or environmental heating. A feedback mechanism regulates this cooling effect by opening sweat glands when blood temperature increases and closing them when the blood returns to normal temperature. This feedback mechanism helps to maintain the body's homeostatic balance. Anhidrosis prevents the sweat glands from operating properly and threatens to upset this homeostatic balance.

WRONG CHOICES EXPLAINED:
(2) It is *not* true that anhidrosis most directly interferes with *an immune system response to harmless antigens.* Such responses are known as allergies. The action of anhidrosis is not known to affect the body's allergic response.

(3) It is *not* true that anhidrosis most directly interferes with *the synthesis of hormones in the circulatory system.* Hormones are synthesized in discrete endocrine tissues embedded in certain glands and body organs, not in the blood. The action of anhidrosis is not known to affect the body's synthesis of hormones.

(4) It is *not* true that anhidrosis most directly interferes with *the enzymatic breakdown of water in cells.* Water molecules cannot be broken down by means of any known enzymatic reaction. This is a nonsense distracter.

11. 2 The hair colors of the sons are most likely a direct result of *heredity*. During sexual reproduction, chromosomes carrying parental genes are transmitted from the parents to their offspring. This transmittal, sometimes referred to as heredity, allows the offspring of the parents to inherit and display traits that may be observed in the parents. Hair color is known to be an observable inheritable trait.

WRONG CHOICES EXPLAINED:
(1), (3) The hair colors of the sons are *not* most likely a direct result of *natural selection in males* or *evolution*. Natural selection is thought to be the major biological mechanism that drives the evolution of Earth's species. These processes occur over vast periods of geological time and are not observable in a single generational change. That being said, hair color in humans is almost certainly a species trait resulting from natural selection and evolution that has occurred over time.

[NOTE: This ambiguity makes these responses possible correct answers to the question as written, although they are not the best answers.]

(4) The hair colors of the sons are *not* most likely a direct result of *environmental influences*. Environmental factors can and often do modify the expression of genetic traits in humans and other organisms. Because the question does not explicitly state that the identified hair colors of the parents and sons are natural, not dyed (an environmental effect), it leaves open the question of whether genetic inheritance or environmental factors are responsible for the observed traits.

[NOTE: This ambiguity makes this response a possible correct answer to the question as written, although it is not the best answer.]

12. 1 The percentage of the bases that will be thymine (T) is approximately *18*. The percentage of cytosine (C) at 32% must match the percentage of guanine (G) in the sample since these two bases are complementary in a DNA strand. The combined total of C and G is 64%. The percentages of thymine (T) and adenine (A), as complementary bases, must also be equal to each other. Their combined total must make up the balance of the 100% of all bases in the sample. T and A, therefore, must make up a combined total of 36% of the sample, or approximately 18% each.

WRONG CHOICES EXPLAINED:
(2), (3), (4) The percentage of the bases that will be thymine (T) is *not* approximately *24*, *32*, or *36*. None of these responses can result from using the appropriate calculation method shown in the correct answer above.

13. **2** This response to carmine is known as *an allergy*. An allergy is an immune response to a usually harmless environmental substance. Common environmental substances, including plant and animal pigments such as carmine, carry foreign proteins known as antigens. When these substances are ingested, the body's immune system recognizes these antigens and may respond to them by producing histamines. Histamines, in turn, cause symptoms such as congestion and swelling of mucous membranes that can interfere with breathing. The symptoms caused by the immune response are known collectively as an allergic reaction.

WRONG CHOICES EXPLAINED:
(1) This response to carmine is *not* known as *a stimulus*. A stimulus is any change in the environment to which an organism responds. The introduction of carmine into the body can be thought of as an environmental change (stimulus) that triggers the immune response in an allergic person. However, the body's allergic reaction to carmine itself is not a stimulus.

(3) This response to carmine is *not* known as *natural selection*. The theory of natural selection postulates that certain members of a species are better adapted to their environment than others and that these organisms are more likely to survive and pass on their favorable adaptations to future generations. The human body's reaction to carmine does not constitute natural selection.

(4) This response to carmine is *not* known as *an adaptation*. An adaptation is any physical, chemical, or behavioral characteristic of an organism that helps it to survive potentially harsh environmental conditions. The allergic reaction (an immune system function) in response to carmine can be thought of as a behavioral characteristic that aids the survival of the allergic person.

[NOTE: This ambiguity makes this response a possible correct answer to the question as written, although it is not the best answer.]

14. **2** *Genetic engineering* is the process that was most likely used to alter these mosquitoes. Genetic engineering is a laboratory technique in which a gene for a desired trait is inserted into the genome of a recipient reproductive cell. When the recipient reproductive cell participates in successful fertilization, the resulting offspring will carry the altered gene and pass it on to future generations. In this case, the *Anopheles* mosquito is the recipient organism and the desired trait is the inability to transmit malaria.

WRONG CHOICES EXPLAINED:
(1) *Cloning studies* is *not* the process that was most likely used to alter these mosquitoes. Cloning is a laboratory technique that involves the removal of a diploid nucleus from a donor organism and inserting it into a denucleated fertilized egg cell of the same or related species. The resulting offspring is a genetic duplicate (clone) of the donor organism, not an artificially altered organism.

(3) *Natural selection* is *not* the process that was most likely used to alter these mosquitoes. The theory of natural selection postulates that certain members of a species are better adapted to their environment than others and that these organisms are more likely to survive and pass on their favorable adaptations to future generations. The question describes a laboratory technique, not a natural process.

(4) *Random mutations* is *not* the process that was most likely used to alter these mosquitoes. Mutations are random events in which genetic material (DNA) is altered through the action of natural mutagenic agents, sometimes in favorable ways and sometimes in unfavorable ways. The question describes a laboratory technique, not a natural process.

15. **4** Row *4* in the chart accurately identifies two causes of mutations and the cells that must be affected in order for the mutations to be passed on to offspring. Mutations are random events in which genetic material (DNA) is altered through the action of mutagenic agents, including chemicals and radiation. When mutations occur in sex cells, they may be passed on to future generations by sexual reproduction, thus perpetuating the mutations and their effects.

WRONG CHOICES EXPLAINED:
(1), (2) Rows *1* and *2* in the chart do *not* accurately identify two causes of mutations and the cells that must be affected in order for the mutations to be passed on to offspring. Mutations occurring in body cells cannot be passed on to future generations. Mutagenic agents do not include infections, antigens, meiosis, or mitosis.

(3) Row *3* in the chart does *not* accurately identify two causes of mutations and the cells that must be affected in order for the mutations to be passed on to offspring. Mutagenic agents, even in sex cells, do not include disease and differentiation.

16. **3** One way this form of reproduction differs from sexual reproduction is *the offspring and the parents are genetically identical.* The reproductive process illustrated in the question is asexual reproduction, in which new organisms are formed from a single parent organism. This process involves mitotic cell division, which does not provide opportunities for the creation of new genetic combinations. For this reason, the offspring of asexual reproduction are genetic duplicates of the parent.

WRONG CHOICES EXPLAINED:

(1) It is *not* true that one way this form of reproduction differs from sexual reproduction is *more genetic variations are seen in the offspring.* In fact, the offspring of asexual reproduction show no genetic variation from each other whatsoever.

(2) It is *not* true that one way this form of reproduction differs from sexual reproduction is *there is a greater chance for mutations to occur.* In fact, the chance of mutations occurring is the same in both sexual and asexual reproduction.

(4) It is *not* true that one way this form of reproduction differs from sexual reproduction is *the new plants possess the combined genes of both parents.* In fact, the offspring of asexual reproduction have only one parent with which they share identical genetic information.

17. **3** If the fish population *decreases,* the most direct effect this will have on the aquatic ecosystem is that *the zooplankton population will increase in size.* According to the diagram, zooplankton serve as a food source for fish. If fewer fish are present in the environment, the zooplankton population will likely increase until some limiting factor (such as a reduction of the phytoplankton population) acts to balance it.

WRONG CHOICES EXPLAINED:

(1) It is *not* true that if the fish population *decreases,* the most direct effect this will have on the aquatic ecosystem is that *the leopard seals will all die from lack of food.* According to the diagram, leopard seals feed on seagulls and penguins in addition to fish. If fewer fish are present in the environment, the leopard seals will likely move to consume more of their alternate food sources.

(2) It is *not* true that if the fish population *decreases,* the most direct effect this will have on the aquatic ecosystem is that *the krill population will only be consumed by seagulls.* According to the diagram, there is no direct nutritional link between krill and seagulls.

(4) It is *not* true that if the fish population *decreases,* the most direct effect this will have on the aquatic ecosystem is that *the phytoplankton population will increase in size.* According to the diagram, phytoplankton serve as a food source for zooplankton. If fewer fish are present in the environment, the zooplankton population will increase and consume more phytoplankton. So, the phytoplankton population will likely decrease, not increase.

18. 1 This sequence of changes is the result of *ecological succession*. Ecological succession refers to the series of changes that occur to the dominant plant community of an area that has been significantly affected by an environmental disruption. The sequence shown in the chart is typical of ecological succession that occurs in abandoned fields in the northeastern United States. Such succession typically begins with grasses, which are then replaced by shrubs, which are replaced in turn by hardwood forest trees.

WRONG CHOICES EXPLAINED:
(2) This sequence of changes is *not* the result of *decreased biodiversity*. Biodiversity is a measure of the number of different, compatible species that inhabit an area. Ecological succession is typically characterized by increased, not decreased, biodiversity.

(3) This sequence of changes is *not* the result of *biological evolution*. Biological evolution is a process characterized by gradual change of existing species over time that results in new or modified species. In this example, new species are not being developed. Rather, the sequence of changes listed in the chart indicates that existing species are replacing other existing species in a modified environment.

(4) This sequence of changes is *not* the result of *environmental trade-offs*. An environmental trade-off is any human decision that takes into consideration the positive and negative aspects of that decision. The sequence of changes listed in the chart describe a natural, not a human-made, phenomenon.

19. 1 *Genetically modified salmon produce more of some proteins than wild salmon* is the statement regarding genetically modified salmon that is correct. Scientists working in this area of science have discovered that adding certain genes to the genome of wild salmon results in larger, faster-growing fish. By modifying the wild salmon in this way, more of certain proteins are synthesized in the modified salmon than in the wild salmon.

WRONG CHOICES EXPLAINED:
(2) *Genetically modified salmon and wild salmon would have identical DNA* is *not* the statement regarding genetically modified salmon that is correct. Although the wild and modified salmon share most genes in common, the fact that scientists added genes to the modified salmon's genome ensures that their DNA is not identical.

(3) *Wild salmon reproduce asexually while genetically modified salmon reproduce sexually* is *not* the statement regarding genetically modified salmon that is correct. Fish of all species and all types reproduce sexually, not asexually. No information is presented to suggest that this characteristic would change under these circumstances.

(4) *Wild salmon have an altered protein sequence, but genetically modified salmon do not* is *not* the statement regarding genetically modified salmon that is correct. In fact, the opposite is true. Scientists have altered the genetic sequence of genetically modified salmon. So, the protein sequence of these salmon is altered by comparison with wild salmon.

20. **2** *Decomposers* is the group of organisms in an ecosystem that fills the niche of recycling organic matter back to the environment. Decomposers, such as bacteria and fungi, secrete enzymes that digest complex organic matter and convert it to simpler materials that can be taken up by green plants. In this way, decomposers aid in the recycling of carbon, hydrogen, oxygen, nitrogen, and other elements in the natural environment.

WRONG CHOICES EXPLAINED:
(1), (4) *Carnivores* and *predators* are *not* the groups of organisms in an ecosystem that fill the niche of recycling organic matter back to the environment. Carnivores and predators represent a nutritional class of animals that consume the bodies of other animals for food. Although carnivores and predators convert organic matter to other forms and release metabolic wastes to the environment, they do not perform the recycling action of decomposers.

(3) *Producers* is *not* the group of organisms in an ecosystem that fills the niche of recycling organic matter back to the environment. Producers represent a nutritional class of organisms that convert inorganic carbon dioxide and water into organic sugars with the input of solar energy. Although producers convert inorganic matter to organic matter and release metabolic wastes to the environment, they do not perform the recycling action of decomposers.

21. **2** A benefit of using solar energy would include *using less fossil fuel to meet energy needs*. Much of the electrical energy produced in the United States comes from power plants that burn fossil fuels, such as coal, oil, and natural gas. Producing electrical energy using solar collectors instead of fossil fuels preserves limited carbon-based resources and reduces pollution of the air and water.

WRONG CHOICES EXPLAINED:
(1), (4) A benefit of using solar energy would *not* include *adding more carbon dioxide to the atmosphere* or *releasing more gases for photosynthesis*. Photosynthesis uses carbon dioxide gas in the atmosphere as a raw material for the manufacture of glucose. Solar energy technology does not produce carbon dioxide for release to the atmosphere.

(3) A benefit of using solar energy would *not* include *using a nonrenewable source of energy*. Solar energy is renewable because the Sun is a stable energy source that will not run out for billions of years into the future.

22. 3 For these microorganisms, the wastewater facility serves as *an ecosystem*. An ecosystem is the basic functional unit of the living environment, including all the communities of organisms that interact with each other and with the nonliving environment. Because the microorganisms in the wastewater facility represent a community of living species interacting with each other and with a limited set of nonliving components, it can be said that the microorganisms inhabit an ecosystem of sorts.

WRONG CHOICES EXPLAINED:
(1) It is *not* true that for these microorganisms, the wastewater facility serves as *its carrying capacity*. The carrying capacity of an environment is the maximum number of individuals of a species that can be supported given limited food and other resources. To the degree that the wastewater facility contains a finite amount of food and other resources, it may, in fact, provide a set of factors that limits the number of microorganisms that can be supported in that facility. However, the facility itself cannot be the species' carrying capacity.

(2) It is *not* true that for these microorganisms, the wastewater facility serves as *a food chain*. Food chains illustrate nutritional relationships among species in terms of the transfer of energy from producers (green plants), which manufacture their own food via photosynthesis, to sequential levels of consumer organisms. A wastewater facility contains only decomposers. So, it cannot represent a true food chain.

(4) It is *not* true that for these microorganisms, the wastewater facility serves as *an energy pyramid*. An energy pyramid is a construct used to illustrate the fact that energy is lost at each trophic level in a food web, with producers containing the most energy. A wastewater facility contains only decomposers. So, it cannot represent a true energy pyramid.

23. 3 The type of organic molecule that is being synthesized is *protein*. The diagram illustrates the manufacture of amino acid chains at the ribosome based on the codon pattern supplied by messenger RNA (mRNA) and transcribed by transfer RNA (tRNA). This amino acid chain is the basis of a specific protein molecule produced in the cell.

WRONG CHOICES EXPLAINED:
(1) It is *not* true that the type of organic molecule that is being synthesized is *DNA*. DNA is synthesized via the process of replication (exact self-duplication) in cells, which involves only DNA, not mRNA, tRNA, or ribosomes.

(2), (4) It is *not* true that the type of organic molecule that is being synthesized is *starch* or *fat*. Starch and fat are organic molecules that are synthesized in cells via the action of enzymatically controlled reactions, not mRNA, tRNA, or ribosomes.

24. 3 This is because, in these new habitats, the nonnative insects might *not have natural predators*. Invasive organisms are organisms native to one ecosystem that are introduced into a foreign ecosystem. Invasive organisms often have the effect of competing with and eliminating native species from their natural habitats. Because invasive species are not native to their new habitat, it is likely that no native predators will be adapted to control their growth.

WRONG CHOICES EXPLAINED:
(1) It is *not* true that this is because, in these new habitats, the nonnative insects might *become food for birds*. Birds represent one type of natural predator that might control an invasive species. However, the birds in a particular area may not be adapted to hunt and consume the nonnative insects.

(2) It is *not* true that this is because, in these new habitats, the nonnative insects might *not survive a cold winter*. Cold winters represent one type of natural abiotic pressure that might control an invasive species. However, the winters may not prove sufficiently cold to eliminate the nonnative insects.

(4) It is *not* true that this is because, in these new habitats, the nonnative insects might *add to the biodiversity*. Invasive species tend to compete with and eliminate native species. So, the typical effect of nonnative insects in their new environment is to reduce, not add to, biodiversity.

25. 1 Maintaining an appropriate temperature is beneficial to an organism because *enzymes need a proper range of temperatures to catalyze vital reactions*. Enzymes are proteins that function as biological catalysts in the cells of all living things and that help to regulate the body's metabolism. If the cell's temperature is too high, enzymes' rate of catalytic action will decline and may even cease permanently.

WRONG CHOICES EXPLAINED:
(2) It is *not* true that maintaining an appropriate temperature is beneficial to an organism because *cellular waste products can only be excreted in cooler temperatures*. Cellular wastes are normally excreted from cells via the processes of passive and active transport. These are physical processes that occur more rapidly in warmer, not cooler, temperatures.

(3) It is *not* true that maintaining an appropriate temperature is beneficial to an organism because *hormones can only produce antibodies if temperatures are not excessive*. Hormones do not produce antibodies in humans, although hormones may help to regulate certain aspects of the immune response.

(4) It is *not* true that maintaining an appropriate temperature is beneficial to an organism because *nutrients diffuse faster into cells when temperatures are lower*. Diffusion is a physical process that occurs more rapidly when temperatures are higher, not lower.

26. 3 Row *3* in the chart correctly identifies where in the female reproductive system these two processes occur. Meiosis, the process in which homologous chromosome pairs and the genes they carry are separated into haploid gametes during gametogenesis, normally occurs in the ovary (structure 3). Fertilization, the process in which the haploid sperm nucleus merges with the haploid egg nucleus, normally occurs within the oviduct (structure 4).

WRONG CHOICES EXPLAINED:
(1) Row *1* in the chart does *not* correctly identify where in the female reproductive system these two processes occur. Structure 1 is the vagina, within which haploid sperm is implanted by the male parent, not where meiosis occurs. Structure 3 is the ovary, within which eggs cells are produced by the process of meiosis, not where fertilization occurs.

(2) Row *2* in the chart does *not* correctly identify where in the female reproductive system these two processes occur. Structure 2 is the cervix, which bounds the lower extremity of the uterus and prevents premature birth of the embryo, not where meiosis occurs. Structure 5 is the uterus, within which fertilized eggs develop into embryos, not where fertilization occurs.

(4) Row *4* in the chart does *not* correctly identify where in the female reproductive system these two processes occur. Structure 4 is the oviduct, within which fertilization of the haploid egg by the haploid sperm occurs, not where meiosis occurs. Structure 5 is the uterus, within which fertilized eggs develop into embryos, not where fertilization occurs.

27. 3 The prevention of browning is likely the result of *slowing the rate of enzyme action*. Enzymes are known to be sensitive to the acidity (pH) of the environment in which they operate. Most enzymes work best in a neutral pH (7.0). Squeezing lemon juice onto the surface of the apple slices lowers the pH of the apple cells to an acidic level (pH < 7.0). This, in turn, slows the catalytic action of enzymes in the apple cells and delays browning of the apple slices.

WRONG CHOICES EXPLAINED:
(1) The prevention of browning is *not* likely the result of *increasing the concentration of enzymes*. Increasing the concentration of enzymes in the system would typically increase, not decrease, the rate of the enzyme's catalytic action.

(2) The prevention of browning is *not* likely the result of *increasing the temperature*. Increasing the temperature slightly would typically increase, not decrease, the rate of the enzyme's catalytic action. If the temperature of the system gets too high, the enzyme molecules may denature, which can end the enzyme's action altogether.

(4) The prevention of browning is *not* likely the result of *maintaining the pH*. If the pH and other factors remain unchanged in the system, the enzyme's action will be neither decreased nor increased.

28. 4 In the presence of a higher amount of estrogen, it would be most likely that *more females would be found because estrogen promotes the development of female characteristics*. Estrogen is a hormone produced in many plant and animal species and by human agricultural activities. Estrogen's effect in most species is to promote the development of female characteristics and to regulate the production of egg cells needed for sexual reproduction. Scientists studying frog populations have noticed a higher ratio of female frogs compared to male frogs in estrogen-rich freshwater environments surrounded by human suburbs than in those surrounded by natural forest environments. The scientists hypothesize that estrogens produced by human activities and released into the frogs' freshwater environments are responsible for this phenomenon.

WRONG CHOICES EXPLAINED:
(1) It is *not* true that in the presence of a higher amount of estrogen, it would be most likely that *fewer males would be found because they are much larger and fewer are produced*. Male frogs are typically smaller than females of the same species. In a balanced natural freshwater environment, male frogs are typically found in numbers nearly equal to that of females.

(2) It is *not* true that in the presence of a higher amount of estrogen, it would be most likely that *fewer females would be found because they are more sensitive to pesticides*. No information is provided in the question that would lead to this conclusion.

(3) It is *not* true that in the presence of a higher amount of estrogen, it would be most likely that *more males would be found because estrogen promotes the development of male characteristics*. Estrogen is known to be associated with the production of female, not male, characteristics.

29. 3 *Use alternate sources of energy such as wind* is an action that humans could take to slow the rate of global warming. Competent scientific evidence around the issue of climate change points to the release of greenhouse gases (such as carbon dioxide) into the atmosphere as being responsible for increases in Earth's average temperature (global warming) sufficient to cause major disruption of long-stable climate patterns. A major source of carbon dioxide has been identified as electrical plants powered by fossil fuels (coal, oil, and natural gas). By encouraging the implementation of clean alternative energy solutions such as wind turbine electrical generators, our dependence on fossil fuel electrical plants and the carbon dioxide they release will decrease.

WRONG CHOICES EXPLAINED:

(1) *Cut down trees for more efficient land use* is *not* an action that humans could take to slow the rate of global warming. Trees and other green plants absorb carbon dioxide from the atmosphere and fix it into stable carbon chain molecules such as cellulose. It is imperative that we maintain our forests, not cut them down, to help reduce the quantity of carbon dioxide in the air.

(2) *Increase the consumption of petroleum products* is *not* an action that humans could take to slow the rate of global warming. Petroleum products include coal, oil, and natural gas, which are carbon-based fuels, the burning of which is thought to be responsible for the increase of carbon dioxide in the atmosphere. It is imperative that we reduce, not increase, our consumption of petroleum products to help reduce the quantity of carbon dioxide in the air.

(4) *Reduce the use of fuel-efficient automobiles* is *not* an action that humans could take to slow the rate of global warming. Cars and trucks burn gasoline and release many tons of carbon dioxide into the atmosphere each day. It is imperative that we increase, not reduce, the fuel efficiency of automobiles to help reduce the quantity of carbon dioxide in the air.

30. **4** The role of antibodies in the human body is to *recognize foreign antigens and mark them for destruction*. Antigens are proteins produced by the cells of all living things, including pathogens, and are specific to the species that produces them. The immune response in humans detects the antigens of foreign pathogens and responds to these antigens by manufacturing antibodies that specifically target and neutralize the infectious agents.

WRONG CHOICES EXPLAINED:

(1) The role of antibodies in the human body is *not* to *stimulate pathogen reproduction to produce additional white blood cells*. Pathogens are infectious agents, including viruses, bacteria, fungi, and parasites, that cause homeostatic disruption in the human body. Antibodies are specialized to slow and stop the reproduction of pathogens in the body, not stimulate them.

(2) The role of antibodies in the human body is *not* to *increase the production of guard cells to defend against pathogens*. Guard cells are specialized structures found in green plant leaves that have no role in the human immune response. This is a nonsense distracter.

(3) The role of antibodies in the human body is *not* to *promote the production of antigens to stimulate an immune response*. Antigens are proteins produced by the cells of all living things, including pathogens. Antibodies are produced in the body in response to the presence of foreign antigens, not to produce antigens.

PART B-1

31. **4** *One individual had larger feet than the other* is the statement that is an accurate observation that can be made based on this trail of footprints. Direct observation of the footprints illustrated shows a size difference of those on the top (left) from those on the bottom (right). A reasonable inference may be drawn that one of the individuals had larger feet than the other.

WRONG CHOICES EXPLAINED:
(1), (2), (3) *The individuals were running from a predator, the volcano was about to erupt again,* and *one individual was much taller than the other* are *not* the statements that are accurate observations that can be made based on this trail of footprints. Insufficient data are provided in the diagram to support any of these inferences.

32. **4** The type of information directly provided by these fossil footprints is useful because it *is a record of some similarities and differences they share with present-day species.* To the extent that these fossil footprints can be compared to the footprints of modern human beings, it is possible to point to similarities of foot structure, walking gait (pattern), and other features that these individuals share with modern humans.

WRONG CHOICES EXPLAINED:
(1), (2), (3) It is *not* true that the type of information directly provided by these fossil footprints is useful because it *offers details about how these individuals changed during their lifetime, offers data regarding their exposure to ultraviolet (UV) radiation,* or *is a record of information about what these individuals ate during their lifetime.* Insufficient data are provided in the diagram to support any of these inferences.

33. **1** *Researchers have been able to collect more data than were available in the 1980s* is the statement that best explains why the safety of PPIs is now in question when clinical experiments in the 1980s provided evidence that they were safe. A characteristic of scientific inquiry is that the results of earlier experiments form the basis of continued study and the discovery of new information. Such continued study commonly results in new information that modifies the inferences drawn from earlier experiments.

WRONG CHOICES EXPLAINED:
(2), (3), (4) *Fewer people had acid reflux in the 1980s compared to today, the medication containing PPIs has changed since the 1980s when tests were done,* and *the original experiments in the 1980s used only test animals and did not use human subjects* are *not* the statements that best explain why the safety of PPIs is now in

question when clinical experiments in the 1980s provided evidence that they were safe. No information is provided in the question that would lead to any of these explanations.

34. **1** The three arrows in the diagram each represent a process known as *mitotic cell division*. The diagram illustrates a zygote (fertilized egg) undergoing the process of cleavage to form the developing embryo. Each new cell shown in the diagram is formed by the process of mitotic cell division, in which a diploid cell divides to produce two genetically identical daughter cells. When this process has occurred several times, a cell mass is formed that is the developing embryo.

WRONG CHOICES EXPLAINED:

(2) The three arrows in the diagram do *not* each represent a process known as *meiotic cell division*. Meiotic cell division is the process in which homologous chromosome pairs and the genes they carry are separated into haploid gametes during gametogenesis. Meiotic cell division results in the formation of haploid gametes, not diploid developing embryos.

(3) The three arrows in the diagram do *not* each represent a process known as *fertilization of gamete cells*. Fertilization is the process in which a haploid sperm nucleus merges with a haploid egg nucleus to produce a diploid zygote. The diagram shows one cell dividing to create two, not two cells merging to create one.

(4) The three arrows in the diagram do *not* each represent a process known as *differentiation of tissues*. Differentiation is the process by which the cells of a developing embryo undergo specialization to become specific body tissues. The cells of the developing embryo illustrated in the diagram have not yet begun to differentiate.

35. **1** Chemical *1* would most likely interfere with the activity of hormone *A*, but not hormone *B*. The cell is illustrated to have receptors that are shaped in a concave circular manner so as to accommodate the circular shape of hormone *A*. Because the shape of chemical *1* is convex circular, it may theoretically bind to the cell's receptors and block hormone *A* from binding to the receptor, thus inhibiting the activity of hormone *A* in the cell's metabolic reactions. Chemical *1* will not, however, be able to bind with the cell's concave square receptors. So, it will not interfere with the activities of hormone *B*.

WRONG CHOICES EXPLAINED:

(2), (3), (4) Chemicals *2*, *3*, and *4* would *not* most likely interfere with the activity of hormone *A*, but not hormone *B*. The shapes of these chemicals are illustrated to be square. As such, they would tend to bind to the cell receptors for hormone *B*, which are shaped as concave squares. Thus, these chemicals would tend to interfere with hormone *B*, but not hormone *A*.

36. 2 The results of this scientific investigation will most likely lead other scientists to hypothesize that *other reptiles may have the TRPV4 protein in their eggs.* Closely related organisms often share genetic traits in common. It would be a logical assumption that other reptile species, such as crocodiles, turtles, lizards, and snakes, would carry the genes for production of TRPV4 in their genomes.

WRONG CHOICES EXPLAINED:
(1) The results of this scientific investigation will *not* most likely lead other scientists to hypothesize that *human sex cells also contain the TRPV4 protein.* Humans and reptiles are both vertebrate animals that share many genes in common. Although it is possible that humans carry the gene to produce TRPV4, it is unlikely that the protein would perform the same function in humans since humans are warm-blooded mammals whose offspring develop in the temperature-controlled environment of the uterus.

(3) The results of this scientific investigation will *not* most likely lead other scientists to hypothesize that *the TRPV4 protein affects the growth of plants.* Plants are only distantly related to alligators and other animal species. It is unlikely, though possible, that plants have the genetic ability to synthesize the TRPV4 protein.

(4) The results of this scientific investigation will *not* most likely lead other scientists to hypothesize that *the TRPV4 protein is present in all of the foods eaten by alligators.* It is possible that some of an alligator's food contains TRPV4, but this TRPV4 would not be the source of the TRPV4 in the alligator's cells. Instead, this food would only be the source of amino acids necessary for the synthesis of TRPV4 in an alligator's cells.

37. 3 The information that was most essential in preparing to carry out this scientific investigation was *the effects of temperature on the incubation of alligator eggs.* Although the investigation carried out is not described in the passage, it is clear from the information given that scientists have been studying the effects of incubation temperature on sex determination in alligators. Scientists have also studied the biochemical basis for this phenomenon and determined that the protein TRPV4 is involved.

WRONG CHOICES EXPLAINED:
(1) It is *not* true that the information that was most essential in preparing to carry out this scientific investigation was *a knowledge of the variety of mutations found in American alligator populations.* There is no mention of studies of mutations provided in the information given in the passage.

(2) It is *not* true that the information that was most essential in preparing to carry out this scientific investigation was *the arrangement of the DNA bases found in the TRPV4 protein*. Proteins such as TRPV4 are constructed of chains of amino acid molecules, not DNA bases.

(4) It is *not* true that the information that was most essential in preparing to carry out this scientific investigation was *a knowledge of previous cloning experiments conducted on alligators and other reptiles*. There is no mention of cloning experiments provided in the information given in the passage.

38. 2 The movement of the calcium ions into certain cells is most likely due to *the action of TRPV4 proteins on the cells involved with sex determination*. This fact is clearly stated in the second sentence of the second paragraph of the passage.

WRONG CHOICES EXPLAINED:
(1), (3), (4) The movement of the calcium ions into certain cells is *not* most likely due to *the destruction of the TRPV4 when it contacts the cell membrane, the sex of the alligator embryo present in that particular egg*, or *the action of receptor proteins attached to the mitochondria in alligator sex cells*. None of these phenomena are described in the passage and are unlikely to be the cause of calcium ion movement into certain alligator cells.

39. 2 Environmental changes such as global warming could affect species such as the American alligator because even slight increases in environmental temperature could *lead to an overabundance of males and few, if any, females*. The passage states in the second sentence of the first paragraph that increasing environmental temperatures as little as 3 degrees Celsius (from 30°C to 33°C) tips the probabilities of sex determination in alligators in favor of the production of all males.

WRONG CHOICES EXPLAINED:
(1) It is *not* true that environmental changes such as global warming could affect species such as the American alligator because even slight increases in environmental temperature could *lead to an overabundance of females and few, if any, males*. According to the passage, just the opposite effect could be expected if environmental temperatures rise.

(3) It is *not* true that environmental changes such as global warming could affect species such as the American alligator because even slight increases in environmental temperature could *cause the breakdown of the TRPV4 protein in female alligators*. No information is provided in the passage that would lead to this inference. In general, proteins are sensitive to environmental temperatures and many begin to denature (deform and become inactive) at temperatures above 40°C.

(4) It is *not* true that environmental changes such as global warming could affect species such as the American alligator because even slight increases in environmental temperature could *increase the rate at which calcium ions exit the sex cells*. No information is provided in the passage that would lead to this inference.

40. 2 *Building dams to control the flow of water in rivers in order to produce electricity* is the human activity that can have a *negative* impact on the stability of a mature ecosystem. The process of hydroelectric dam construction typically involves the removal of vegetation from, and the flooding of, mature forested areas and freshwater ecosystems. This activity destroys natural habitats and disrupts animal breeding and migration cycles, often permanently. These are generally considered to be negative consequences of such projects and must be weighed against their potential benefits.

WRONG CHOICES EXPLAINED:
(1), (3), (4) *Replanting trees in areas where forests have been cut down for lumber*; *preserving natural wetlands, such as swamps, to reduce flooding after heavy rainfalls*; and *passing laws that limit the dumping of pollutants in forests* are *not* the human activities that can have a *negative* impact on the stability of a mature ecosystem. Each of these responses represents a human activity that can have a positive effect on the ecosystem.

41. 4 Energy in this ecosystem passes directly from the Sun to *autotrophs*. Autotrophs (self-feeders) found in this pond ecosystem (such as grass, sedge, cattail, pickerelweed, water lily, duckweed, and algae) are all capable of absorbing solar energy and using it to convert carbon dioxide and water into glucose and oxygen. Most autotrophs employ the process of photosynthesis to absorb solar energy and incorporate it into the chemical bonds of glucose.

WRONG CHOICES EXPLAINED:
(1), (2), (3) Energy in this ecosystem does *not* pass directly from the Sun to *herbivores*, *consumers*, or *heterotrophs*. Each of these categories includes animal species that consume preformed nutrients, either plant or animal tissue, for their energy. The term *heterotroph* means "other feeder" and is the category that includes herbivores and consumers.

42. 4 The risk to the horses during the roundups compared to the entire loss of plants and soils is considered *a trade-off*. The death of 1% of the mustangs, though regrettable, can be considered a small price to pay for the preservation of the prairie lands that could potentially become overgrazed and irreparably damaged by these animals. Weighing such alternatives in decision making is known as a trade-off between two alternatives and their outcomes.

WRONG CHOICES EXPLAINED:
(1) The risk to the horses during the roundups compared to the entire loss of plants and soils is *not* considered *selective breeding*. Selective breeding is a technique used by animal and plant breeders in which organisms with desirable traits are cross-bred with the hope of producing offspring that also display those traits. This is not the process described in the question.

(2) The risk to the horses during the roundups compared to the entire loss of plants and soils is *not* considered *a technological fix*. A technological fix is a general term that usually refers to a method of solving a problem using a computer program or other modern technology. This is not the process described in the question.

(3) The risk to the horses during the roundups compared to the entire loss of plants and soils is *not* considered *direct harvesting*. Direct harvesting is a human practice in which plants or animals are removed (harvested) from a natural environment for their economic value. This is not the process described in the question.

43. 3 The number of organisms that an area of land can sustain over a long period of time is known as *its carrying capacity*. The carrying capacity of an environment is the maximum number of individuals of a species that can be supported in an area given limited food and other resources. In this example, the amount of food (grass and other vegetation) available on the prairie environment would provide a check on the mustang population, limiting the number of horses that could be naturally maintained in the area over time.

WRONG CHOICES EXPLAINED:
(1) The number of organisms that an area of land can sustain over a long period of time is *not* known as *its finite resources*. A finite resource is a general term that refers to any biotic or abiotic factor that is considered nonrenewable. This description does not match the biological definition of carrying capacity.

(2) The number of organisms that an area of land can sustain over a long period of time is *not* known as *ecological succession*. Ecological succession refers to the series of changes that occur to the dominant plant community of an area that has been significantly affected by an environmental disruption. This description does not match the biological definition of carrying capacity.

(4) The number of organisms that an area of land can sustain over a long period of time is *not* known as *evolutionary change*. Evolutionary change is a term that refers to the process of natural selection that results in new or altered species characteristics leading to species change. This description does not match the biological definition of carrying capacity.

PART B-2

44. One credit is allowed for correctly marking an appropriate scale, without any breaks in the data, on the axis labeled "Number of Bats." [1]

45. One credit is allowed for correctly plotting the data for big brown bats on the grid, connecting the points, and surrounding each point with a small circle. [1]

46. One credit is allowed for correctly plotting the data for little brown bats on the grid, connecting the points, and surrounding each point with a small triangle. [1]

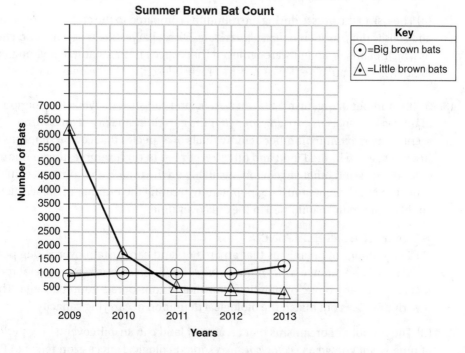

Summer Brown Bat Count

47. **1** *A few of the bats possessed an immunity to the WNS disease and produced offspring that were immune* is the statement that best explains this increased survival rate. This statement is consistent with the theory of natural selection, which is thought to be the major biological mechanism that drives the evolution of Earth's species. It hypothesizes that some bats carried genetic mutations that gave them a natural immunity to WNS and therefore provided them with an adaptive advantage over other bats that lacked this adaptation. Surviving bats passed on this favorable trait to future generations. If this hypothesis is true, after several generations, the majority of the bats will be immune to WNS.

[NOTE: The experiment as described in the passage included no genetic testing to determine whether or not a mutation providing immunity against WNS was present in the healthy bats but not in the diseased bats. Without this information, it is impossible to say with certainty that this statement is accurate with respect to the surviving bats' resistance to WNS.]

WRONG CHOICES EXPLAINED:
(2) *The bats needed to reproduce in greater numbers, otherwise they would have died out completely* is *not* the statement that best explains this increased survival rate. This statement implies that the bats possess reasoning ability common to humans but not to the vast majority of Earth's species. It is more likely that the reproductive process in bats is an instinctive behavior that cannot be altered by the bats' free will.

(3) *The people that performed the recent counts did not identify the bats correctly and were counting bats of a different species* is *not* the statement that best explains this increased survival rate. This statement implies that the scientists conducting this count were not competent to recognize the similarities and differences between these two bat species and to record scientific data properly. Such an implication is highly unlikely to be the case.

(4) *The original decline of the bat population due to WNS was a natural occurrence and is part of a natural cycle* is *not* the statement that best explains this increased survival rate. This statement makes a case for the concept that fungal infections and other diseases come and go within a species depending on varying environmental conditions, the immune response of the bats, and the life cycle of the infectious agent.

[NOTE: Insufficient information is given in the passage to eliminate this concept as a possible contributing factor in the WNS infection in bats. For this reason, this response may be considered a potential correct answer to the question as written, although it is not the best answer.]

48. One credit is allowed for correctly explaining why this may have a positive
effect on the control of WNS in bats. Acceptable responses include but are not
limited to: [1]

- *The bat houses are more easily monitored than the natural bat habitats, and
  the sick bats could be removed or treated.*

- *The factors of temperature and humidity could be more easily controlled in the
  constructed bat houses.*

- *The bat houses can be sterilized, and the disease will be less likely to be transmitted.*

- *The amount of contact among members of the bat population would be less,
  slowing the spread of the disease.*

- *Bat houses would keep bats away from infected caves/mines/bats.*

49. **2** *The starfish is important in maintaining the biodiversity of this ecosystem* is the
conclusion that can be made regarding the role of the starfish in this ecosystem.
Biodiversity is a measure of the number of different, compatible species that
inhabit an area. In 1963, prior to the start of the experiment, scientists counted
approximately 17 species in the experimental tract. However, only about two
species were present ten years after the removal of starfish. In the control tract
that included starfish, the number of species present after ten years rose from
17 to 20 species. Assuming all other factors were identical in these two tracts, it is
likely that the loss of the starfish had a negative effect on the biodiversity in the
experimental tract. It can therefore be concluded that starfish have an important
role in maintaining biodiversity in this ecosystem.

WRONG CHOICES EXPLAINED:
(1), (3) *The biodiversity of this ecosystem increased within ten years as organisms
adjusted to the loss of the starfish* and *when the starfish were removed, the ecosystem decreased in stability and increased in biodiversity* are *not* the conclusions
that can be made regarding the role of the starfish in this ecosystem. The data in
the graph show that the biodiversity of the ecosystem decreased, not increased,
over the ten-year study period. A decrease in biodiversity is known to result in a
decrease of stability in an environment.

(4) *Biodiversity in this ecosystem is not dependent on the presence of starfish* is *not*
the conclusion that can be made regarding the role of the starfish in this ecosystem. On the contrary, the data in the graph demonstrate that the loss of the starfish had a significant impact on the biodiversity of this ecosystem.

50. 3 The ecologist would most likely be asked about *what effect the production of the fuel will have on the environment.* An ecologist is a scientist who specializes in the science of ecology (study of the environment). Ecologists understand and can speak knowledgeably about the impacts of an industry's actions on the living things that inhabit the area where the fuel production facility will be built and operate.

WRONG CHOICES EXPLAINED:
(1) The ecologist would *not* most likely be asked about *the cost of producing the fuel compared with the profit when the fuel is sold.* This question would most probably be asked of a business accounting expert, not an ecologist.

(2) The ecologist would *not* most likely be asked about *whether the fuel will be widely accepted by customers.* This question would most probably be asked of a business marketing expert, not an ecologist.

(4) The ecologist would *not* most likely be asked about *the time it will take to produce large amounts of the fuel.* This question would most probably be asked of a chemical engineering expert, not an ecologist.

51. One credit is allowed for correctly explaining why biomass is considered a renewable energy source. Acceptable responses include but are not limited to: [1]

- *Biomass is continually being produced by plants, animals, and humans.*
- *More plants or trees can be grown to replace those used for fuel.*
- *Humans will always be generating food wastes and garbage.*
- *Biomass is an energy source that is quickly replaced by natural processes.*
- *When available supplies of biomass fuels are exhausted, more can be produced relatively quickly.*

52. One credit is allowed for correctly stating *one* specific advantage and *one* specific disadvantage of the use of biofuels as an energy source. Acceptable responses include but are not limited to: [1]

Advantage:
- *Fossil fuel use will decrease.*
- *Less waste/garbage will end up in landfills.*
- *Less gasoline will be used.*
- *Biofuels are renewable.*
- *Biofuels can be used to power cars and generate electricity.*

Disadvantage:

- *There will be fewer farm crops available as food.*
- *Burning biofuels will still produce some pollution.*
- *Burning wood for heat produces carbon dioxide and particulate matter.*
- *Collecting and transporting biofuels can be costly.*
- *Harvesting native plants/trees as biofuel will alter natural ecosystems.*

53. One credit is allowed for correctly identifying *two* raw materials necessary for photosynthesis. Acceptable responses include: [1]

- *Carbon dioxide, CO_2, $O=C=O$*
- *Water, H_2O, $H–O–H$*

54. One credit is allowed for correctly stating *one* reason why photosynthesis is necessary for animals to survive. Acceptable responses include but are not limited to: [1]

- *Photosynthesis provides the raw materials/glucose and oxygen for cell respiration.*
- *A product of photosynthesis is glucose, which can be converted to other energy-containing compounds that animals use for food.*
- *Animals are heterotrophic, so they are dependent on plants to manufacture food.*
- *Photosynthesis produces oxygen gas needed by animals for respiration.*
- *Only green plants can use photosynthesis to convert solar energy into food energy.*

55. One credit is allowed for correctly identifying *one* abiotic factor in the pond eco-system and explaining how this abiotic factor would affect the frogs in the pond. Acceptable responses include but are not limited to: [1]

- *Sunlight—The Sun provides energy to the plants so they can perform photosynthesis/produce food/release oxygen that the frogs need to survive.*
- *Oxygen—Frogs use oxygen for aerobic respiration.*
- *Temperature—If the average environmental temperature gets too low or too high, it could cause the frogs to die out.*
- *Soil—Soil anchors the plants that the frogs use to hide from predators.*
- *Water—Frogs depend on abundant water in their environment for their survival/breeding/habitat.*
- *Acidity/pH—The frog's enzymes function best in a neutral acidity (pH = 7) environment.*

PART C

56. One credit is allowed for correctly stating how the TR4 fungus threatens homeostasis within the banana plant. Acceptable responses include but are not limited to: [1]

- *The TR4 fungus interferes with the transport of water and other materials within the banana plant.*

- *The fungus that attacks the banana plant interferes with the plant's normal functions, and the plant basically dies of thirst.*

- *The fungus prevents water from reaching the leaves, preventing photosynthesis.*

- *The plant cannot get nourishment from water and nutrients.*

57. One credit is allowed for correctly explaining why the entire Cavendish banana crop worldwide is particularly vulnerable to the TR4 fungus. Acceptable responses include but are not limited to: [1]

- *All of the Cavendish banana plants are genetically identical, so none of them have adaptations to fight off the disease.*

- *There is no genetic diversity among the Cavendish bananas to provide immunity against TR4.*

- *Without genetic variation that provides defense against TR4, the banana plants are more likely to be killed by the fungus.*

- *Once the fungus is transferred to an uninfected banana grove, it spreads rapidly through all the trees.*

- *The crop is grown as a monoculture, so all the banana plants are susceptible to TR4.*

- *Soil infected by TR4 can easily be transferred to uninfected groves on the boots of workers.*

58. One credit is allowed for correctly describing *one* other possible way that the growers may be able to combat the disease if the fungus cannot be stopped by chemical treatment of the soil. Acceptable responses include but are not limited to: [1]

- *Look for a biological control that would attack the fungus.*

- *Genetically engineer the bananas so that they are not affected by the fungus.*

- *Do not allow people to bring contaminated boots to uninfected banana fields.*

- *Develop a "clean room" protocol to ensure that workers and others are completely decontaminated before entering and after leaving any Cavendish banana grove.*

59. One credit is allowed for correctly explaining why having children eat foods high in calcium and iron is a scientifically valid recommendation. Acceptable responses include but are not limited to: [1]

- *If there is a high amount of calcium in the diet, it is less likely that lead will be used in the formation of enzymes.*

- *Having a lot of calcium and iron available in the cells will make it more likely that these will be incorporated into enzymes when they are synthesized.*

- *A greater concentration of iron and calcium will make it more likely that they will diffuse from the blood vessels, through cell membranes, and into the cells.*

60. One credit is allowed for correctly describing how the presence of lead in the body could interfere with the ability of enzymes to function. Acceptable responses include but are not limited to: [1]

- *Lead could give the enzyme a different shape/molecular structure, which could change the enzyme's function.*

- *If lead replaces calcium or iron in the enzyme molecule, the enzyme will not have the right shape to do its job.*

- *The enzyme changes shape, which interferes with its function in the cell.*

- *If the enzyme is changed and can no longer catalyze key cell reactions, the health of the child/person may be adversely affected.*

61. One credit is allowed for correctly identifying *one* type of cell that would be expected to have numerous calcium channels and supporting the answer. Acceptable responses include but are not limited to: [1]

- *Nerve cells—Nerve/brain cells are known to be damaged by lead in the body. Therefore, nerve cells likely contain numerous calcium channels that admit lead ions instead of calcium ions.*

- *Brain cells—Lead poisoning in adults/children often leads to learning/memory/ concentration/hearing/speech problems.*

- *Bone/connective tissue—Calcium is needed for healthy bone/joint develop- ment. However, lead interferes with that development/slows bone growth/ causes joint pain in adults/children.*

- *Muscle—Lead poisoning can result in poor muscular coordination/muscular spasms.*

62. One credit is allowed for correctly stating *one* reason why the mutation in Manx cat embryos causes them to have such short tails. Acceptable responses include but are not limited to: [1]

- *Manx genes cause abnormalities in the number and/or shape/size of bones in the spine.*

- *The mutation causes the spine to be shorter in Manx cats, so there could be too few or smaller tail bones formed.*

- *The mutation interferes with the development of the spine.*

- *The Manx mutation in the embryo is replicated so that all cells, including those in the spine, have a copy of the mutation.*

63. One credit is allowed for correctly explaining how the genes that the kittens inherit determine whether they will have a normal tail or a short tail. Acceptable responses include but are not limited to: [1]

- *If a kitten gets one mutated gene from one of the parents, it will have a short tail or no tail.*

- *If a kitten receives a normal gene from each parent, it will develop a normal tail.*

- *Each parent has a Manx gene and a normal gene, so kittens will be born with either a Manx tail or a normal tail depending on whether they inherit one or two normal genes.*

- *Heterozygous kittens will develop Manx tails, and homozygous normal kittens will develop normal tails. Homozygous Manx embryos will die before birth because that allelic combination is lethal.*

64–66. Three credits are allowed for correctly discussing how sweet sensitivity in hummingbirds has developed. In your answer, be sure to:

- Identify the initial event responsible for the new sweet-sensing gene. [1]

- Explain how the presence of the sweet-sensing gene increased in the hummingbird population over time. [1]

- Describe how the fossil record of hummingbird ancestors might be used to learn more about the evolution of food preferences in hummingbirds. [1]

Acceptable responses include but are not limited to: [3]

The initial appearance of the hummingbird sweet-sensing gene was most likely the result of a random mutation of its savory-sensing gene. [1] It is likely that hummingbirds that were able to detect and feed on sweeter nectar were more successful and produced more offspring than hummingbirds that did not have this ability. [1] A study of the shapes of fossil hummingbird beaks might reveal a series of physical changes in the beaks that made sweet-sensing hummingbirds increasingly efficient or successful at gathering food. [1]

The most likely reason for the appearance of this trait was an alteration of the base sequence in one or more DNA strands in the hummingbird genome. [1] Hummingbirds that were able to sense and ingest sweeter nectars gained more energy than those that could not, increasing the sweet-sensing birds' ability to survive and reproduce. [1] By studying fossil evidence of ancestral hummingbirds and other species, scientists might be able to learn more about the environment that hummingbirds lived in and any changes in food types that may have occurred and selected for the sweet-sensing gene. [1]

67. One credit is allowed for correctly explaining why taking folic acid early in pregnancy is important to the prevention of neural tube defects. Acceptable responses include but are not limited to: [1]

   - *Early in pregnancy, folic acid promotes the normal development of the brain and spinal cord.*

   - *Folic acid is essential for normal development of the central nervous system.*

   - *The brain and spinal cord form early in pregnancy. Folic acid helps them to develop normally.*

   - *Folic acid reduces the risk of an embryo developing neural tube defects.*

68. One credit is allowed for correctly describing how a fetus receives folic acid and other essential materials directly from its mother for its development. Acceptable responses include but are not limited to: [1]

   - *Materials diffuse from the mother's bloodstream into the bloodstream of the fetus.*

   - *Materials are transported from mother to fetus across the tissues of the placenta.*

   - *Essential materials, such as folic acid, are exchanged between the mother and the fetus within the structure of the placenta.*

   - *The placenta allows the exchange of materials, including folic acid, between the mother's blood and her developing fetus.*

69. One credit is allowed for correctly identifying *one* factor, other than the lack of folic acid, that may interfere with the proper development of essential organs during pregnancy. Acceptable responses include but are not limited to: [1]

- *Genetic mutations in the fetal tissues*
- *Mother's use of alcohol/drugs/tobacco during pregnancy*
- *Mother's exposure to environmental toxins*
- *Bacterial/viral infections that occur in the mother during pregnancy*
- *Poor diet/nutritional deficiency while the fetus is developing*

70. One credit is allowed for correctly explaining why adding folic acid to foods is an advantage to people other than pregnant women. Acceptable responses include but are not limited to: [1]

- *Folic acid is essential for normal growth and development of cells in children and adults.*
- *The body is continually producing new cells for the purpose of tissue repair/ growth, and folic acid is essential to these processes.*
- *A deficiency of folic acid may interfere with a person's ability to repair damaged tissues/organs.*
- *Children use folic acid to add new cells that are needed to help their bodies grow normally.*

71. One credit is allowed for correctly circling *Yes* and supporting the answer. Acceptable responses include but are not limited to: [1]

- A *and* B *share a more recent common ancestor (H) than do* A *and* D.
- A *and* B *evolved from* H, *while* D *evolved from* G.
- *The most recent common ancestor of* A *and* B *is* H; *the most recent common ancestor of* A *and* D *is* K.
- D *is on an evolutionary line that split from that of* A *in the distant past (at* K). *The long period of time since that split occurred makes it unlikely that* A *and* D *share as much DNA in common as do* A *and* B.

72. One credit is allowed for correctly stating *one* possible cause for the extinction of species *E*. Acceptable responses include but are not limited to: [1]

- *An environmental change occurred that did not favor the survival of species* E.

- *Species* E *could not successfully compete against other similar species in a changed environment.*

- *The environment changed, and species* E *was not adapted to this change.*

- *An extinction level event such as an asteroid strike changed the environment drastically, and members of species* E *could not survive the change.*

- *Plant species that species* E *used for food became extinct, so all the members of species* E *died off from starvation because they could not adapt to new food sources.*

PART D

73. 3 *No. Additional tests should be done to test for other chemical similarities* is the conclusion that is correct, based on the results of this comparison alone, to determine if there is enough information to conclude which of the other three species is most closely related to *Botana curus*. Although the banding patterns of *B. curus* and species *Z* appear to be nearly identical, enough differences can be seen in the electrophoresis results (middle band) to warrant additional tests to determine the degree of relationship between these species.

WRONG CHOICES EXPLAINED:

(1) *Yes. Only species X has the same bands as* Botana curus is *not* the conclusion that is correct, based on the results of this comparison alone, to determine if there is enough information to conclude which of the other three species is most closely related to *Botana curus*. The banding pattern of species *X* is actually quite different from that of *B. curus*, so they are not likely to be closely related.

(2) *Yes. Species Z has only two of the bands that* Botana curus *has* is *not* the conclusion that is correct, based on the results of this comparison alone, to determine if there is enough information to conclude which of the other three species is most closely related to *Botana curus*. The banding pattern of species *Z* shows three, not two, bands that are similar though not identical to those of *B. curus*. So, it is inconclusive as to whether or not they are closely related.

(4) *No. Other rainforest plant species should be tested* is *not* the conclusion that is correct, based on the results of this comparison alone, to determine if there is enough information to conclude which of the other three species is most closely related to *Botana curus*. The conclusion called for in this experiment involves only these four species and not additional, unidentified species.

74. 1 Among the birds living there, the finches most likely to experience a drastic *decrease* in population size would be the *warbler finches*. The information provided in the diagram shows that warbler finches are 100% dependent on animal (insect) food for their nutritional requirements. If all the insects on this island are killed by the infection, warbler finches will have no food source remaining and will have to migrate to another island that still has insects or die of starvation.

WRONG CHOICES EXPLAINED:
(2), (3), (4) Among the birds living there, the finches most likely to experience a drastic *decrease* in population size would *not* be the *cactus finches*, the *large ground finches*, or the *medium ground finches*. The information provided in the diagram shows that these finch species are mainly dependent on plant food for their nutritional requirements. If all the insects on this island are killed by the infection, these finch species will still have adequate plant food sources remaining and so will survive in the short term.

75. 3 A variety of species of Galapagos finches evolved from one original species long ago through the process of *natural selection*. Natural selection, sometimes referred to as survival of the fittest, postulates that certain members of a species are better adapted to their environment than others and that these organisms are more likely to survive and pass on their favorable adaptations to future generations. It has long been thought that an original, ancestral finch species arrived on the Galapagos from the South American mainland and that genetic variation in that ancestral finch promoted the development of new finch species via natural selection as time and environmental selection pressures allowed.

WRONG CHOICES EXPLAINED:
(1) It is *not* true that a variety of species of Galapagos finches evolved from one original species long ago through the process of *asexual reproduction*. Asexual reproduction is a type of reproductive activity in which a single parent organism produces genetically identical offspring by mitotic cell division. Asexual reproduction minimizes genetic change, which limits evolutionary change. So, asexual reproduction is unlikely to be the force behind the speciation of finches in the Galapagos.

(2) It is *not* true that a variety of species of Galapagos finches evolved from one original species long ago through the process of *ecological succession*. Ecological succession refers to the series of changes that occur to the dominant plant community of an area that has been significantly affected by an environmental disruption. This description does not explain the biological evolution of finch species in the Galapagos.

(4) It is *not* true that a variety of species of Galapagos finches evolved from one original species long ago through the process of *selective breeding*. Selective breeding is a technique used by animal and plant breeders in which organisms with desirable traits are cross-bred with the hope of producing offspring that also display those traits. This description does not explain the biological evolution of finch species in the Galapagos.

76. **2** If scientists want to determine the similarities in the DNA fragments in several plant species, they should *analyze electrophoresis results*. Electrophoresis is a laboratory technique in which DNA fragments from an organism are placed in and drawn through an electrified gel to produce a unique banding pattern of the DNA for that organism. Based on the reference to DNA fragments in the question, electrophoresis is the only reasonable methodology available to perform this analysis.

WRONG CHOICES EXPLAINED:
(1), (3), (4) It is *not* true that if scientists want to determine the similarities in the DNA fragments in several plant species, they should *add salt water to cells from each plant, compare seed structures of the plants*, or *examine their chromosomes with a microscope*. These techniques are inadequate to analyze accurately the similarities and differences among DNA fragments in several plant species.

77. One credit is allowed for correctly identifying the dependent variable in this experiment. Acceptable responses include but are not limited to: [1]

- *The time it takes each team to row a specific distance on a specific course under identical conditions*

- *The speed at which the teams complete the same course under the same conditions*

- *Which team finishes in the fastest time in a head-to-head competition*

78. One credit is allowed for correctly stating *one* biological reason for the decrease in the number of squeezes during the second trial. Acceptable responses include but are not limited to: [1]

- *The student squeezed the clothespin less the second time because the muscles in his or her hand began to fatigue.*

- *The student squeezed the clothespin fewer times because her or his hand muscles had less oxygen.*

- *The student squeezed the clothespin less the second time because waste products were building up in his or her cells.*

- *As more squeezes occurred, cellular respiration switched over from aerobic respiration to less efficient anaerobic respiration, so less ATP energy was available to the muscles.*

- *The buildup of lactic acid in the student's hand muscles caused a burning sensation and made continued squeezing more difficult and painful.*

79. One credit is allowed for correctly stating *one* reason why the heart rate increased during exercise. Acceptable responses include but are not limited to: [1]

- *The faster heart rate helped to lower the level of carbon dioxide in the bloodstream.*

- *Increased blood flow helped to maintain homeostasis throughout the body.*

- *Muscles required more nutrients and oxygen, which were delivered by the circulatory system.*

- *Carbon dioxide entered the bloodstream and was detected by the central nervous system, which stimulated the heart to beat faster.*

- *The CNS sent a command to the heart muscle to increase its rate of contraction based on elevated CO_2 concentration in the blood.*

80. One credit is allowed for correctly explaining why a paramecium would require contractile vacuoles while a similar protozoan living in salt water would *not*. Acceptable responses include but are not limited to: [1]

- *Excess water would diffuse into the freshwater paramecium, but the saltwater organism would lose water.*

- *The saltwater organism would lose water to its environment/dehydrate instead.*

- *In freshwater organisms, the higher water concentration outside causes water to enter cells. This is the opposite of what happens in saltwater organisms.*

- *For a freshwater paramecium, the net diffusion of water molecules is from the environment into the cell. For a saltwater protozoan, the net diffusion of water molecules is out of the cell to the environment.*

81. **1** Row *1* in the chart correctly describes what would be expected to occur in the potato cells with regard to both the starch and water molecules. Diffusion of water (osmosis) is the movement of water molecules across a membrane with the concentration gradient from a region of high relative concentration of water to a region of low relative concentration of water. Distilled water is the most highly concentrated form of water. So, more water will enter the potato cell through its cell membrane than will leave the cell. Starch molecules are too large to pass through a cell membrane and will therefore be retained inside the cell.

WRONG CHOICES EXPLAINED:
(2), (3), (4) Rows *2, 3,* and *4* in the chart do *not* correctly describe what would be expected to occur in the potato cells with regard to both the starch and water molecules. Each of these rows contains one or more descriptions that are inconsistent with the correct answer above.

82. **2** *The indicator would remain amber in color if starch molecules were not present in the water in the beaker* is the statement that correctly describes a possible result if starch indicator is added to the water in the beaker one hour after the potato cube was added. Starch molecules are too large to pass through a cell membrane from inside the cell to the beaker water. So, no starch will be present in the beaker water with which the indicator can react. For this reason, the indicator in the beaker water will retain its amber color.

WRONG CHOICES EXPLAINED:
(1) *The indicator solution would turn to an amber color in the water if starch molecules were present in the water in the beaker* is *not* the statement that correctly describes a possible result if starch indicator is added to the water in the beaker one hour after the potato cube was added. Starch indicator is typically amber in color. If starch molecules were present in the beaker water, the indicator would change to a blue-black color, not remain an amber color.

(3) *The indicator would change to a black color if starch molecules were not present in the water in the beaker* is *not* the statement that correctly describes a possible result if starch indicator is added to the water in the beaker one hour after the potato cube was added. Starch indicator is typically amber in color. If starch molecules were *not* present in the beaker water, the indicator would remain an amber color, not change to a blue-black color.

(4) *The indicator solution would remain black in color if starch molecules were present in the water in the beaker* is *not* the statement that correctly describes a possible result if starch indicator is added to the water in the beaker one hour after the potato cube was added. Starch indicator is typically amber in color. If starch molecules were present in the beaker water, the indicator would change from an amber color to a blue-black color.

83. One credit is allowed for correctly describing how this information could specifically be used to determine if water moved during the investigation. Acceptable responses include but are not limited to: [1]

- *If the cube's mass increased, it would indicate that water had moved into the potato cells.*

- *A positive change in mass would indicate a change in the water content of the potato cells.*

- *If the mass increased, water had probably moved into the cells.*
- *The cube soaked up water and became heavier.*

84. One credit is allowed for correctly selecting *one* row in the chart and explaining how the systems in that row work together during exercise. Acceptable responses include but are not limited to: [1]

Row 1:

- *These systems work together to take in and move oxygenated blood to the muscles for their use.*
- *The respiratory system takes in oxygen, which is passed into the circulatory system, which then takes the oxygen to the muscles.*
- *The respiratory and circulatory systems work together to remove carbon dioxide from the muscles.*

Row 2:

- *These systems work together to move wastes away from muscle tissues to be released into the environment.*
- *Muscle cells produce wastes and circulatory organs transport these wastes to excretory organs to be excreted.*

Row 3:

- *These systems work together to take in and digest food and move soluble nutrients to the muscles for their use.*
- *The digestive system breaks down food into nutrients, and the circulatory system transports nutrients to muscle cells for energy.*

85. One credit is allowed for correctly stating *one* way the bag or its contents will have changed by the next day and supporting the answer. Acceptable responses include but are not limited to: [1]

- *The bag will become smaller because the water will diffuse from inside the bag to outside the bag.*
- *Water will move from higher concentration inside the bag to lower concentration outside the bag.*
- *The membrane bag will decrease in size due to osmosis of the water out of the bag.*
- *The size of the bag will become smaller due to water loss.*
- *The salt concentration inside the bag will have increased as water moved out of the bag.*

Standards/Key Ideas	August 2019 Question Numbers	Number of Correct Responses
Standard 1		
Key Idea 1: The central purpose of scientific inquiry is to develop explanations of natural phenomena in a continuing and creative process.	33, 58, 59	
Key Idea 2: Beyond the use of reasoning and consensus, scientific inquiry involves the testing of proposed explanations involving the use of conventional techniques and procedures and usually requiring considerable ingenuity.	36	
Key Idea 3: The observations made while testing proposed explanations, when analyzed using conventional and invented methods, provide new insights into natural phenomena.		
Laboratory Checklist	31, 32, 44, 45, 46	
Standard 4		
Key Idea 1: Living things are both similar to and different from each other and from nonliving things.	4, 7, 8, 17, 20, 22, 23, 35, 38, 61, 70	
Key Idea 2: Organisms inherit genetic information in a variety of ways that result in continuity of structure and function between parents and offspring.	11, 12, 14, 19, 37, 39, 62, 63	
Key Idea 3: Individual organisms and species change over time.	47, 57, 64, 65, 66, 71, 72	
Key Idea 4: The continuity of life is sustained through reproduction and development.	16, 26, 28, 34, 67, 68, 69	
Key Idea 5: Organisms maintain a dynamic equilibrium that sustains life.	5, 10, 13, 15, 25, 27, 30, 53, 54, 56, 60	
Key Idea 6: Plants and animals depend on each other and their physical environment.	1, 2, 3, 9, 18, 41, 43, 49, 55	
Key Idea 7: Human decisions and activities have a profound impact on the physical and living environment.	6, 21, 24, 29, 40, 42, 48, 50, 51, 52	
Required Laboratories		
Lab 1: "Relationships and Biodiversity"	73, 76	
Lab 2: "Making Connections"	77, 78, 79, 84	
Lab 3: "The Beaks of Finches"	74, 75	
Lab 5: "Diffusion Through a Membrane"	80, 81, 82, 83, 85	

Examination June 2021
Living Environment

PART A

Answer all questions in this part. [30]

Directions (1–30): For *each* statement or question, record in the space provided the *number* of the word or expression that, of those given, best completes the statement or answers the question.

1. The respiratory system of an elephant functions in a similar way to which organelle in a single-celled organism?

    (1) cell membrane

    (2) nucleus

    (3) vacuole

    (4) chloroplast   1 _____

2. The carrying capacity of an environment may be *decreased* by

    (1) maintaining biodiversity

    (2) replacing lost minerals

    (3) removing dead organisms

    (4) preventing deforestation   2 _____

3. The offspring of a species of bird known as the European roller possess an effective defense mechanism. When they sense a threat by predators, the young birds vomit and cover themselves with a foul-smelling liquid.

European roller
Source: http:/www.hbw.com/species/

Which two systems work together to alert the young birds of danger and help produce the vomit?

(1) respiratory and excretory
(2) circulatory and immune
(3) nervous and digestive
(4) reproductive and muscular

3 _____

4. A *decrease* in the biodiversity of an ecosystem usually leads to

(1) an increase in predator and prey populations
(2) the elimination of material cycling
(3) a decrease in stability
(4) an increase in dynamic equilibrium

4 _____

5. Down syndrome occurs when an individual has an extra copy of chromosome 21. This additional genetic material alters development and results in Down syndrome. This genetic abnormality is an example of

(1) a mutation   (3) a substitution
(2) fertilization   (4) differentiation

5 _____

6. Most of the reactions by which energy from carbohydrates is released for use by the cell take place within the

    (1)   mitochondria   (3)   ribosomes

    (2)   nuclei   (4)   vacuoles   6 _____

7. Which human activity best represents a method for recycling nutrients?

    (1)   mixing lawn clippings with vegetable waste to produce compost used to fertilize gardens

    (2)   raking and bagging lawn clippings in plastic bags for disposal in landfills

    (3)   collecting lawn and garden wastes for burning

    (4)   clearing a forested area to provide open land for cattle   7 _____

8. Rabbit populations vary in size over time. An increase in which factor would likely prevent the rabbit population from steadily increasing?

    (1)   food   (3)   predators

    (2)   mates   (4)   prey   8 _____

9. The diagrams below represent two reproductive processes used by different organisms.

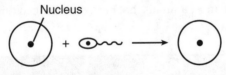

Process A

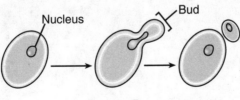

Process B

(Not drawn to scale)

When compared to organisms that utilize process *A*, organisms that utilize process *B* would most likely produce offspring with

(1)   a greater variety of genetic combinations

(2)   fewer genetic differences

(3)   more genetic combinations

(4)   more DNA within each nucleus   9 _____

10. A weightlifter has spent years building his muscular strength. His newborn daughter has normal strength for a baby. Which statement best explains this situation?

(1)   A daughter inherits most of her traits from her mother. The daughter's muscles are unlikely to resemble her father's.

(2)   The weightlifter's wife probably did not lift weights. Both parents must have this trait before the baby can inherit it.

(3)   Babies do not have strong muscles. The daughter's muscles will be unusually strong in a few more months.

(4)   The weightlifter's highly developed muscles resulted from exercise. A characteristic such as this will not be inherited.   10 _____

11. When it is disturbed, the bombardier beetle is able to produce and release a hot spray of irritating chemicals from the end of its body, as shown in the photo below. As a result, most animals that have experienced this defense avoid the beetles in the future.

Source: http://www.bbc.com/news/uk-england-leeds-11959381

The beetle's defense mechanism has developed as a result of

(1) the need for an effective protection against its enemies

(2) competition with its predators

(3) natural selection over many generations

(4) ecological succession over hundreds of years   11 _____

12. Rejection of a newly transplanted organ is caused by

(1) the immune system reacting to the presence of the organ

(2) antibiotics that stimulate the immune system to attack the organ

(3) inheritance of genetic disorders from infected individuals

(4) development of cancerous cells in the organ   12 _____

13. One of the largest and oldest organisms on Earth is located in Fishlake National Forest in Utah. Pando is an 80,000-year-old grove of aspen trees that covers 100 acres. Although it looks like a forest, DNA analysis of several of the "trees" has confirmed it is really just one huge organism. Therefore, the "trees" must have been reproduced

(1) sexually and have genetic variability

(2) asexually and have genetic variability

(3) sexually and are genetically identical

(4) asexually and are genetically identical   13 _____

14. A female giraffe has 62 chromosomes in each of her skin cells.

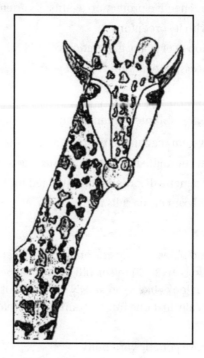

How many chromosomes will be in the skin cells of her offspring?

(1)   124   (3)   31
(2)   62   (4)   30   14 _____

15. Many female mammals, such as dogs, give birth to litters consisting of multiple offspring. All of the characteristics described below are reproductive adaptations that female dogs have for giving birth and caring for several offspring at once, *except*

(1)   a specialized structure for internal development of several young
(2)   several pairs of mammary glands that provide milk for their pups
(3)   ovaries capable of releasing many gametes at one time for fertilization
(4)   a pancreas that produces excess insulin to trigger the release of eggs

15 _____

16. As blood glucose levels increase, hormones are released to return glucose levels to normal. This is an example of

(1) a nervous system disorder

(2) the synthesis of antibodies

(3) a stimulus and a response

(4) an antigen and antibody reaction 16 _____

17. BRCA genes are human genes that normally work to help shut down cancer cells before they can harm the body. Scientists have learned that individuals inheriting a damaged form of a BRCA gene are at greater risk of developing breast or ovarian cancer. This discovery is an important first step in

(1) preventing the uncontrolled meiotic division of cells in humans

(2) identifying individuals at risk and recommending preventive treatment

(3) being able to detect all the genes that regulate meiosis

(4) helping to eliminate all BRCA genes 17 _____

18. In humans, embryonic development during the first two months is more sensitive to environmental factors than during the remaining months. The best explanation for this statement is that

(1) during the first two months, organs are being formed and any unusual change during cell division can interfere with normal development

(2) the genes that control development function only during the first two months of development

(3) no changes occur in a developing fetus after the second month

(4) organ development is not affected by environmental factors after the second month 18 _____

19. Gene editing can be used to swap out an unwanted gene for a desirable one from the same species. Which statement best explains why the desired gene will be found in all cells that come from the genetically edited cell?

(1) The original cell will reproduce by meiosis and a mutation will occur.

(2) The altered DNA in the edited cell will be replicated and passed on to each new cell during mitosis.

(3) DNA replication in body cells will result in sperm and egg cells with the edited gene.

(4) The desired gene will be inserted into each new cell by using restriction enzymes.   19 _____

20. Which sequence of events best represents ecological succession?

(1) A squirrel eats acorns, and a hawk eats the squirrel.

(2) Grass grows on a sand dune and is slowly replaced by shrubs.

(3) After many years of planting corn in the same field, minerals present in the soil are used up.

(4) The decomposition of plant material releases nutrients, and other plants use these nutrients.   20 _____

21. Which human activity has the potential to greatly affect the equilibrium of an ecosystem?

(1) cutting down a few small evergreen trees and using them to make holiday decorations

(2) mowing the playing fields in a city park

(3) washing a car with a detergent-based cleaner

(4) emptying an aquarium containing many nonnative fish of several species into a local lake   21 _____

22. Which statement describes a failure of homeostasis in humans?

    (1)   When activity in an individual increases, the body temperature rises and the individual sweats.

    (2)   As the concentration of carbon dioxide increases in the human body, the lungs begin to expel more carbon dioxide.

    (3)   A viral infection leads to a decrease in the number of white blood cells being produced in the body.

    (4)   After an individual gets a cut, certain chemical changes begin the healing process.   22 _____

23. Some environmental engineering companies have recently designed "manufactured wetlands" to serve as natural sewage treatment plants. Utilizing the ability of wetland organisms to reduce human wastes makes use of naturally occurring

    (1)   nutrient cycles   (3)   limiting factors

    (2)   energy cycles   (4)   finite resources   23 _____

24. A hummingbird may need to consume up to 50% of its body weight in sugar each day, just to meet its energy needs. Some of this energy is stored and some is used for metabolic activities, but much of the energy is

    (1)   converted into amino acids needed for the production of starch

    (2)   released as heat energy back into the hummingbird's environment

    (3)   changed into radiant energy, which can be used by plants for photosynthesis

    (4)   used to synthesize inorganic compounds necessary for cellular respiration   24 _____

25. Sustainable development occurs when people use their resources without depleting them. Which human activity is the best example of sustainable development?

    (1)   draining a wetland to build houses

    (2)   loggers planting a tree for each one cut down

    (3)   using nets to quickly capture large numbers of fish

    (4)   building coal-burning power plants to provide electricity   25 _____

26. Ringworm is a skin infection common among school-aged children. Although the name suggests that a worm causes the disease, it is actually caused by a fungus that lives and feeds on the dead outer layer of the skin. The relationship between ringworm and humans can be described as

(1) predator/prey   (3) parasite/prey

(2) predator/host   (4) parasite/host   26 _____

27. Genetically identical yarrow plants were grown at different altitudes. Even though their genetic makeup was identical, the plants grew to different heights. One likely explanation for the different heights of the plants at each altitude is that

(1) gene expression was influenced by the environment

(2) genes mutated when the plants were grown at higher elevations

(3) chromosomes increased in number with elevation change

(4) the sequence of DNA bases was altered at different altitudes   27 _____

28. Which biological process is represented in the diagram below?

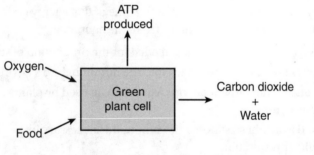

(1) photosynthesis   (3) digestion

(2) respiration   (4) replication   28 _____

29. The diagram below shows specialized plant cells that control openings called stomates.

The proper function of these cells is vital to the survival of the plant because they regulate the

(1) rate of glucose use by root cells

(2) absorption of sunlight by leaf cells

(3) products of photosynthesis in the stem

(4) exchange of gases in leaves   29 _____

30. Substance *X* directly supplies energy for various life functions, as shown in the diagram below.

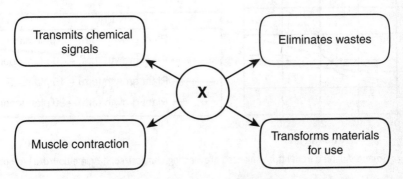

Which substance is represented by *X* in the diagram?

(1) ATP   (3) starch

(2) DNA   (4) glucose   30 _____

PART B–1

Answer all questions in this part.   [13]

Directions (31–43): For *each* statement or question, record in the space provided the *number* of the word or expression that, of those given, best completes the statement or answers the question.

Base your answers to questions 31 through 33 on the information below and on your knowledge of biology.

A student set up an experiment to test the effect of the number of seedlings planted in one pot on the rate of growth. All conditions in the experiment were the same, except for the number of plants in each pot. The results are shown in the graph below.

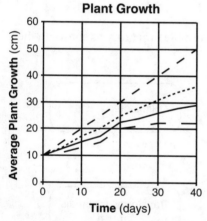

Source: Adapted from http://science.halleyhosting.com/sci/soph/scimethod/q/q1/q9.htmthod

31. The most likely reason for the differences in plant growth in the different pots was

   (1) cyclic changes in the plants' ecosystems
   (2) ecological succession over time
   (3) the amount of light available for each setup
   (4) competition for resources in each setup

31 _____

32. According to the graph, which statement is true concerning the growth of the plants?

(1)   The plants in the pot with only 5 plants grew to be an average of 40 cm tall in 30 days.

(2)   The plants in the pot with only 10 plants grew to be an average of 30 cm tall in 20 days.

(3)   The plants in the pot with 15 plants grew an average of 20 cm taller after a period of 10 days.

(4)   The plants in the pot with 20 plants grew an average of 20 cm taller after a period of 40 days.   32 _____

33. The dependent variable for this experiment was

(1)   the number of plants per pot

(2)   time in days

(3)   average plant growth

(4)   the amount of water per pot   33 _____

Base your answers to questions 34 and 35 on the diagram below and on your knowledge of biology. The diagram represents interactions between organisms in an ecosystem.

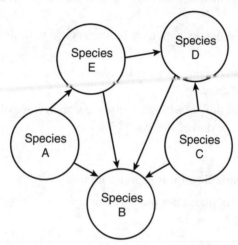

34. Which statement correctly identifies a possible role of *one* organism in this ecosystem?

(1) Species *A* may carry out autotrophic nutrition.

(2) Species *B* may be a producer that synthesizes nutrients.

(3) Species *C* carries out heterotrophic nutrition.

(4) Species *D* can recycle energy from the Sun.   34 _____

35. Which statement correctly describes an interaction that contributes to the stability of this ecosystem?

(1) Species *E* is not affected by the activity of species *A*.

(2) Species *B* returns compounds to the environment that may later be used by species *C*.

(3) Species *C* recycles nutrients from species *B* and *D* to obtain energy.

(4) Species *D* is directly dependent on the autotrophic activity of species *B*.   35 _____

Base your answers to questions 36 and 37 on the information below and on your knowledge of biology.

The Venus flytrap is a plant that uses specialized leaves in order to capture and digest small insects.

Source: https://www.britannica.com/plant/Venus-flytrap

36. Although the Venus flytrap uses its prey to obtain certain molecules that it needs, it is still classified as a producer because it

  (1)   uses its prey to produce food

  (2)   consumes the prey to produce energy

  (3)   synthesizes energy by using oxygen and releasing carbon dioxide

  (4)   synthesizes glucose by using carbon dioxide and water   36 _____

37. Enzymes secreted by cells in the leaves of the Venus flytrap can digest

  (1)   proteins into amino acids

  (2)   sugars into starches

  (3)   amino acids into fats

  (4)   proteins into sugars   37 _____

Base your answers to questions 38 and 39 on the information below and on your knowledge of biology.

Ulcers: Mystery Solved

Stomach ulcers are painful sores that develop in the stomach. Doctors once thought that ulcers were caused by stress. In the 1980s, a pair of physicians, Barry J. Marshall and J. Robin Warren, questioned the cause of ulcers. They found the bacterium *Helicobacter pylori* in the ulcer tissue of their patients. Even though they repeatedly presented their findings to colleagues, they were ignored until Marshall performed an astonishing experiment: He drank broth containing the bacteria and made himself sick with an ulcer! He then cured himself by taking an antibiotic.

The results were published in 1985, but it took another 10 years for doctors to regularly use antibiotics to treat ulcers. Marshall and Warren received a Nobel Prize in 2005 for this discovery.

38. Which choice represents a possible hypothesis for Marshall's experiment?

(1) Does *Helicobacter pylori* cause stomach ulcers in people?

(2) If a person takes an antibody, then they will not develop an ulcer.

(3) Does exposure to infectious bacterial cells make people sick?

(4) If a patient is infected by *Helicobacter pylori*, then they will get an ulcer.

38 _____

39. The work of Marshall and Warren shows that

(1) hypotheses made by physicians are always correct

(2) scientific explanations are revised based on new evidence

(3) peer review always leads to the immediate acceptance of results

(4) conclusions must always be consistent with those made by other scientists

39 _____

40. The structural formulas shown below represent parts of two different complex carbohydrate molecules composed of glucose subunits. Molecules 1 and 2 differ in their overall structure.

Molecule 1

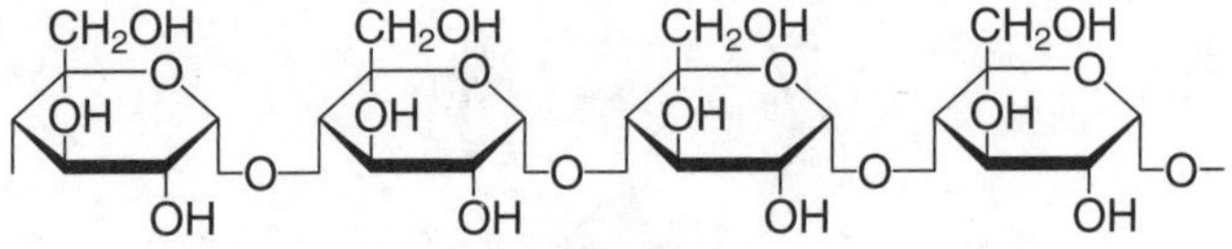

Molecule 2

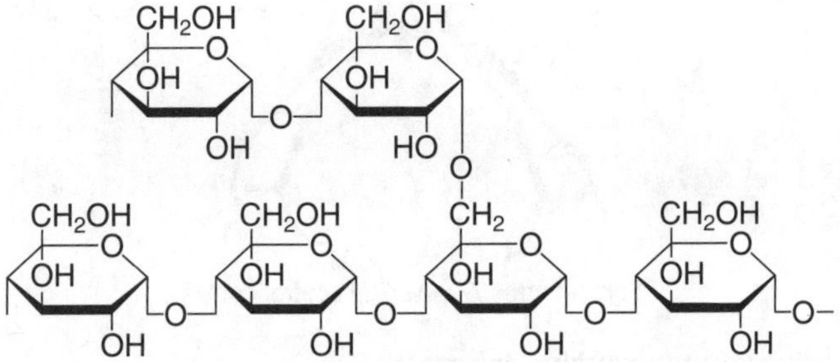

Source: Adapted from http://www.rsc.org/Education/Teachers/
Resources/cfb/carbohydrates.htm

Due to the differences in structure, each of these molecules most likely

(1) is composed of different molecular bases

(2) forms a different protein

(3) contains different elements

(4) performs a different function   40 _____

41. To capture their prey, spiders have fangs, which pierce the body wall of insects and inject venom. Spider venoms usually contain specific proteins that attack the cell membranes of the prey. The membranes and most of the contents of the insect's body turn into a liquid that the spider then ingests for food.

Source: https://www.pest-control.com/

These specific venom proteins are most likely

(1)   ATP molecules

(2)   DNA molecules

(3)   biological catalysts

(4)   regulatory hormones   41 _____

42. Lymphatic capillaries are found throughout the body. Both the lymphatic and circulatory systems transport substances between the bloodstream and body tissues. These two systems are also involved in fighting infections.

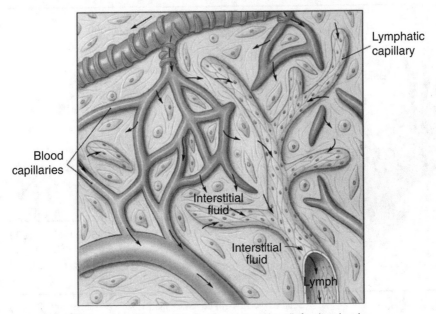

Source: Adapted from http://droualb.faculty.mjc.edu

The arrows shown in the diagram that go from the blood capillaries to the interstitial fluid most likely represent the

(1) release of red blood cells, so that they can diffuse into body cells and fight bacteria

(2) movement of materials from the circulatory system that will eventually enter lymphatic capillaries

(3) transport of digestive enzymes from the blood to help with the digestion of glucose in muscle cells

(4) transport of glucose molecules from the blood to be used by cells to attack proteins and fats

42 _____

43. The graph below shows how the introduction of the opossum shrimp, as a food source for salmon, affected a lake ecosystem in Montana.

Changes in Montana Lake Species

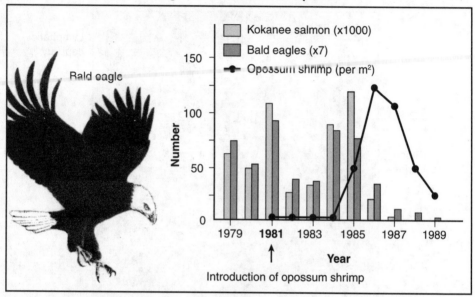

Source: *Biology*, 9th Ed. Sylvia Mader, McGraw-Hill, Boston, 2007, p. 929

Based on the data in this graph, one likely conclusion that can be made is that over approximately ten years

(1) the lake ecosystem stabilized after the introduction of the new species

(2) competition between organisms was reduced as more producers were introduced into the lake

(3) more predators moved into the lake ecosystem once the opossum shrimp were added

(4) the introduction of the opossum shrimp into the lake ecosystem disrupted the food webs that were present

43 _____

PART B-2

Answer all questions in this part.   [12]

Directions (44–55): For those questions that are multiple choice, record your answers in the spaces provided. For all other questions in this part, record your answers in accordance with the directions.

Base your answers to questions 44 through 47 on the information and data table below and on your knowledge of biology.

> Peregrine falcons are an endangered species in New York State. This crow-sized predator feeds primarily on birds. Starting in the 1940s, exposure to the pesticide DDT in their prey caused declines in the peregrine falcon population. These pesticides caused eggshell thinning, which drastically lowered breeding success. By the early 1960s, peregrine falcons no longer nested in New York State. After the United States banned DDT in 1972, efforts were made to reintroduce peregrine falcons into the Northeast. Since the 1980s, the peregrine falcons are once again breeding in many areas of New York State.

Source: http://www.dailymail.co.uk/news/article-1018309/Peregrine-falcons-return-breed-time-200-years.html

The table below shows the number of peregrine falcon offspring produced in New York State over a 20-year period.

**Number of Peregrine Falcon Offspring
Produced in New York State From 1992 to 2012**

Year	Number of Offspring Produced
1992	30
1996	48
2000	75
2004	79
2008	129
2012	148

Directions **(44–45): Using the information in the data table, construct a line graph on the grid provided, following the directions below.**

44. Mark an appropriate scale, without any breaks in the data, on each labeled axis. [1]

45. Plot the data on the grid. Connect the points and surround each point with a small circle. [1]

Example:

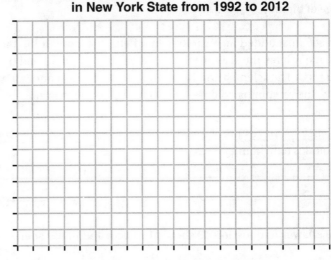

Number of Peregrine Falcon Offspring Produced in New York State from 1992 to 2012

Number of Young Produced

Years

46. Identify a body system in the falcon that was directly affected by DDT and led to the loss of nesting peregrine falcons from New York State in the early 1960s. Support your answer. [1]

Body system: _____

Support: _____

Note: The answer to question 47 should be recorded in the space provided.

47. Which conclusion is best supported by the information presented in the graph?
- (1)   The greatest decrease was during the time period of 1992 and 1996.
- (2)   The greatest increase was during the time period of 2004 and 2008.
- (3)   There has been a steady decline since the banning of DDT in 1972.
- (4)   The population reached carrying capacity in 2004.   47 _____

Base your answers to questions 48 and 49 on the information below and on your knowledge of biology.

A scientist added an antibiotic to a Petri dish containing bacterial colonies. A day later, the scientist noticed that many colonies had died, but a few remained. The scientist continued to observe the dish and noted that, eventually, the remaining colonies of bacteria increased in size.

48. Explain why the results of this study may indicate *one disadvantage* of using antibiotics to fight infections.   [1]

_____

_____

Note: The answer to question 49 should be recorded in the space provided.

49. The survival of some bacterial colonies was most likely due to
- (1)   the bacterial cells changing so that they could live
- (2)   a resistance to the antibiotic
- (3)   meiotic cell division in the bacteria
- (4)   a DNA change caused by the antibiotic   49 _____

Base your answers to questions 50 and 51 on the information below and on your knowledge of biology.

The diagrams below provide information about two separate species of tree frogs found in the United States. The shaded areas represent the habitats of each of the two species.

Tree Frogs of the United States

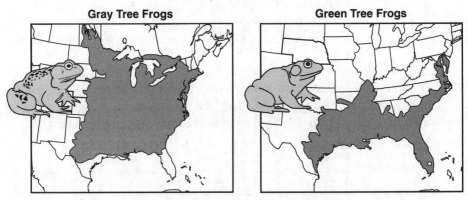

Source: Adapted from Roger Conant and Joseph T Collins. 1998. *A Field Guide to Reptiles & Amphibians of Eastern & Central North America* (Peterson Field Guide Series).

Note: The answer to question 50 should be recorded in the space provided.

50. One likely reason that the gray tree frog occupies a larger environmental area than the green tree frog is that the gray tree frog species

   (1) eats only prey found in central areas in the United States

   (2) is adapted to live in any environment in the United States

   (3) has adaptations that enable survival in a wider variety of habitats

   (4) outcompetes the green tree frogs in Florida and any state where they both live   50 _____

51. Identify a biological process that led to the presence of 90 different species of frogs throughout the United States. Support your answer. [1]

Biological process: _____

_____

_____

_____

Base your answer to question 52 on the information below and on your knowledge of biology. The diagram below represents the human female reproductive system.

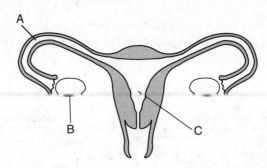

52. Select *one* of the lettered parts from the diagram. Circle the letter of the part that you selected, and identify the part. State how a malfunction in the structure that you identified could interfere with an individual's ability to reproduce.   [1]

Part selected (circle one)   A   B   C

Identification: _____

Explanation: _____

_____

_____

53. The diagram below represents a cell nucleus. Complete the diagram so that it shows the arrangement of the genetic material in the two new cells that are produced by mitosis.   [1]

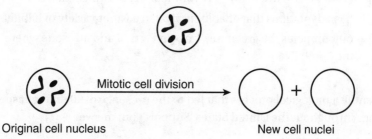

54. Sometimes a hypothesis is not supported. Yet, scientists consider the findings valuable. State *one* reason scientists would value an experiment that does *not* support the initial hypothesis.   [1]

_____

_____

55. The sequence below represents different organizational levels within the human body, from the simplest to more complex. Complete the sequence by correctly filling in the missing levels.   [1]

organelles → _____ → tissues → _____ → organ systems → organism

_____

PART C

Answer all questions in this part.   [17]

Directions (56–72): **Record your answers in the spaces provided.**

Base your answers to questions 56 through 58 on the passage below and on your knowledge of biology.

Indian Ocean Ecosystem in Danger

The Indian Ocean is under increasing environmental pressures. Until recently, this ocean was considered to have the least ecologically disrupted coastline. However, as the surface water temperatures have increased, there has been a reduction in the phytoplankton population (microscopic producers). This reduction in phytoplankton has been linked to a decline in some fish populations.

Also affecting the fish populations is the urbanization of coastal areas. As the human population grows in this area, more of the coastline region is being developed. In addition, the mining of natural resources has led to oil spills, the destruction of mangrove forests, and an increase in the area's acidity level.

Countries along the coast are trying to encourage development while, at the same time, trying to maintain a healthy coastal ecosystem.

56. Explain how a reduction in phytoplankton can lead to a reduction in fish populations in the Indian Ocean.   [1]

_____

_____

57. Describe how *one* specific human activity mentioned in the passage could negatively affect the Indian Ocean ecosystem.   [1]

Human activity: _____

_____

_____

58. State *one* specific reason why it is important to maintain a healthy ecosystem in the Indian Ocean.   [1]

_____

_____

_____

Base your answers to questions 59 through 61 on the photo and reading passage below, and on your knowledge of biology.

Invasive Water Chestnuts Challenge Environmentalists

Environmental scientists are troubled by the rapid spread of the water chestnut plant. This invasive plant is a freshwater species with leaves that blanket the surface of water. The leaves grow so densely, they stop people from swimming and prevent boats from moving.

Invasive water chestnut leaves prevent 95% of the sunlight from reaching the water below. Local animals and insects cannot eat this plant. New York ecosystems infested by the water chestnut are quickly disrupted. Water chestnut seeds can survive more than ten years under water in the sediments.

The most effective way to kill the water chestnut is to pull out each plant by hand. This can be done in a small pond, but for rivers and lakes that are blocked by huge numbers of water chestnut plants, other methods are needed. Chemical herbicides kill the leaves, but, after several weeks, the water chestnut plants grow back. Large machines have been used to clear these plants and seeds from the water and sediments of ecosystems, but the machines remove many other organisms too.

Water chestnut
seed

Source: estuarylive.pbworks.com

59. State *one* way that the presence of water chestnut plants affects the other organisms in the freshwater ecosystem. [1]

_____

_____

60. Some scientists recommend bringing in biological controls, such as introducing a new species of insect to eat the water chestnut leaves and stop its growth. State *one* advantage and *one disadvantage* of using biological controls in this situation. [1]

Advantage: _____

_____

Disadvantage: _____

_____

61. Harvesting machines are used to scrape water chestnut plants and seeds from the bottom of lakes and rivers. State *one disadvantage* of this method of controlling water chestnuts. [1]

_____

_____

Base your answers to questions 62 and 63 on the information below and on your knowledge of biology.

Rising CO_2 [Carbon Dioxide] Levels in Ocean Block Sharks' Ability to Smell Prey

...Changes in the chemistry of the world's oceans expected by century's end could impact the hunting ability of sharks, which depend heavily on their sense of smell to locate prey, researchers say.

As ocean waters turn increasingly acidic from absorbing atmospheric CO_2 created by human activities, the odor-detecting ability of sharks to locate prey could diminish, they say. ...

Source: Jim Algar, *Tech Times*, 9/9/14

62. Identify *one* human activity and describe how it contributes to increasing levels of carbon dioxide in the environment. [1]

Human activity: _____

_____

_____

63. Describe how the inability of sharks to detect their prey could affect an ocean ecosystem. [1]

_____

_____

Base your answers to questions 64 through 66 on the information and photo below and on your knowledge of biology. The photo shows an adult female weasel.

Weasels Are Built for the Hunt

Weasels are fierce and quick-witted carnivores that must compete for food with larger predators. Their slender, elongated body plan allows them to pursue prey in tight spaces that other carnivores can't enter, a key factor in controlling rodent and rabbit populations. This body plan is important to the success of weasels. Female weasels have evolved to give birth to fetuses that have not fully completed development. The fetuses complete their development externally. In this way, there is no baby bump to limit the mother's access to tight feeding locations.

A high energy level is key to the weasel's success in capturing prey, but it comes at a price. To survive, weasels need to eat a third of their body weight per day. This need can make them unpopular with poultry farmers, because they can enter through the smallest opening and consume large numbers of chickens.

Source: NY Times 6/13/16

64. State how the body plan of the weasel is effective for successfully competing with other organisms.   [1]

_____

_____

65. If the weasels are so successful, explain why they do *not* completely overpopulate the areas where they live.   [1]

_____

_____

66. Indicate whether the weasels' relationship with humans is positive or negative by circling the appropriate term below. Support your answer.   [1]

Relationship (circle one):   positive   negative

Support: _____

_____

Base your answers to questions 67 and 68 on the information and diagram below and on your knowledge of biology.

HIV Infection

The human immunodeficiency virus (HIV), which can lead to AIDS, is a type of virus that adds its genetic material to the DNA of the host cell. HIV reproduces within the host cell and exits through a process called budding.

In the process of budding, the newly forming virus merges with the host cell membrane and pinches off, taking with it a section of the host-cell membrane. It then enters into circulation.

HIV Budding

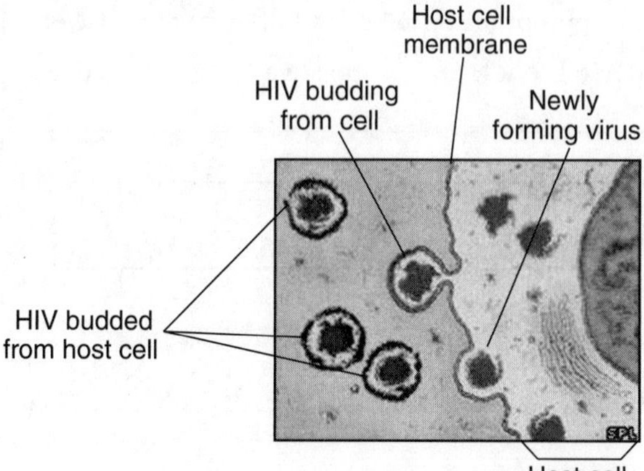

Source: Adapted from http://news.bbc.
co.uk/2/hi/health/5221744.stm

67. Explain how an outer covering composed of a section of a cell membrane from the host would protect HIV from attack by the host's immune system.   [1]

_____

_____

68. Describe *one* specific way that HIV makes the body unable to deal with other pathogens and cancer.   [1]

_____

_____

Base your answers to questions 69 through 72 on the information below and on your knowledge of biology.

Snakes Used to Have Legs and Arms Until These Mutations Happened

The ancestors of today's slithery snakes once sported full-fledged arms and legs, but genetic mutations caused the reptiles to lose all four of their limbs about 150 million years ago, according to two new studies. ...

Both studies showed that mutations in a stretch of snake DNA called ZRS (the Zone of Polarizing Activity Regulatory Sequence) were responsible for the limb-altering change. But the two research teams used different techniques to arrive at their findings. ...

...According to one study, published online today (Oct. 20, 2016) in the journal *Cell,* the snake's ZRS anomalies [differences] became apparent to researchers after they took several mouse embryos, removed the mice's ZRS DNA, and replaced it with the ZRS section from snakes. ...

...The swap had severe consequences for the mice. Instead of developing regular limbs, the mice barely grew any limbs at all, indicating that ZRS is crucial for the development of limbs, the researchers said. ...

Looking deeper at the snakes' DNA, the researchers found that a deletion of 17 base pairs within the snakes' DNA appeared to be the reason for the loss of limbs.

Source: http://www.livescience.com/56573-mutation-caused-snakes-to-lose-legs.html

69. State *one* possible advantage for a snake to have no limbs instead of four limbs.   [1]

_____

_____

70. Identify the technique that the scientists used to remove the ZRS DNA from mice and replace it with the ZRS section from snakes.   [1]

_____

71. Identify the type of mutation responsible for the loss of limbs in snakes.   [1]

_____

72. Without having DNA samples from snakes 150 million years ago, state how scientists could know that snakes once actually had legs.   [1]

_____

_____

_____

PART D

Answer all questions in this part. [13]

Directions (73–85): For those questions that are multiple choice, record your answers in the spaces provided. For all other questions in this part, record your answers in accordance with the directions.

Base your answers to questions 73 and 74 on the information and chart below and on your knowledge of biology.

Finding Relationships Between Organisms

Organisms living in the same environment may have similar body structures, but this does not always indicate a close biological relationship. The chart below provides information about four organisms that live in an Antarctic Ocean ecosystem.

Body Structures of Four Antarctic Marine Organisms				
Organism	Killer whale	Adélie penguin	Leopard seal	Baleen whale
Skin covering	Very little hair	Feathers	Thick hair	Very little hair
Diagram* *Pictures are not drawn to scale.				

Note: The answer to question 73 should be recorded in the space provided.

73. Two features that would be the most useful in determining which of these organisms are most closely related are

(1) presence of hair and similar proteins

(2) presence of feathers and similar body structures

(3) habitat and diet

(4) body size and color   73 _____

Note: The answer to question 74 should be recorded in the space provided.

74. Which lab procedure can be done to find molecular evidence for relationships between these Antarctic marine organisms?

    (1)   Compare slides of cell organelles.

    (2)   Examine fossils and ocean sediments.

    (3)   Set up and perform gel electrophoresis.

    (4)   Use a dichotomous key and test for pH.   74 _____

Note: The answer to question 75 should be recorded in the space provided.

75. As an extension of the lab activity *Making Connections*, a biology teacher asked students to brainstorm variables other than exercise that would affect heart rate. The students hypothesized that eating a lunch high in protein would decrease heart rates. They recorded resting heart rates of 20 students, had them eat high-protein meals, and then recorded their heart rates again. The heart rates of 15 students were lower while the heart rates for 5 students were higher after lunch.

The best explanation for the observation that the heart rates of 5 students were higher after lunch is

    (1)   the heart rates of female students are not affected by a high-protein meal

    (2)   the students all participated in physical education class immediately before lunch

    (3)   the students all had varying physical fitness levels and consumed different amounts of protein

    (4)   the students were all the same gender and age   75 _____

Base your answers to questions 76 and 77 on the passage below and on your knowledge of biology.

A recent study of Darwin's finches in the Galapagos Islands identified the gene, HMGA2, that is involved in beak size. It played a role in which finches feeding on smaller seeds survived a severe drought in 2004–2005. Following the drought, the average size of the medium ground finch beak decreased. This change was traced directly to changes in the frequency of the HMGA2 gene. Previous studies have shown that HMGA2 affects body size in animals, including dogs and horses, and even humans.

Note: The answer to question 76 should be recorded in the space provided.

76. One possible reason that such diverse species could be affected by the HMGA2 gene is that

    (1)   they all lived on the Galapagos Islands

    (2)   they share a common ancestor

    (3)   the drought caused the formation of the gene

    (4)   the gene allowed all these species to grow larger   76 _____

77. State *one* possible reason the medium ground finches with a smaller beak were able to survive during the 2004–2005 drought. Support your answer.   [1]

_____

_____

Base your answer to question 78 on the information and the Universal Genetic Code Chart below and on your knowledge of biology.

Universal Genetic Code Chart
Messenger RNA Codons and the Amino Acids for Which They Code

		U	C	A	G	
SECOND BASE						
FIRST BASE	**U**	UUU UUC **PHE** UUA UUG **LEU**	UCU UCC UCA UCG **SER**	UAU UAC **TYR** UAA UAG **STOP**	UGU UGC **CYS** UGA **STOP** UGG **TRP**	U C A G
	C	CUU CUC CUA CUG **LEU**	CCU CCC CCA CCG **PRO**	CAU CAC **HIS** CAA CAG **GLN**	CGU CGC CGA CGG **ARG**	U C A G
	A	AUU AUC **ILE** AUA AUG **MET or START**	ACU ACC ACA ACG **THR**	AAU AAC **ASN** AAA AAG **LYS**	AGU AGC **SER** AGA AGG **ARG**	U C A G
	G	GUU GUC GUA GUG **VAL**	GCU GCC GCA GCG **ALA**	GAU GAC **ASP** GAA GAG **GLU**	GGU GGC GGA GGG **GLY**	U C A G

(THIRD BASE)

Original DNA for protein *X*:   TAC-GGC-TTA-GCT-CCC-GCG-CTA-AAA

Mutated DNA for protein *X*:   TAC-GGC-TTG-GCT-CCT-GCG-CTA-AAA

78. Would the mutated DNA strand affect the functioning of protein *X*? Support your answer.   [1]

_____

_____

_____

Base your answers to questions 79 and 80 on the diagram below and on your knowledge of biology. The diagram represents a hypothetical result of a technique used in a lab.

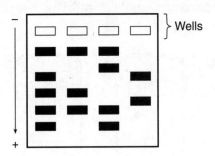

79. State where on the diagram the largest fragments of DNA would be located.   [1]

_____

80. Identify the factor that caused the fragments to move through the gel rather than remaining in the wells.   [1]

_____

_____

Base your answer to question 81 on the diagram below and on your knowledge of biology.

The diagram represents a sugar cube being dropped into an undisturbed beaker of water at room temperature. One sugar molecule is labeled.

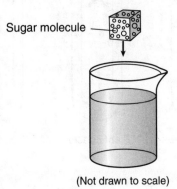

(Not drawn to scale)

Note: The answer to question 81 should be recorded in the space provided.

81. Which diagram below represents the distribution of sugar molecules in the water a day later?

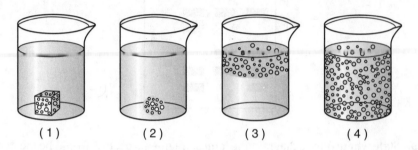

(1)   (2)   (3)   (4)

81 _____

Note: The answer to question 82 should be recorded in the space provided.

82. In an effort to determine how closely related several plant species are, a student performed the laboratory test shown below.

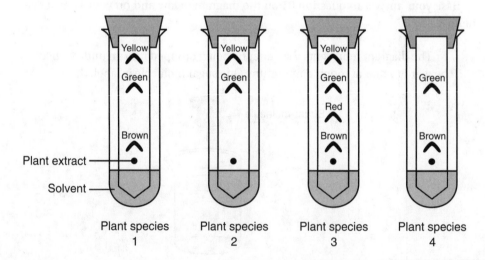

The method used by the student to compare plant extracts from the different species is

(1) gel electrophoresis
(2) DNA banding
(3) a staining technique
(4) paper chromatography   82 _____

Base your answer to question 83 on the graph below and on your knowledge of biology. The graph shows the average heart rate data for a group of students before, during, and after exercise.

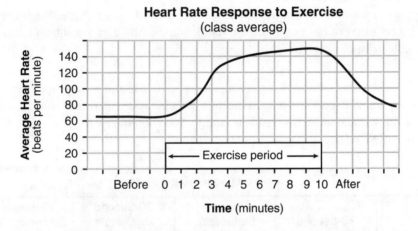

Heart Rate Response to Exercise
(class average)

83. State *one* benefit of the increase in average heart rate during exercise.   [1]

_____

_____

_____

Base your answers to questions 84 and 85 on the information below and on your knowledge of biology.

A Clothespin Experiment

A student in a Living Environment class designed an experiment to investigate if the number of times a student squeezes a clothespin varies with the hand used. Her hypothesis was that students could squeeze a clothespin more times in a minute when they used their dominant hand than when they used their nondominant hand.

During her investigation, she first squeezed and released a clothespin as often as possible for 20 seconds with her dominant hand. She recorded the number of squeezes in a chart.

She performed three trials before resting. After that, she repeated the entire procedure with her nondominant hand. Some of the data are shown in the table below.

84. Calculate the clothespin-squeezing rates per minute and average for the dominant hand. Record the data in the data table below for all three trials, as well as the average squeezing rate per minute. You should have four numbers recorded. [1]

Clothespin Squeezing Activity

Trial	20-Second Clothespin Squeezing (Dominant Hand)	Clothespin-Squeezing Rate Per Minute (Dominant Hand)	20-Second Clothespin Squeezing (Nondominant Hand)	Clothespin-Squeezing Rate Per Minute (Nondominant Hand)
Trial 1	26	_____	18	54
Trial 2	33	_____	28	84
Trial 3	24	_____	29	87
Average		_____		75

85. After performing the experiment, the student's laboratory write-up indicated that the hypothesis was supported. Do you agree with this student? Support your answer. [1]

Agree (circle one):   Yes   No

Support: _____

Answers June 2021
Living Environment

Answer Key

PART A

1. 1	6. 1	11. 3	16. 3	21. 4	26. 4
2. 3	7. 1	12. 1	17. 2	22. 3	27. 1
3. 3	8. 3	13. 4	18. 1	23. 1	28. 2
4. 3	9. 2	14. 2	19. 2	24. 2	29. 4
5. 1	10. 4	15. 4	20. 2	25. 2	30. 1

PART B-1

31. 4	34. 1	36. 4	38. 4	40. 4	42. 2
32. 1	35. 2	37. 1	39. 2	41. 3	43. 4
33. 3					

PART B-2

44. *See* Answers Explained.

45. *See* Answers Explained.

46. *See* Answers Explained.

47. 2

48. *See* Answers Explained.

49. 2

50. 3

51. *See* Answers Explained.

52. *See* Answers Explained.

53. *See* Answers Explained.

54. *See* Answers Explained.

55. *See* Answers Explained.

PART C. *See* **Answers Explained**.

PART D

73. 1

74. 3

75. 3

76. 2

77. *See* Answers Explained.

78. *See* Answers Explained.

79. *See* Answers Explained.

80. *See* Answers Explained.

81. 4

82. 4

83. *See* Answers Explained.

84. *See* Answers Explained.

85. *See* Answers Explained.

Answers Explained

PART A

1. 1 The *cell membrane* is the organelle in a single-celled organism that functions in a similar way to the respiratory system of an elephant. Within the elephant's lung, oxygen is absorbed and carbon dioxide is released through cell membranes of the alveoli. In a like manner, oxygen and carbon dioxide are exchanged through the cell membrane of the single-celled organism.

WRONG CHOICES EXPLAINED:
(2) The *nucleus* is *not* the organelle in a single-celled organism that functions in a similar way to the respiratory system of an elephant. The nucleus is an organelle in a single-celled organism that contains the genetic material of the cell and that controls the cell's biochemical processes.

(3) The *vacuole* is *not* the organelle in a single-celled organism that functions in a similar way to the respiratory system of an elephant. The vacuole is an organelle in a single-celled organism that collects, stores, and excretes some of the cell's biochemical wastes.

(4) The *chloroplast* is *not* the organelle in a single-celled organism that functions in a similar way to the respiratory system of an elephant. The chloroplast is an organelle in a single-celled organism that contains chlorophyll and that operates the biochemical process known as photosynthesis.

2. 3 The carrying capacity of an environment may be *decreased* by *removing dead organisms*. The carrying capacity of an environment is the number of individuals of a particular species that may be sustainably supported by the natural resources present in the environment. Any action that diminishes those resources may have the effect of reducing the number of individuals that can be supported. By removing dead organisms of a species from the environment, the biochemicals making up those organisms are effectively diminished, thereby decreasing the environment's carrying capacity for that species.

WRONG CHOICES EXPLAINED:
(1), (2), (4) It is *not* true that the carrying capacity of an environment may be *decreased* by *maintaining biodiversity*, by *replacing lost minerals*, or by *preventing deforestation*. These actions help to preserve or reestablish the resources present in the environment upon which species depend for their homeostatic balance. These actions would tend to stabilize, not decrease, the environment's carrying capacity.

3. 3 The two systems that work together to alert young birds of danger and help produce the vomit are the *nervous and digestive*. The young birds are able to detect the presence of the predator through their senses of sight, smell, and hearing. These senses are taken in by the bird's central nervous system, which then sends a command to the muscles of the bird's digestive organs to reverse its peristaltic direction to produce the vomit.

WRONG CHOICES EXPLAINED:

(1) The two systems that work together to alert young birds of danger and help produce the vomit are *not* the *respiratory and excretory*. The respiratory system regulates the intake of oxygen and the release of carbon dioxide. The excretory system is responsible for releasing metabolic wastes to the environment. Neither system is directly involved in the stimulus-response behavior described.

(2) The two systems that work together to alert young birds of danger and help produce the vomit are *not* the *circulatory and immune*. The circulatory system distributes dissolved materials throughout the body. The immune system is responsible for protecting the body from foreign invaders such as pathogens. Neither system is directly involved in the stimulus-response behavior described.

(4) The two systems that work together to alert young birds of danger and help produce the vomit are *not* the *reproductive and muscular*. The reproductive system produces gametes and secretes hormones that result in the production of new members of the species. The muscular system, with the skeletal system, enables movement such as flapping of the wings in birds. Neither system is directly involved in the stimulus-response behavior described.

4. 3 A *decrease* in the biodiversity of an ecosystem usually leads to *a decrease in stability*. Biodiversity is a measure of the number of different species inhabiting an environment (ecosystem). It is well established that different species interact with each other in a stable, healthy environment. Each species has a specific role to play (niche to fill) in such an ecosystem. When even a single species is removed from that ecosystem, the niche it filled is vacated and the interactive role it played in the environment no longer exists, effectively decreasing the stability of that ecosystem.

WRONG CHOICES EXPLAINED:

(1) It is *not* true that a *decrease* in the biodiversity of an ecosystem usually leads to *an increase in predator and prey populations*. Although this may be an outcome of decreasing biodiversity in specific cases, it is not a common result of this action.

(2) It is *not* true that a *decrease* in the biodiversity of an ecosystem usually leads to *the elimination of material recycling*. Materials such as oxygen, carbon, hydrogen, and nitrogen are constantly being recycled in any natural ecosystem. Such recycling is unlikely to be eliminated as a result of this action.

(4) It is *not* true that a *decrease* in the biodiversity of an ecosystem usually leads to *an increase in dynamic equilibrium*. Dynamic equilibrium is synonymous with stability in an ecosystem. Loss of biodiversity will likely decrease, not increase, dynamic equilibrium,

5. 1 The genetic abnormality leading to Down syndrome is an example of *a mutation*. Mutations are broadly defined as any alterations of the normal genetic structure of the cell. In this case, the extra chromosome 21 represents such an alteration. Also known as trisomy 21, this condition results in the Down syndrome phenotype.

WRONG CHOICES EXPLAINED:
(2) This genetic abnormality is *not* an example of *fertilization*. Fertilization is a life process by which male and female gametes fuse to form a zygote. This is a life process that maintains the normal genetic structure of the species.

(3) This genetic abnormality is *not* an example of *a substitution*. Substitution is a type of mutation in which a normal genetic code is altered by substituting an abnormal code sequence for a normal sequence. Substitution does not result in an extra chromosome 21.

(4) This genetic abnormality is *not* an example of *differentiation*. Differentiation is a process by which cells in a mass resulting from cleavage begin to specialize into ectodermic, endodermic, and mesodermic tissues that eventually form specific organs and systems within the body of a new organism. Cells undergoing differentiation do not normally alter their genetic structure in the process.

6. 1 Most of the reactions by which energy from carbohydrates is released for use by the cell occur within the *mitochondria*. The mitochondria absorb glucose and oxygen and, using respiratory enzymes, break the chemical bonds of the glucose to release energy and to form energy-carrying molecules of ATP. Carbon dioxide and water are by-products of this process and are carried out of the mitochondria for excretion from the cell.

WRONG CHOICES EXPLAINED:
(2) It is *not* true that most of the reactions by which energy from carbohydrates is released for use by the cell occur within the *nuclei*. Nuclei are the control centers of the cells of which they are a part. They contain the genetic codes of the cells and initiate the process of protein synthesis in those cells. Nuclei are not directly involved in energy release in cells.

(3) It is *not* true that most of the reactions by which energy from carbohydrates is released for use by the cell occur within the *ribosomes*. Ribosomes act as the sites of protein synthesis in the cell. Ribosomes are not directly involved in energy release in cells.

(4) It is *not* true that most of the reactions by which energy from carbohydrates is released for use by the cell occur within the *vacuoles*. Vacuoles function by absorbing and storing insoluble metabolic waste for release to the environment outside the cell. Vacuoles are not directly involved in energy release in cells.

7. **1** *Mixing lawn clippings with vegetable waste to produce compost used to fertilize gardens* is the human activity that best represents a method for recycling nutrients. By taking this action, humans can take advantage of natural processes that slowly break down the complex biochemicals that make up these wastes into simpler units that can be absorbed by plants and then be used to produce new plant tissues.

WRONG CHOICES EXPLAINED:
(2) *Raking and bagging lawn clippings in plastic bags for disposal in landfills* is *not* the human activity that best represents a method for recycling nutrients. Doing so would lock the grass clippings away from bacteria and fungi that otherwise could break down the grasses into simple, absorbable components.

(3) *Collecting lawn and garden wastes for burning* is *not* the human activity that best represents a method for recycling nutrients. Although burning the grass clippings would release recyclable carbon dioxide and water as by-products and leave ash containing nitrogen, phosphorus, and other elements that could be absorbed by plants, it would also release harmful particulates into the atmosphere that pollute the air and water in the environment.

(4) *Clearing a forested area to provide open land for cattle* is *not* the human activity that best represents a method for recycling nutrients. This action reduces biodiversity in an ecosystem and sacrifices valuable plant and animal species and environmental stability to benefit an unsustainable economic enterprise.

8. **3** An increase in *predators* would likely prevent the rabbit population from steadily increasing. Predators of rabbits, such as foxes and hawks, would take a proportion of the rabbit population as food, eliminating those rabbits from breeding activity. This situation would help to limit the reproduction of new rabbits.

WRONG CHOICES EXPLAINED:
(1) An increase in *food* would *not* likely prevent the rabbit population from steadily increasing. Increasing food supplies would likely have the effect of increasing, not decreasing, the rabbit population.

(2) An increase in *mates* would *not* likely prevent the rabbit population from steadily increasing. Increasing the availability of mates would likely have the effect of increasing, not decreasing, the rabbit population.

(4) An increase in *prey* would *not* likely prevent the rabbit population from steadily increasing. Increasing the availability of prey, such as mice and voles, in the environment would reduce predatory pressure on the rabbits and would likely have the effect of increasing, not decreasing, the rabbit population.

9. **2** When compared to organisms that utilize process *A*, organisms that utilize process *B* would most likely produce offspring with *fewer genetic differences*. Process *B* illustrates the asexual reproductive process known as budding. The offspring (buds) resulting from this process are normally genetically identical to the parent organism. By contrast, process *A* illustrates the process of sexual reproduction, in which new genetic combinations are common in offspring and result in more, not fewer, genetic differences.

WRONG CHOICES EXPLAINED:
(1), (3), (4) It is *not* true that, when compared to organisms that utilize process *A*, organisms that utilize process *B* would most likely produce offspring with *a greater variety of genetic combinations, more genetic combinations,* or *more DNA within each nucleus.* Offspring produced by asexual means (process *B*) are normally genetically identical to the parent organism.

10. **4** *The weightlifter's highly developed muscles resulted from exercise. A characteristic such as this will not be inherited* is the statement that best explains the situation described. It is well established through scientific study that characteristics acquired during an organism's lifetime cannot be passed on to that organism's offspring. Only traits that are genetically determined may be inherited in this manner.

WRONG CHOICES EXPLAINED:

(1) *A daughter inherits most of her traits from her mother. The daughter's muscles are unlikely to resemble her father's* is *not* the statement that best explains the situation described. The daughter's genetic traits are provided equally by her father and her mother. Acquired characteristics such as muscle strength due to exercise are determined by physical-training habits.

(2) *The weightlifter's wife probably did not lift weights. Both parents must have this trait before the baby can inherit it* is *not* the statement that best explains the situation described. Acquired characteristics such as muscle strength due to exercise cannot be passed on to the daughter from either or both parents but, rather, are determined by the daughter's own physical-training habits.

(3) *Babies do not have strong muscles. The daughter's muscles will be unusually strong in a few more months* is *not* the statement that best explains the situation described. Acquired characteristics such as muscle strength due to exercise cannot be passed on to the daughter from either or both parents but, rather, are determined by the daughter's own physical-training habits.

11. **3** The beetle's defense mechanism has developed as a result of *natural selection over many generations*. The scientific theory of evolution by natural selection posits that such traits occur randomly within a species population through the process of mutation in gametes. Those mutations that prove useful to the species and that prolong the mutant organism's breeding life are passed on to future generations via sexual reproduction. Over many generations, such mutations may become common in a majority, or even all, of the individuals making up the population of that species.

WRONG CHOICES EXPLAINED:

(1) The beetle's defense mechanism has *not* developed as a result of *the need for an effective protection against its enemies*. The traits of a species do not change as a function of need but only through the processes of genetic mutation, reproduction, and natural selection.

(2) The beetle's defense mechanism has *not* developed as a result of *competition with its predators*. The traits of a species do not change as a function of predatory pressures but only through the processes of genetic mutation, reproduction, and natural selection.

(4) The beetle's defense mechanism has *not* developed as a result of *ecological succession over hundreds of years*. Ecological succession is a term used to describe the replacement of one plant community by another until a stable, self-perpetuating plant community is established. This term is not used to describe the establishment of new genetic traits in any species.

12. 1 Rejection of a newly transplanted organ is caused by *the immune system reacting to the presence of the organ*. An organ transplanted from a donor organism contains protein molecules known as antigens that are unique to that donor. When the donated organ is placed into the body of a recipient organism, the recipient's immune system detects the donor's antigens in the donated organ and initiates an immune response to protect the recipient from the foreign tissues. If this immune response is not muted by the use of antirejection medications, the recipient's body may damage the donated organ, causing it to cease functioning (be rejected).

WRONG CHOICES EXPLAINED:
(2) Rejection of a newly transplanted organ is *not* caused by *antibiotics that stimulate the immune system to attack the organ*. Antibiotics are biochemical agents produced by certain fungal molds that have an antibacterial effect. Antibiotics act independently of the immune system to kill bacterial infections in the body. They are not involved in rejecting donated organs.

(3) Rejection of a newly transplanted organ is *not* caused by *inheritance of genetic disorders from infected individuals*. The genetic traits of an organ donor cannot be inherited by the recipient. Inheritance is a function of the reproductive process.

(4) Rejection of a newly transplanted organ is *not* caused by *development of cancerous cells in the organ*. Doctors are very careful to ensure that donated organs are free of major defects or disorders such as cancer so as to protect the life and health of the recipient.

13. 4 The "trees" must have been reproduced *asexually and are genetically identical*. All somatic (body) cells of an individual organism are genetically identical. The fact that researchers have confirmed that this aspen grove is "just one huge organism" indicates that all the "trees" in the grove must have identical DNA. This phenomenon is possible only if all the "trees" in the grove were reproduced asexually, probably through the use of runners that grow underground and produce new "trees" through the asexual budding at nodes located on the runners. Each of the resulting "trees" would appear to be an independent organism but would actually be one node of a single organism.

WRONG CHOICES EXPLAINED:
(1) It is *not* true that the "trees" must have been reproduced *sexually and have genetic variability*. The process of sexual reproduction ensures that genetic traits are constantly reshuffled to produce new variants. Researchers have confirmed that the aspen "trees" are parts of a single organism with identical DNA, so reproduction must be asexual, not sexual.

(2) It is *not* true that the "trees" must have been reproduced *asexually and have genetic variability*. The process of asexual reproduction ensures that genetic traits are identical in the parent and offspring. Researchers have confirmed that the aspen "trees" are parts of a single organism with identical DNA, so no genetic variability was detected.

(3) It is *not* true that the "trees" must have been reproduced *sexually and are genetically identical*. The process of sexual reproduction ensures that genetic traits are constantly reshuffled to produce new variants. Researchers have confirmed that the aspen "trees" are parts of a single organism with identical DNA, which is possible only through asexual reproductive processes.

14. **2** The number of chromosomes in the skin cells of her offspring will be *62*. Prior to mating, cells in the female giraffe's ovaries undergo the process of meiotic cell division to produce haploid ($n = 31$) gametes, or eggs. Meanwhile, using the same process in the testes, a male giraffe produces haploid ($n = 31$) gametes known as sperm cells. When the female giraffe mates with a male giraffe, a haploid ($n = 31$) egg fuses with a haploid ($n = 31$) sperm to produce a diploid ($2n = 62$) zygote. In this process, the species number of 62 chromosomes is reestablished. As the offspring giraffe develops using mitotic cell division, each of its body cells will contain the species number of 62 chromosomes.

WRONG CHOICES EXPLAINED:
(1), (3), (4) The number of chromosomes in the skin cells of her offspring will *not* be *124*, *31*, or *30*. These answers are inconsistent with the process known as sexual reproduction. See the correct answer described above.

15. **4** All of the characteristics described below are reproductive adaptations that female dogs have for giving birth and caring for several offspring at once except *a pancreas that produces excess insulin to trigger the release of eggs*. The ability of the pancreas to produce and secrete insulin is a correct statement. However, the hormone insulin has no direct effect on the release of eggs in either dogs or other mammals. The pancreas has no direct role in any aspect of reproduction in mammals.

WRONG CHOICES EXPLAINED:
(1) Dogs have *a specialized structure for internal development of several young*, which is a reproductive adaptation that female dogs have for giving birth and caring for several offspring at once. The uterus is the reproductive structure that houses and nourishes the developing pup embryos.

(2) Dogs have *several pairs of mammary glands that provide milk for their pups*, which is a reproductive adaptation that female dogs have for giving birth to and caring for several offspring at once. The mammary glands provide nutrition for young pups during their first weeks of life after birth.

(3) Dogs have *ovaries capable of releasing many gametes at one time for fertilization*, which is a reproductive adaptation that female dogs have for giving birth and caring for several offspring at once. The ovaries produce eggs and manage their release during the female's estrous cycle.

16. 3 This is an example of *a stimulus and a response*. It is also known as a feedback loop or reflex arc. This involuntary behavior depends on the body's ability to detect elevated glucose concentrations in the blood (stimulus) and to release the hormone insulin in order to counteract it (response) and return blood glucose concentrations to acceptable levels. As glucose concentrations fall below the acceptable range (stimulus), the body releases the hormone glucagon to increase the amount of glucose in the blood (response).

WRONG CHOICES EXPLAINED:
(1) This is *not* an example of *a nervous system disorder*. In the case described, the nervous and endocrine systems are working exactly as they should, so no nervous system disorder is indicated.

(2), (4) This is *not* an example of *the synthesis of antibodies* or *an antigen and antibody reaction*. Antibodies are synthesized and released by the immune system in direct response to the presence of a foreign antigen in the body. In the case described, no such antigen-antibody reaction is indicated.

17. 2 This discovery is an important step in *identifying individuals at risk and recommending preventive treatment*. By understanding the scientific information available from legitimate research on the BRCA gene, it is possible for doctors to apply this information in the diagnosis of human patients. Evidence of a damaged BRCA gene in a patient allows the doctor to be on the alert for other symptoms of developing cancer and to provide early interventions that could save the patient's life.

WRONG CHOICES EXPLAINED:
(1), (3), (4) It is *not* true that this discovery is an important step in *preventing the uncontrolled meiotic division of cells in humans, being able to detect all the genes that regulate meiosis,* or *helping to eliminate all BRCA genes*. No information provided is in the question that would lead to any of these outcomes. Cancer is a disorder characterized by uncontrolled mitotic, not meiotic, cell division.

18. 1 The best explanation for this statement is that *during the first two months, organs are being formed and any unusual change during cell division can interfere with normal development.* The process of fetal development begins immediately following fertilization. Any environmental factor, including alcohol, tobacco smoke, or drugs in the system of the mother, can enter the fetal environment and cause changes to the genetic structure of rapidly dividing fetal cells. Any such genetic changes will be inherited by every new cell that arises from the affected cell in the developing fetus. This can lead to abnormal tissues that can result in birth defects or fetal death. After two months of fetal development, the new organs are sufficiently developed to resist such environmental changes, although it is normally recommended that pregnant mothers avoid alcohol, tobacco, and drugs for the full nine months of fetal development.

WRONG CHOICES EXPLAINED:
(2), (3) It is *not* true that the best explanation for this statement is that *the genes that control development function only during the first two months of development* or that *no changes occur in a developing fetus after the second month.* Genetic control of development functions throughout the entire period of fetal development and childhood. Some body tissues never stop developing for the entire life of a human being.

(4) It is *not* true that the best explanation for this statement is that *organ development is not affected by environmental factors after the second month.* After two months of fetal development, the new organs are sufficiently developed to resist such environmental changes, although it is normally recommended that pregnant mothers avoid alcohol, tobacco, and drugs for the full nine months of fetal development.

19. 2 *The altered DNA in the edited cell will be replicated and passed on to each new cell during mitosis* is the statement that best explains why the desired gene will be found in all cells that come from the genetically edited cell. When DNA replicates during mitotic cell division, all genes—including the inserted desired gene—will be accurately copied and passed on to the two daughter cells. When these daughter cells in turn divide by mitosis, their four daughter cells will also carry the desired gene. This process will continue and will eventually produce substantial tissue made up of cells carrying the desired gene.

WRONG CHOICES EXPLAINED:
(1) *The original cell will reproduce by meiosis and a mutation will occur* is *not* the statement that best explains why the desired gene will be found in all cells that come from the genetically edited cell. Meiosis is a specialized form of cell division that results in the formation of haploid gametes. Because of the way meiotic cell division operates, these gametes will not all contain the inserted desired gene.

(3) *DNA replication in body cells will result in sperm and egg cells with the edited gene* is *not* the statement that best explains why the desired gene will be found in all cells that come from the genetically edited cell. Sperm and egg cells are formed from specialized cells in germ tissues, not body cells, of humans.

(4) *The desired gene will be inserted into each new cell by using restriction enzymes* is *not* the statement that best explains why the desired gene will be found in all cells that come from the genetically edited cell. The use of restriction enzymes is a laboratory technique by which scientists insert a desired gene into the genome of a living cell. This technique cannot be used to produce cells arising from the altered cell.

20. **2** *Grass grows on a sand dune and is slowly replaced by shrubs* is the sequence of events that best represents ecological succession. Ecological succession is a term used to describe the replacement of one plant community by another until a stable, self-perpetuating plant community is established. In this example, grass represents the pioneer community, while the shrubs represent the first succession community,

WRONG CHOICES EXPLAINED:
(1) *A squirrel eats acorns, and a hawk eats the squirrel* is *not* the sequence of events that best represents ecological succession. This is an example of a food chain, not ecological succession.

(3) *After many years of planting corn in the same field, minerals present in the soil are used up* is *not* the sequence of events that best represents ecological succession. This is an example of a negative human activity known as monocropping, not ecological succession.

(4) *The decomposition of plant material releases nutrients, and other plants use these nutrients* is *not* the sequence of events that best represents ecological succession. This is an example of recycling of materials or nutrient cycling, not ecological succession.

21. **4** *Emptying an aquarium containing many nonnative fish of several species into a local lake* is a human activity that has the potential to greatly affect the equilibrium of an ecosystem. These nonnative fish may reproduce rapidly in the lake and eventually crowd out native species, significantly altering the balance of nature in that environment and potentially endangering native fish populations in connecting waters.

WRONG CHOICES EXPLAINED:
(1), (2), (3) *Cutting down a few small evergreen trees and using them to make holiday decorations, mowing the playing fields in a city park,* and *washing a car with a detergent-based cleaner* are *not* human activities that have the potential to greatly affect the equilibrium of an ecosystem. These actions, by themselves, represent relatively minor environmental disruptions that are unlikely to have major negative consequences for an ecosystem. However, if actions such as these become widespread, they could result in deforestation, monocropping, or chemical pollution of soil and water in the ecosystem.

22. **3** *A viral infection leads to a decrease in the number of white blood cells being produced in the body* is the statement that describes a failure of homeostasis in humans. Homeostasis is a term that refers to the total of all metabolic activities that serve to maintain physical and chemical balance, or steady state, in the body. In order to function properly and defend itself from foreign invaders, a certain concentration of white blood cells is necessary to maintain this balance. When the concentration of white blood cells is diminished, the body may experience a weakened immune response that can lead to a failure of homeostasis, also known as disease.

WRONG CHOICES EXPLAINED:
(1), (2), (4) *When activity in an individual increases, the body temperature rises and the individual sweats; as the concentration of carbon dioxide increases in the human body, the lungs begin to expel more carbon dioxide;* and *after an individual gets a cut, certain chemical changes begin the healing process* are *not* the statements that describe a failure of homeostasis in humans. In fact, each of these statements describes a different example of the maintenance of homeostasis, or steady state, in the human body.

23. **1** Utilizing the ability of wetland organisms to reduce human wastes makes use of naturally occurring *nutrient cycles*. Some of these wetland organisms (e.g., bacteria, fungi) are able to break down complex molecules in the sewage into absorbable products that can be taken up by other wetland organisms (e.g., plants, algae) to produce new living tissues that can be used by other wetland organisms (e.g., worms, mollusks, insects) to produce still other living tissues that are consumed by other wetland organisms (e.g., amphibians, fish, birds, mammals). Eventually, the tissues of all these organisms are returned to the wetland as biological wastes, where they reenter the food web. In this way, nutrients such as carbon, hydrogen, and nitrogen are cycled within the environment.

WRONG CHOICES EXPLAINED:
(2) Utilizing the ability of wetland organisms to reduce human wastes *does not* make use of naturally occurring *energy cycles*. Energy does not cycle in the environment but is instead used and diminished at each level of a food pyramid until it is dissipated into the environment as waste heat.

(3) Utilizing the ability of wetland organisms to reduce human wastes *does not* make use of naturally occurring *limiting factors*. Limiting factors are the abiotic and biotic conditions of an ecosystem that serve to cap the number of an individual species that can be supported by that ecosystem. Limiting factors are not described in this example, although they are always present in any natural ecosystem.

(4) Utilizing the ability of wetland organisms to reduce human wastes *does not* make use of naturally occurring *finite resources*. Finite resources is a concept that describes the fixed amount of natural resources, primarily mineral, that are available on Earth for human extraction and exploitation. Finite resources are not described in this example, although they are always present in any natural ecosystem.

24. **2** Much of the energy is *released as heat energy back into the hummingbird's environment*. In fact, scientists estimate that as much of 90% of the energy that supports a trophic (feeding) level of a food pyramid is dissipated into the environment as waste heat.

WRONG CHOICES EXPLAINED:
(1) It is *not* true that much of the energy is *converted into amino acids needed for the production of starch*. Energy cannot be converted directly into substances such as amino acids. Amino acids are the building blocks of proteins, not starch.

(3) It is *not* true that much of the energy is *changed into radiant energy, which can be used by plants for photosynthesis*. Heat is a form of radiant energy, but plants cannot absorb it for use in photosynthesis. Green plants absorb light energy for this process.

(4) It is *not* true that much of the energy is *used to synthesize inorganic compounds necessary for cellular respiration*. Energy cannot be converted directly into substances such as inorganic compounds. Inorganic compounds are not used in respiration, although water and carbon dioxide are the inorganic compounds that are by-products of the process. Molecular oxygen is an element that is used in aerobic respiration.

25. **2** *Loggers planting a tree for each one cut down* is the human activity that is the best example of sustainable development. By taking this action, the loggers help to ensure that the forest environment they are harvesting will be returned to its healthy state and will be available for reharvesting in the future. The tree species planted should represent the same species as those harvested in order to ensure biodiversity.

WRONG CHOICES EXPLAINED:
(1), (3) *Draining a wetland to build houses* and *using nets to quickly capture large numbers of fish* are *not* the human activities that are the best examples of sustainable development. By draining the wetland habitat for human development or removing large numbers of fish from their water habitat, the environment will experience a net decrease in biodiversity. Such a loss in biodiversity will not promote sustainability in the environment.

(4) *Building coal-burning power plants to provide electricity* is *not* the human activity that is the best example of sustainable development. By building a power plant that burns coal, the entire environment will be negatively impacted. The resulting air and water pollution will be harmful to all living things, including humans. This action will not promote sustainability in the environment.

26. **4** The relationship between ringworm and humans can be described as *parasite/host*. A parasite is an organism that lives in or on another organism and derives some benefit (usually nutritional) from the host, harming the host in the process. The relationship described is a type of symbiosis that fits the definition given.

WRONG CHOICES EXPLAINED:
(1) It is *not* true that the relationship between ringworm and humans can be described as *predator/prey*. A predator is an animal that hunts and consumes other animals (prey) for food. This is a type of symbiosis that exists in nature but does not fit the description given.

(2), (3) It is *not* true that the relationship between ringworm and humans can be described as *predator/host* or *parasite/prey*. These relationships do not exist in nature. These are nonsense distracters.

27. **1** *Gene expression was influenced by the environment* is one likely explanation for the different heights of the plants at each altitude. Because the yarrow plants are genetically identical, it can be assumed that each plant has the potential to reach the same height, all other factors being equal. It is known that the expression of certain genetic traits may be affected by a variety of environmental factors. The independent variable of altitude is the only environmental factor mentioned in this observation. It can therefore be inferred, based on the information given, that altitude (or elevation) is an environmental factor that may affect expression of the genetic trait of height in yarrow plants.

WRONG CHOICES EXPLAINED:

(2) *Genes mutated when the plants were grown at higher elevations* is *not* one likely explanation for the different heights of the plants at each altitude. Mutations are random events that may or may not alter the genetic structure of yarrow plants. At the relatively low altitudes at which yarrow plants grow, it is unlikely that mutagenic agents such as ultraviolet radiation would significantly differ and cause this type of genetic change.

(3), (4) *Chromosomes increased in number with elevation change* or *the sequence of DNA bases was altered at different altitudes* are *not* likely explanations for the different heights of the plants at each altitude. There is no scientific basis for either of these explanations. Extreme genetic alterations of these types would not occur as a function of altitude (or elevation).

28. **2** *Respiration* is the biological process that is represented in the diagram. The diagram clearly shows that oxygen and food are taken in, that carbon dioxide and water are released, and that ATP is produced in this process occurring inside a green plant cell. This is the chemical definition of cellular respiration.

WRONG CHOICES EXPLAINED:

(1) *Photosynthesis* is *not* the biological process that is represented in the diagram. Photosynthesis is a biochemical process that occurs in green plant cells in which carbon dioxide and water are combined into molecules of glucose (food) using the energy of light. Oxygen gas is a by-product of this process.

(3) *Digestion* is *not* the biological process that is represented in the diagram. Digestion is a cellular process by which complex foods are broken down into simple molecules with the aid of digestive enzymes.

(4) *Replication* is *not* the biological process that is represented in the diagram. Replication is a cellular process by which molecules of DNA synthesize identical duplicate DNA molecules during cell division.

29. **4** The proper function of these cells is vital to the survival of the plant because they regulate the *exchange of gases in leaves*. During daylight hours the specialized cells, known as guard cells, that surround the stomate are rigid. This rigidity forces the stomate to remain open. This opening allows the free exchange of oxygen, carbon dioxide, and water vapor through the opening and promotes efficiency in the photosynthetic process.

WRONG CHOICES EXPLAINED:

(1) It is *not* true that the proper function of these cells is vital to the survival of the plant because they regulate the *rate of glucose use by root cells*. Glucose is used by root cells as a function of the amount of energy needed to operate cell processes such as growth and active transport.

(2) It is *not* true that the proper function of these cells is vital to the survival of the plant because they regulate the *absorption of sunlight by leaf cells*. The rate of sunlight absorption by leaf cells varies as a function of both biotic and abiotic factors, including sunlight intensity, ambient temperature, and water concentration.

(3) It is *not* true that the proper function of these cells is vital to the survival of the plant because they regulate the *products of photosynthesis in the stem*. The products of photosynthesis are the same no matter where in the plant the photosynthesis is occurring.

30. 1 *ATP* is the substance that is represented by X in the diagram. ATP, also known as adenosine triphosphate, is a chemical produced in the cell as a result of energy release from molecules such as glucose during the process of cellular respiration. ATP releases its energy as needed to enable other vital biochemical reactions that occur in the cell.

WRONG CHOICES EXPLAINED:

(2) *DNA* is *not* the substance that is represented by X in the diagram. DNA, also known as deoxyribonucleic acid, is a complex biochemical that carries the genes that determine traits in most living things. DNA does not directly supply energy used to operate cell processes.

(3) *Starch* is *not* the substance that is represented by X in the diagram. Starch is a complex carbohydrate that stores food in most living things. Starch does not directly supply energy used to operate cell processes.

(4) *Glucose* is *not* the substance that is represented by X in the diagram. Glucose is a simple sugar that provides the energy for ATP formation. Glucose does not directly supply energy used to operate cell processes.

PART B-1

31. **4** The most likely reason for the differences in plant growth in the different pots was *competition for resources in each setup.* Since the experimenter was careful to keep all variables constant other than the number of plants, it is likely that resources such as soil nutrients and water were used up more rapidly in the pots containing more than five plants. This was likely the cause of the variation in growth, indicating a level of competition among the seedlings for limited available resources.

WRONG CHOICES EXPLAINED:
(1), (3) It is *not* true that the most likely reason for the differences in plant growth in the different pots was *cyclic changes in the plants' ecosystems* or *the amount of light available for each setup.* These responses imply that variables were present in the experiment that the experimenter reportedly ensured were not present. The description clearly states, "All conditions in the experiment were the same, except for the number of plants in each pot."

(2) It is *not* true that the most likely reason for the differences in plant growth in the different pots was *ecological succession over time.* Ecological succession is a term used to describe the replacement of one plant community by another until a stable, self-perpetuating plant community is established. Nothing in the description of this experiment would suggest that this as a variable that needs to be considered.

32. **1** *The plants in the pot with only 5 plants grew to be an average of 40 cm tall in 30 days* is the statement that is true concerning the growth of plants according to the graph. This can be determined by first tracing on the graph from the x-axis (horizontal axis) at the value of 30 days until it intersects the data line for 5 plants per pot and then tracing from this intersection left to the y-axis (vertical axis), where the value 40 cm can be read.

WRONG CHOICES EXPLAINED:
(2) *The plants in the pot with only 10 plants grew to be an average of 30 cm tall in 20 days* is *not* the statement that is true concerning the growth of plants according to the graph. Following the tracing method described above for 10 plants per pot yields a y-axis value of about 25 cm, not 30 cm.

(3) *The plants in the pot with 15 plants grew an average of 20 cm taller after a period of 10 days* is *not* the statement that is true concerning the growth of plants according to the graph. By following the data line for 15 plants per pot from 0 days to 10 days, it can be determined that the average increase during this period was about 5 cm, not 20 cm.

(4) *The plants in the pot with 20 plants grew an average of 20 cm taller after a period of 40 days* is *not* the statement that is true concerning the growth of plants according to the graph. By following the data line for 20 plants per pot from 0 days to 40 days, it can be determined that the average increase during this period was about 12 cm, not 20 cm.

33. **3** The dependent variable for this experiment was *average plant growth.* The dependent variable in a scientific experiment is the measured value that changes as a result of the independent variable manipulated by the experimenter. In this experiment, the independent variable is the number of plants per pot, and the value that changes over a period of 40 days (dependent variable) is the average plant growth.

WRONG CHOICES EXPLAINED:
(1) It is *not* true that the dependent variable for this experiment was *the number of plants per pot.* This value represents the independent variable manipulated by the experimenter, not the dependent variable.

(2), (4) It is *not* true that the dependent variable for this experiment was *time in days* or *the amount of water per pot.* These values are constants in this experiment. A constant is a condition that is maintained equally in all experimental setups by the experimenter.

34. **1** *Species* A *may carry out autotrophic nutrition* is the statement that correctly identifies a possible role of *one* organism in this ecosystem. The diagram represents a food web and the interactions among organisms in it. By convention, the arrows in this diagram indicate the direction of food energy flow from one organism to another. Green plants capable of autotrophic nutrition (producers) would show only arrows leading from those organisms to other, heterotrophic organisms (consumers). In this diagram, organisms *A* and *C* both demonstrate this condition.

WRONG CHOICES EXPLAINED:
(2) *Species* B *may be a producer that synthesizes nutrients* is *not* the statement that correctly identifies a possible role of *one* organism in this ecosystem. In this diagram, species *B* is shown to be a consumer, not a producer, of food energy.

(3) *Species* C *carries out heterotrophic nutrition* is *not* the statement that correctly identifies a possible role of *one* organism in this ecosystem. In this diagram, species *C* is shown to be an autotrophic organism (producer), not a heterotrophic organism (consumer).

(4) *Species* D *can recycle energy from the Sun* is *not* the statement that correctly identifies a possible role of *one* organism in this ecosystem. In this diagram, species *D* is shown to be a consumer, not a producer, of food energy. Also, energy is not recycled in a food web but only converted from one form to another until it dissipates into the environment as waste heat.

35. 2 *Species* B *returns compounds to the environment that may later be used by species* C is the statement that correctly describes an interaction that contributes to the stability of this ecosystem. In this diagram, species *B* represents a consumer of food energy also known as a decomposer. Decomposers break down complex organic waste materials into simple compounds that may be absorbed and used by autotrophs such as species *C*. This recycling of materials helps to ensure the long-term stability of the ecosystem.

WRONG CHOICES EXPLAINED:
(1) *Species* E *is not affected by the activity of species* A is *not* the statement that correctly describes an interaction that contributes to the stability of this ecosystem. In the diagram, species *E* is shown to be directly dependent on species *A* for food. If species *A* were to cease its autotrophic activity, species *E* would be greatly affected.

(3) *Species* C *recycles nutrients from species* B *and* D *to obtain energy* is *not* the statement that correctly describes an interaction that contributes to the stability of this ecosystem. In the diagram, species *C* represents an autotrophic organism. As such, species *C* does not obtain food energy from other organisms in the ecosystem but, rather, produces food using the process of photosynthesis.

(4) *Species* D *is directly dependent on the autotrophic activity of species* B is *not* the statement that correctly describes an interaction that contributes to the stability of this ecosystem. In the diagram, species *B* represents a decomposer organism, not an autotrophic organism.

36. 4 Although the Venus flytrap uses its prey to obtain certain molecules that it needs, it is still classified as a producer because it *synthesizes glucose by using carbon dioxide and water*. The Venus flytrap is a green plant whose cells contain chloroplasts capable of carrying out the autotrophic nutritional process known as photosynthesis. This process uses the energy of light to convert molecules of carbon dioxide and water into molecules of glucose. For this reason, the Venus flytrap is classified as a producer organism.

WRONG CHOICES EXPLAINED:

(1) It is *not* true that, although the Venus flytrap uses its prey to obtain certain molecules that it needs, it is still classified as a producer because it *uses its prey to produce food*. Producers typically do not trap and consume prey for food. This is a nutritional adaptation found in Venus flytraps and a few other insectivorous plant species.

(2), (3) It is *not* true that, although the Venus flytrap uses its prey to obtain certain molecules that it needs, it is still classified as a producer because it *consumes the prey to produce energy* or because it *synthesizes energy by using oxygen and releasing carbon dioxide*. Energy is neither produced nor synthesized in any nutritional process but is only converted from one form to another (e.g., light, chemical bond, heat).

37. **1** Enzymes secreted by cells in the leaves of the Venus flytrap can digest *proteins into amino acids*. Amino acids are the building blocks of proteins. When digestive enzymes known as proteases encounter proteins, these enzymes break the peptide bonds holding the amino chain together and release free amino acids for absorption.

WRONG CHOICES EXPLAINED:

(2), (3), (4) It is *not* true that enzymes secreted by cells in the leaves of the Venus flytrap can digest *sugars into starches*, *amino acids into fats*, or *proteins into sugars*. None of these examples represents a biochemical reaction involving digestion of a complex molecule by enzymes.

38. **4** *If a patient is infected by* Helicobacter pylori, *then they will get an ulcer* is the choice that represents a possible hypothesis for Marshall's experiment. A hypothesis in a scientific experiment is the investigator's educated guess as to the outcome of that experiment. Hypotheses are usually based on some prior experience or scientific work by the researcher him/herself or by other scientists, and so is informed by what is already known about the subject.

WRONG CHOICES EXPLAINED:

(1), (3) *Does* Helicobacter pylori *cause stomach ulcers in people?* and *Does exposure to infectious bacterial cells make people sick?* are *not* the choices that represent possible hypotheses for Marshall's experiment. These are scientific questions that may be asked by a researcher to begin the experimental method, leading the researcher to form a hypothesis, develop and execute an experimental procedure, analyze experimental results, and form a conclusion that answers the question.

(2) *If a person takes an antibody, then they will not develop an ulcer* is *not* the choice that represents a possible hypothesis for Marshall's experiment. The passage mentions that Marshall took an antibiotic, not an antibody. The antibiotic was taken at the conclusion of the experiment specifically to treat and reverse the ulcerating effects of the bacterium *Helicobacter pylori* on Marshall's stomach lining.

39. **2** The work of Marshall and Warren shows that *scientific explanations are revised based on new evidence*. This statement summarizes a basic tenet of scientific inquiry that should guide any experimental procedure any time research is conducted. The passage describes the way these researchers went about disproving a widely held theory in favor of a new theory supported by scientific evidence.

WRONG CHOICES EXPLAINED:
(1) The work of Marshall and Warren does *not* show that *hypotheses made by physicians are always correct*. No researcher is ever correct all the time. Failure in scientific research should not be viewed negatively but as an opportunity to do further research until the truth is found.

(3), (4) The work of Marshall and Warren does *not* show that *peer review always leads to the immediate acceptance of results* or that *conclusions must always be consistent with those made by other scientists*. Very often, the peer review of scientific results by other scientists does not support the conclusions drawn from the original research. Failure in scientific research should not be viewed negatively but as an opportunity to do further research until the truth is found.

40. **4** Due to the differences in structure, each of these molecules most likely *performs a different function*. These complex carbohydrates may serve as foods, as structural features, or as any one of several other functions in living things.

WRONG CHOICES EXPLAINED:
(1) It is *not* true that, due to the differences in structure, each of these molecules most likely *is composed of different molecular bases*. Molecular bases are subcomponents of complex nucleic acids including DNA and RNA, not of carbohydrates.

(2) It is *not* true that, due to the differences in structure, each of these molecules most likely *forms a different protein*. Proteins are made up of amino acid, not glucose, subunits.

(3) It is *not* true that, due to the differences in structure, each of these molecules most likely *contains different elements*. An examination of the molecules illustrated shows that both are composed only of the elements carbon, hydrogen, and oxygen. This chemical composition is common to all carbohydrates.

41. 3 These specific venom proteins are most likely *biological catalysts*. Biological catalysts are also known as enzymes. These enzymes convert complex proteins in the prey animal's body to amino acids that are suspended in the fluids released from cells destroyed by the venom.

WRONG CHOICES EXPLAINED:
(1) These specific venom proteins are *not* most likely *ATP molecules*. ATP, also known as adenosine triphosphate, is a chemical that is produced in the cell as a result of energy release from molecules such as glucose during the process of cellular respiration. ATP is not a protein.

(2) These specific venom proteins are *not* most likely *DNA molecules*. DNA, also known as deoxyribonucleic acid, is a complex biochemical that carries the genes that determine traits in most living things. DNA is not a protein.

(4) These specific venom proteins are *not* most likely *regulatory hormones*. Regulatory hormones are beneficial proteins secreted by endocrine tissues in the bodies of many living things. These hormones exert chemical control over various metabolic processes that aid the maintenance of homeostasis in the body.

42. 2 The arrows shown in the diagram that go from the blood capillaries to the interstitial fluid most likely represent the *movement of materials from the circulatory system that will eventually enter lymphatic capillaries*. These materials may include dissolved oxygen, nutrients, and other dissolved or suspended substances necessary for the maintenance of homeostatic balance in body tissues.

WRONG CHOICES EXPLAINED:
(1) The arrows shown in the diagram that go from the blood capillaries to the interstitial fluid do *not* most likely represent the *release of red blood cells, so that they can diffuse into body cells and fight bacteria*. Red blood cells are too large to pass easily through capillary walls or cell membranes. Red blood cells function to carry oxygen bound to hemoglobin, not to fight bacteria.

(3) The arrows shown in the diagram that go from the blood capillaries to the interstitial fluid do *not* most likely represent the *transport of digestive enzymes from the blood to help with the digestion of glucose in muscle cells*. Digestive enzymes are not normally carried in the bloodstream. Glucose is not digested but, rather, oxidized in muscle cells.

(4) The arrows shown in the diagram that go from the blood capillaries to the interstitial fluid do *not* most likely represent the *transport of glucose molecules from the blood to be used by cells to attack proteins and fats*. Glucose is used by cells as a source of energy-rich food. Glucose is not used by cells to attack proteins or fats.

43. **4** Based on the data in this graph, one likely conclusion that can be made is that over approximately ten years *the introduction of the opossum shrimp into the lake ecosystem disrupted the food webs that were present.* Some research indicates that certain species of invasive opossum shrimp introduced into lake environments may have the effect of competing with fish for the supply of plankton (microscopic organisms) found in these lakes. When the shrimp population blooms, they may deplete the plankton population below normal levels. If severe enough, this depletion may interfere with the survival rates of young salmon, which depend on plankton for food. This situation can be one cause of population decline of mature salmon in the lake ecosystem. Because eagles depend on salmon for their food, the eagle population may decline rapidly as well. An examination of the data presented in the graph appears to support this inference, although not enough information is provided in the question for the student to easily arrive at this conclusion.

WRONG CHOICES EXPLAINED:
(1) It is *not* true that, based on the data in this graph, one likely conclusion that can be made is that over approximately ten years *the lake ecosystem stabilized after the introduction of the new species.* On the contrary, beginning with the introduction of opossum shrimp to the lake ecosystem in 1981, we can see significant fluctuations in the populations of salmon and eagles, with no salmon in evidence in 1988 and 1989. The data do not indicate a stabilized ecosystem.

(2) It is *not* true that, based on the data in this graph, one likely conclusion that can be made is that over approximately ten years *competition between organisms was reduced as more producers were introduced into the lake.* There is no information provided in the question that would indicate the introduction of more producers into the lake ecosystem at any time.

(3) It is *not* true that, based on the data in this graph, one likely conclusion that can be made is that over approximately ten years *more predators moved into the lake ecosystem once the opossum shrimp were added.* There is no information provided in the question that would indicate the migration of more predators into the lake ecosystem at any time.

PART B-2

44. One credit is allowed for correctly marking an appropriate scale, without any breaks in the data, on each labeled axis. [1]

45. One credit is allowed for correctly plotting the data on the grid, connecting the points, and surrounding each point with a small circle. [1]

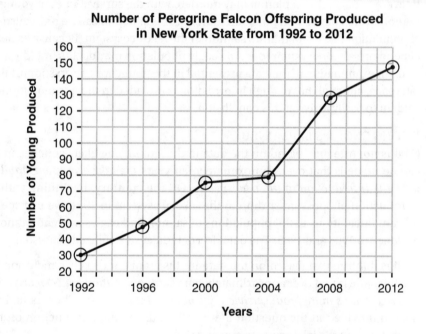

46. One credit is allowed for correctly identifying a body system in the falcon that was directly affected by DDT and led to the loss of nesting peregrine falcons from New York State in the early 1960s and supporting the answer. Acceptable responses include but are not limited to: [1]

- Body system:
  - *Reproductive*
- Support:
  - *Egg laying/egg shells are part of the reproduction of falcons.*
  - *The eggs had thin shells, which drastically lowered breeding success.*
  - *DDT has the effect of thinning the egg shells, which would break before the chicks could develop.*

47. 2 *The greatest increase was during the time period of 2004 and 2008* is the conclusion that is best supported by the information presented in the graph. A review of the graphed data shows a steep increase in the production of peregrine falcon offspring between these two survey years.

WRONG CHOICES EXPLAINED:
(1) *The greatest decrease was during the time period of 1992 and 1996* is the *not* conclusion that is best supported by the information presented in the graph. A review of the graphed data shows an increase, not a decrease, in the production of peregrine falcon offspring between these two survey years.

(3) *There has been a steady decline since the banning of DDT in 1972* is the *not* conclusion that is best supported by the information presented in the graph. A review of the graphed data shows an increase, not a decline, in the production of peregrine falcon offspring since 1992. No information is provided for the period 1972 to 1992.

(4) *The population reached carrying capacity in 2004* is the *not* conclusion that is best supported by the information presented in the graph. The carrying capacity of an environment is the maximum number of individuals of a particular species that may be sustainably supported by the natural resources present in the environment. This phenomenon would be shown on the graph as a leveling of the production of new peregrines. A review of the graphed data shows an increase, not a leveling, in the production of peregrine falcon offspring during this period.

48. One credit is allowed for correctly explaining why the results of this study may indicate *one disadvantage* of using antibiotics to fight infections. Acceptable responses include but are not limited to: [1]

- *The antibiotics a person receives may not kill all of the bacteria causing the infection.*

- *The overuse of antibiotics may cause antibiotics to become less effective.*

- *If bacteria develop resistance to an antibiotic, the antibiotic will be less effective for humans infected by the bacteria.*

- *The bacteria may develop strains that are resistant to the antibiotic.*

- *Antibiotics can be selective in their action, killing some types of bacteria but encouraging the growth of others.*

49. **2** The survival of some bacterial colonies was most likely due to *a resistance to the antibiotic.* The surviving colonies were probably resistant strains of one type of bacteria or colonies of a different type of bacteria not sensitive to the antibiotic.

WRONG CHOICES EXPLAINED:
(1) The survival of some bacterial colonies was *not* most likely due to *the bacterial cells changing so that they could live.* The traits of a species do not change as a function of need but only through the processes of genetic mutation, reproduction, and natural selection.

(3) The survival of some bacterial colonies was *not* most likely due to *meiotic cell division in the bacteria.* Bacteria are simple unicellular organisms that reproduce by mitotic, not meiotic, cell division.

(4) The survival of some bacterial colonies was *not* most likely due to *a DNA change caused by the antibiotic.* Antibiotics are not known to be mutagenic agents that could alter the genetic code of DNA.

50. **3** One likely reason that the gray tree frog occupies a larger environmental area than the green tree frog is that the gray tree frog species *has adaptations that enable survival in a wider variety of habitats.* These adaptations may include an enhanced tolerance for cold climates that allows it to survive the harsh temperatures and deep frost common to winters in the northern United States and southern Canada.

WRONG CHOICES EXPLAINED:
(1) It is *not* true that one likely reason that the gray tree frog occupies a larger environmental area than the green tree frog is that the gray tree frog species *eats only prey found in central areas of the United States.* No information is provided concerning the diets of these tree frog species. However, the diagram shows that there is considerable overlap in the ranges of both of these species. It is likely that they seek and consume some of the same insect prey in areas that they inhabit in common but inhabit different niches that minimize competition for food.

(2) It is *not* true that one likely reason that the gray tree frog occupies a larger environmental area than the green tree frog is that the gray tree frog species *is adapted to live in any environment in the United States.* The diagram shows that gray tree frogs are not found in a substantial portion of the United States, especially humid subtropical and arid western plains and mountain areas.

(4) It is *not* true that one likely reason that the gray tree frog occupies a larger environmental area than the green tree frog is that the gray tree frog species *outcompetes the green tree frogs in Florida and any state where they both live.* There is no information provided in the diagram that indicates the level of interspecies competition that may or may not occur. However, the fact that both species inhabit some of the same states indicates that they each inhabit slightly different niches and have slightly different requirements to be successful.

51. One credit is allowed for correctly identifying a biological process that led to the presence of 90 different species of frogs throughout the United States and supporting the answer. Acceptable responses include but are not limited to: [1]

- Biological process:
  - *Sexual reproduction*
  - *Genetic recombination*
  - *Natural selection*
  - *Evolution*
  - *Mutations*
  - *Survival of the fittest*
  - *Adaptation*
  - *Geographic isolation*
- Support:
  - *Sexual reproduction results in the production of offspring with many variations that can lead to the evolution of new species.*
  - *Genetic recombination promotes new gene arrangements that give rise to new adaptations that gradually lead to new varieties/species of frogs.*
  - *Natural selection gives an advantage to certain members of a frog species that have favorable genetic adaptations. Eventually, the descendants of these successful frogs give rise to a new species of frogs.*
  - *Some frogs are better able to survive in certain environments, reproduce, and pass on their successful traits, eventually resulting in new species of frogs.*
  - *Mutations can make the frogs better fit to their environments. They survive and pass on their traits to future generations.*

— *Some frogs have characteristics that make them better adapted to certain environments. They can pass these genetic characteristics to their offspring.*

— *Separating populations of a species of frog by physical/geographic barriers makes it easier for new varieties to arise. After many generations of separation, these varieties may evolve into distinct species.*

52. One credit is allowed for selecting *one* lettered part of the diagram and circling it, correctly identifying the part, and correctly stating how a malfunction in the structure that you identified could interfere with an individual's ability to reproduce. Acceptable responses include but are not limited to: [1]

- Part selected: *A*
- Identification:
  - *Oviduct*
  - *Fallopian tube*
- Explanation:
  - *Fertilization might not occur.*
  - *Eggs may not be able to get to the uterus.*
- Part selected: *B*
- Identification:
  - *Ovary*
- Explanation:
  - *Hormones that regulate the menstrual cycle might not be synthesized.*
  - *Eggs may not be produced/released.*
- Part selected: *C*
- Identification:
  - *Uterus*
- Explanation:
  - *The embryo might not implant.*
  - *The placenta might not form correctly.*
  - *The fetus may not be able to develop.*

53. One credit is allowed for correctly completing the diagram so that it shows the arrangement of the genetic material in the two new cells that are produced by mitosis. Acceptable responses include but are not limited to: [1]

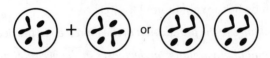

54. One credit is allowed for correctly stating *one* reason scientists would value an experiment that does *not* support the initial hypothesis. Acceptable responses include but are not limited to: [1]

- *It helps scientists design new experiments or develop new hypotheses.*

- *The new information can be used for future investigations.*

- *An unsupported hypothesis provides scientists with information that is important to understanding the scientific concepts being studied.*

- *It tells scientists they are incorrect and that more study is needed.*

55. One credit is allowed for completing the sequence by correctly filling in the missing levels. Acceptable responses include: [1]

- Organelles → *cells* → tissues → *organs* → organ systems → organism

PART C

56. One credit is allowed for correctly explaining how a reduction in phytoplankton can lead to a reduction in fish populations in the Indian Ocean. Acceptable responses include but are not limited to: [1]

- *Phytoplankton are the foundation of the food web in the ocean.*

- *If the amount of phytoplankton decreases in the ocean, there will be less food/ oxygen available for the fish.*

- *Producers provide energy in the form of food for all other organisms in the ocean. A reduction in the number of producers will lead to a reduction in the number of fish that can be supported.*

- *It disrupts the stability of the ecosystem because there is less food for the fish to eat.*

- *It would lead to more competition for food among the ocean fish species.*

57. One credit is allowed for correctly describing how *one* specific human activity mentioned in the passage could *negatively* affect the Indian Ocean ecosystem. Acceptable responses include but are not limited to: [1]

- Human activity: *urbanization/coastline development*
  - *As towns develop/cities grow, land is taken away from the organisms that live along the coast.*
  - *This can increase water pollution by human waste/sewage/trash.*

- Human activity: *mining of natural resources*
  - *This activity could deplete natural resources.*
  - *Mining runoff could pollute coastal waters/increase acidity levels.*

- Human activity: *oil spills*
  - *Oil spills from pipelines/storage sites/tankers can kill plants and animals in the area, disrupting food webs.*
  - *Oil is less dense than water, so it floats on the surface and cuts off oxygen absorption for gill-breathing marine life.*

- Human activity: *destruction of mangrove forests*
  - *Deforestation decreases photosynthesis and increases CO_2 levels.*
  - *Deforestation removes/destroys animal habitats.*
  - *Removing mangroves decreases natural water purification/increases coastline erosion.*

58. One credit is allowed for correctly stating *one* specific reason why it is important to maintain a healthy ecosystem in the Indian Ocean. Acceptable responses include but are not limited to: [1]

- *The Indian Ocean provides food for a large number of people.*
- *The more biodiversity in the ocean, the healthier/more stable it will be.*
- *If all the resources in the ocean are used/greatly reduced, there won't be enough food or other ocean resources for future generations.*
- *The world's ocean environments are interconnected, so the health of the Indian Ocean environment affects the health of all other oceans, including those that bound the United States.*

59. One credit is allowed for correctly stating *one* way that the presence of water chestnut plants affects the other organisms in the freshwater ecosystem. Acceptable responses include but are not limited to: [1]

- *The water chestnut blocks out the light, so other plants die out because they can't compete.*
- *There is less food for animals and insects.*
- *The local food web is disrupted.*
- *The plants block 95% of the sunlight.*

60. One credit is allowed for correctly stating *one* advantage and *one disadvantage* of using biological controls in this situation. Acceptable responses include but are not limited to: [1]

- Advantage:
  - *No chemicals are added to disrupt the ecosystem.*
  - *Biological control will not poison the water.*
  - *It may eliminate water chestnut infestations.*
  - *This process may restore the balance of the freshwater ecosystems.*
- Disadvantage:
  - *Native plants in the ecosystem may be eaten/eliminated by the new insects, not just the water chestnuts.*
  - *The new species may spread to other ecosystems, where they could cause damage by competing with native insects.*
  - *The introduced species may become invasive.*
  - *The new insects could overpopulate due to the absence of natural predators.*

61. One credit is allowed for correctly stating *one disadvantage* of this method of controlling water chestnuts. Acceptable responses include but are not limited to: [1]

- *Many other plants and some animals are also removed from the sediment, not just the water chestnut plants.*
- *Other kinds of plants/animals are killed also.*
- *The harvester removes many local species/organisms together with the water chestnuts.*
- *The water chestnut plants may grow back.*
- *The machine could pollute the water.*
- *The machine may disturb the bottom sediments, releasing particulates/toxins from the mud that could harm fish and other aquatic animals.*

62. One credit is allowed for correctly identifying *one* human activity and describing how it contributes to increasing levels of carbon dioxide in the environment. Acceptable responses include but are not limited to: [1]

- Human activity: *burning fossil fuels*
  - *Burning coal/oil/natural gas releases carbon dioxide into the environment.*
  - *Cars/trucks expel carbon dioxide into the air.*
- Human activity: *cutting down forests/deforestation*
  - *This activity removes trees/plants that absorb carbon dioxide from the atmosphere.*
  - *Crops that are planted on deforested plots are not as efficient at removing CO_2 from the air as the native trees/plants that were destroyed.*
- Human activity: *industrial/commercial/residential development*
  - *Power plants/factories/homes that burn carbon-based fuels release carbon dioxide from chimneys into the air.*
  - *Development requires that natural habitats be removed and replaced with buildings/parking lots, reducing the number of plants available to scrub the air of CO_2.*

63. One credit is allowed for correctly describing how the inability of sharks to detect their prey could affect an ocean ecosystem. Acceptable responses include but are not limited to: [1]

- *The sharks might not detect enough prey to feed on and might not survive.*

- *Since sharks lack the ability to detect their food source, the population of their prey would sharply increase.*

- *There could be an increase in prey and a decrease in the populations that the prey themselves consume.*

- *Without the shark population holding its prey in check, the prey population will increase.*

64. One credit is allowed for correctly stating how the body plan of the weasel is effective for successfully competing with other organisms. Acceptable responses include but are not limited to: [1]

- *The body plan of the weasel allows it to get food/prey by entering small spaces that other carnivores cannot.*

- *Their long, slender body allows weasels to travel very close to the ground, making it easier for them to approach prey and avoid predators/humans.*

- *Female weasels are adapted to give birth to immature young, making it possible for them to hunt for food soon afterward.*

65. One credit is allowed for correctly explaining why weasels do *not* completely overpopulate the areas where they live. Acceptable responses include but are not limited to: [1]

- *Limiting factors such as available food, space, and nesting habitats can limit the size of weasel populations.*

- *Predators of the weasel take a certain proportion of the weasel population, including unguarded young, reducing the weasel's reproductive capacity.*

- *Since they need to eat 1/3 of their body weight every day, they may starve if prey populations decline.*

- *The weasels may have reached the carrying capacity for their species in the environment.*

66. One credit is allowed for indicating whether the weasels' relationship with humans is positive or negative by circling the appropriate choice and supporting your answer. Acceptable responses include but are not limited to: [1]

- Relationship: *positive*
    - *The weasels are able to control the number of rodents and wild rabbits that could otherwise consume farm or garden crops.*
    - *The weasels are able to control the number of rodents that could carry diseases that can affect human health.*
    - *The weasel fills an important ecological niche that helps to maintain the balance of nature, which is essential to human survival.*
- Relationship: *negative*
    - *The weasels can consume small farm animals, such as chickens and domestic rabbits, that are raised for human food.*
    - *Sealing entrances to chicken coops to prevent the weasels' entry can use up valuable time and resources.*

67. One credit is allowed for correctly explaining how an outer covering composed of a section of a cell membrane from the host would protect HIV from attack by the host's immune system. Acceptable responses include but are not limited to: [1]

- *The host will not recognize the virus as an invader.*
- *The cell membrane has antigens that indicate that it is a normal part of the host, so it won't be attacked by the immune system.*
- *The virus is enclosed by a cell membrane that the host's immune system won't recognize as a pathogen.*
- *The immune system would recognize the virus as being part of the individual since it would have receptors and other chemicals that identify it as not being a pathogen.*

68. One credit is allowed for correctly describing *one* specific way that HIV makes the body unable to deal with other pathogens and cancers. Acceptable responses include but are not limited to: [1]

- *HIV destroys white blood cells/helper T cells/B cells.*
- *HIV weakens the immune system, allowing other diseases to enter the body and cause damage.*
- *Once HIV enters the human body and weakens it, the body can no longer defend itself from other pathogens.*

69. One credit is allowed for correctly stating *one* possible advantage for a snake to have no limbs instead of four limbs. Acceptable responses include but are not limited to: [1]

- *Snakes without limbs could escape predators/capture prey better than those with limbs.*
- *Legless snakes could seek shelter more easily in smaller spaces than those with legs.*
- *Snakes display a number of adaptations in addition to the absence of limbs that operate together to make snakes successful in their environments.*

70. One credit is allowed for correctly identifying the technique that the scientists used to remove the ZRS DNA from mice and replace it with the ZRS section from snakes. Acceptable responses include but are not limited to: [1]

- *Genetic engineering*
- *Gene editing*
- *Gene splicing*
- *Genetic manipulation*
- *Restriction enzyme–catalyzed excision/insertion*

71. One credit is allowed for correctly identifying the type of mutation responsible for the loss of limbs in snakes. Acceptable responses include: [1]

- *Deletion*

72. One credit is allowed for correctly stating how scientists could know that snakes once actually had legs. Acceptable responses include but are not limited to: [1]

- *They may have had fossils of snake ancestors that showed the presence of leg bones.*
- *Fossils may have been discovered showing snake ancestors with four limbs.*
- *They might have examined the fossil record and found fossils of snake ancestors with legs.*
- *They could have examined the characteristics of common ancestors of modern snake species.*
- *They might have studied modern snake skeletons and found vestigial leg bones.*

PART D

73. **1** Two features that would be most useful in determining which of these organisms are most closely related are *presence of hair and similar proteins*. The presence of hair is characteristic of the taxonomic class Mammalia (mammals), which includes killer whales, leopard seals, and baleen whales. This would exclude the Adélie penguin, with a skin covering of feathers, as a close relative of the others. An examination of the proteins produced by the three mammals would indicate the relative relatedness among them by revealing their genetic similarities. The more similar their proteins are, the more closely they are related.

WRONG CHOICES EXPLAINED:
(2) It is *not* true that two features that would be most useful in determining which of these organisms are most closely related are *presence of feathers and similar body structures*. The only organism identified as having a skin covering composed of feathers is the Adélie penguin, a member of the taxonomic class Aves (birds). Having similar body structures, such as eyes or limbs/flippers, would be less useful in determining relative relatedness in this case.

(3) It is *not* true that two features that would be most useful in determining which of these organisms are most closely related are *habitat and diet*. All of the organisms illustrated share the same Antarctic habitat. The killer whale, Adélie penguin, and leopard seal consume fish for food, while the baleen whale consumes plankton. These features are not useful in determining relative relatedness.

(4) It is *not* true that two features that would be most useful in determining which of these organisms are most closely related are *body size and color*. All four of the organisms listed vary considerably in terms of body size. The killer whale and Adélie penguin share a similar body color pattern but are not closely related. These features are not useful in determining relative relatedness.

74. **3** *Set up and perform gel electrophoresis* is the lab procedure that can be done to find molecular evidence for relationships among these Antarctic marine organisms. This laboratory procedure involves using restriction enzymes to cut the study organisms' DNA molecules into fragments. This fragmented DNA is then placed into separate wells in a block of specially prepared gel and is subjected to an electric current that draws the DNA fragments through the gel at different rates. The bands of DNA fragments that are created by this process are compared for the patterns they produce. The more similar the patterns of these molecular fragments are, the more closely the organisms are related.

WRONG CHOICES EXPLAINED:

(1), (2), (4) *Compare slides of cell organelles, examine fossils and ocean sediments,* and *use a dichotomous key and test for pH* are *not* the lab procedures that can be done to find molecular evidence for relationships among these Antarctic marine organisms. None of these activities would reveal useful information concerning molecular similarities and differences among these organisms.

75. **3** *The students all had varying physical fitness levels and consumed different amounts of protein* is the best explanation for the observation that the heart rates of 5 students were higher after lunch. There is no discussion of any controls that may or may not have been used in this experiment. Controls should be implemented in scientific experiments of this type to minimize the effects of variables such as physical fitness, grams of protein consumed, gender, prior exercise, or age on experimental results. It can be assumed that no such controls were in place and that the results of the experiment could have been skewed by differences in these factors. Therefore, it is likely that the students differed in their physical fitness levels and were permitted to consume different amounts of protein.

WRONG CHOICES EXPLAINED:

(1) *The heart rates of female students are not affected by a high-protein meal* is *not* the best explanation for the observation that the heart rates of 5 students were higher after lunch. No information is provided concerning the genders of students included in the experimental group.

(2) *The students all participated in physical education class immediately before lunch* is *not* the best explanation for the observation that the heart rates of 5 students were higher after lunch. No information is provided concerning the prior exercise of the students included in the experimental group.

(4) *The students were all the same gender and age* is *not* the best explanation for the observation that the heart rates of 5 students were higher after lunch. No information is provided concerning the genders or ages of students included in the experimental group.

76. **2** One possible reason that such diverse species could be affected by the HMGA2 gene is that *they share a common ancestor*. Finches, horses, dogs, and humans are all vertebrate animals and therefore share a large proportion of their genetic structures in common. It is possible that the HMGA2 gene has been present in the genome of vertebrates for about 650 million years, ever since the first vertebrates evolved from their common ancestor. Since then, it has been passed on intact from generation to generation and from ancestral species to new species, all through the evolutionary history of the vertebrate phylum.

WRONG CHOICES EXPLAINED:

(1) It is *not* true that one possible reason that such diverse species could be affected by the HMGA2 gene is that *they all lived on the Galapagos Islands*. The Galapagos finches migrated from the South American mainland long after the HMGA2 gene was present in its genome. The different geographic locations of diverse vertebrate species have little to do with the genes they share in common.

(3) It is *not* true that one possible reason that such diverse species could be affected by the HMGA2 gene is that *the drought caused the formation of the gene*. Environmental conditions such as drought may affect the survival of better-fit organisms, but it is not recognized as a mutagenic agent capable of causing mutations in reproductive tissues.

(4) It is *not* true that one possible reason that such diverse species could be affected by the HMGA2 gene is that *the gene allowed all these species to grow larger*. The passage mentions that the HMGA2 gene is involved in controlling body size in some vertebrate species, but it does not claim that the gene acts the same way in these species.

77. One credit is allowed for correctly stating *one* possible reason the medium ground finches with a smaller beak were able to survive during the 2004–2005 drought and supporting the answer. Acceptable responses include but are not limited to: [1]

- *Medium ground finches with smaller beaks had less competition for food than the medium ground finches with larger beaks.*

- *A smaller beak was better for obtaining the food that was still available during the drought.*

- *Their smaller beaks provided them with the ability to obtain more seeds.*

- *They were better adapted to eat smaller seeds than were other finches.*

- *Smaller beaks proved to be an adaptive advantage under the severe selection pressure provided by the drought and limited food type available.*

78. One credit is allowed for correctly stating that the mutated DNA strand would *not* affect the functioning of protein *X* and supporting the answer. Acceptable responses include but are not limited to: [1]

- *Protein X would not be affected because the mRNA codons resulting from the change code for the same amino acids as the unchanged codons.*

- *The two changes in the code result in the same amino acids, so the protein produced is the same.*

- *Both changes in the code still result in the same amino acids.*
- *The original amino acids will still be in the same location in the protein, even with the changes.*

79. One credit is allowed for correctly stating where on the diagram the largest fragments of DNA would be located. Acceptable responses include but are not limited to: [1]

    - *The largest fragments are those that have migrated the least distant from the wells.*
    - *The bands closest to the wells contain the largest fragments.*
    - *The most complex DNA fragments would be at the top, near where the DNA is put into the gel.*
    - *Since the DNA fragments migrated through the gel from the negative to the positive end, the larger pieces would have migrated most slowly and would have stayed closer to the negative end.*

80. One credit is allowed for correctly identifying the factor that caused the fragments to move through the gel rather than remaining in the wells. Acceptable responses include but are not limited to: [1]

    - *Electricity*
    - *Electric charge*
    - *Electric current*
    - *Positive and negative charges*

81. **4** Diagram *4* represents the distribution of sugar molecules in the water a day later. Sugar is soluble in water at room temperature. As molecules dissolve from the surface of the sugar cube, they begin to move through the water in the beaker by the process of diffusion until they are evenly distributed. After 24 hours, the entire sugar cube would have dissolved and the molecules would have become evenly distributed through the water in the beaker, as shown in diagram *4*.

    WRONG CHOICES EXPLAINED:
    (1) Diagram *1* does *not* represent the distribution of sugar molecules in the water a day later. This diagram represents the condition in the beaker just after the sugar cube was placed into the beaker.

    (2), (3) Diagrams *2* and *3* do *not* represent the distribution of sugar molecules in the water a day later. These diagrams do not represent a condition that would occur in the beaker at any time during the experiment.

82. 4 The method used by the student to compare plant extracts from the different species is *paper chromatography*. Paper chromatography is a laboratory technique used to study the pigments present in plant leaves. Because the pigment molecules are different shapes and sizes, the dissolved molecules move through a filter paper strip at different rates, allowing them to separate and concentrate in distinct color bands. In this experiment, the technique is being used to demonstrate the similarities and differences in leaf pigments, and therefore relative relatedness, among four separate plant species.

WRONG CHOICES EXPLAINED:

(1) It is *not* true that the method used by the student to compare plant extracts from the different species is *gel electrophoresis*. Gel electrophoresis is a laboratory technique used to demonstrate similarities and differences in the DNA of separate organisms. It is typified by the use of an electrically charged block of agar gel, DNA samples, and restriction enzymes, not solvent, test tubes, plant extract, and filter paper strips.

(2) It is *not* true that the method used by the student to compare plant extracts from the different species is *DNA banding*. DNA banding is the phenomenon that results from the differential migration of DNA fragments in a gel electrophoresis setup.

(3) It is *not* true that the method used by the student to compare plant extracts from the different species is *a staining technique*. Staining is a laboratory technique used in light microscopy that adds contrast or color to cell organelles, making them more visible to the researcher. It is typified by the use of prepared stains such as Lugol's solution and methylene blue, not solvent, test tubes, plant extract, and filter paper strips.

83. One credit is allowed for correctly stating *one* benefit of the increase in average heart rate during exercise. Acceptable responses include but are not limited to: [1]

- *Wastes/carbon dioxide are removed more quickly.*
- *The increased heart rate results in more glucose/oxygen being delivered to the cells.*
- *The increase in heart rate helps maintain homeostasis.*

84. One credit is allowed for correctly calculating the clothespin-squeezing rates per minute and average for the dominant hand. Acceptable responses include: [1]

Clothespin Squeezing Activity

Trial	20-Second Clothespin Squeezing (Dominant Hand)	Clothespin-Squeezing Rate Per Minute (Dominant Hand)	20-Second Clothespin Squeezing (Nondominant Hand)	Clothespin-Squeezing Rate Per Minute (Nondominant Hand)
Trial 1	26	78	18	54
Trial 2	33	99	28	84
Trial 3	24	72	29	87
Average	■■■	83	■■■	75

85. One credit is allowed for indicating either yes or no and supporting the answer in a manner consistent with the data provided in the table. Acceptable responses include: [1]

- *Yes:*
  - *The average of the student's squeezing number with her dominant hand is higher than the student's squeezing number with her nondominant hand.*
  - *In two trials out of three, the student's dominant hand squeezing rate was higher than her nondominant hand squeezing rate.*
- *No:*
  - *The conclusion is based on the work of only one student.*
  - *Not enough trials were run in this experiment.*
  - *The results have not been independently tested by other researchers.*
  - *The experimenter should not be the experimental subject.*

STUDENT SELF-APPRAISAL GUIDE
Living Environment June 2021

Standards/Key Ideas	June 2021 Question Numbers	Number of Correct Responses
Standard 1		
Key Idea 1: The central purpose of scientific inquiry is to develop explanations of natural phenomena in a continuing and creative process.	39, 54	
Key Idea 2: Beyond the use of reasoning and consensus, scientific inquiry involves the testing of proposed explanations involving the use of conventional techniques and procedures and usually requiring considerable ingenuity.	38	
Key Idea 3: The observations made while testing proposed explanations, when analyzed using conventional and invented methods, provide new insights into natural phenomena.	47	
Laboratory Checklist	31, 32, 33, 44, 45	
Standard 4		
Key Idea 1: Living things are both similar to and different from each other and from nonliving things.	1, 3, 6, 8, 34, 35, 36, 37, 55, 59, 63	
Key Idea 2: Organisms inherit genetic information in a variety of ways that result in continuity of structure and function between parents and offspring.	7, 9, 10, 14, 17, 19, 70, 71	
Key Idea 3: Individual organisms and species change over time.	5, 11, 48, 49, 50, 51, 69, 72	
Key Idea 4: The continuity of life is sustained through reproduction and development.	13, 15, 18, 46, 52, 53	
Key Idea 5: Organisms maintain a dynamic equilibrium that sustains life.	12, 16, 22, 28, 29, 30, 40, 41, 42, 67, 68	
Key Idea 6: Plants and animals depend on each other and their physical environment.	2, 4, 20, 24, 26, 56, 64, 65, 66	
Key Idea 7: Human decisions and activities have a profound impact on the physical and living environment.	21, 23, 25, 27, 43, 57, 58, 60, 61, 62	
Required Laboratories		
Lab 1: "Relationships and Biodiversity"	73, 74, 78, 79, 80, 82	
Lab 2: "Making Connections"	75, 83, 84, 85	
Lab 3: "The Beaks of Finches"	76, 77	
Lab 5: "Diffusion Through a Membrane"	81	

Examination June 2022
Living Environment

PART A

Answer all questions in this part. [30]

Directions (1–30): For *each* statement or question, record in the space provided the *number* of the word or expression that, of those given, best completes the statement or answers the question.

1. When the carrying capacity for a species in a habitat is reached, the population of the species levels off. This slowing of the rate of growth is most likely due to
   (1) limited resources
   (2) renewable energy
   (3) an increase in decomposers
   (4) a lack of competition   1 _____

2. On a hot day, dogs sweat through their paw pads and pant, which helps keep them cool. Both the sweating and panting are
   (1) due to a loss of oxygen
   (2) a failure of cell communication
   (3) due to a lack of adaptation to the environment
   (4) a response to a stimulus   2 _____

3. The formation of the many kinds of body cells that make up an embryo begins with
   (1) chemical changes in the cell membranes
   (2) the clumping together of proteins within the cells
   (3) specific genes being activated
   (4) the rapid metabolism of sugar molecules   3 _____

4. Some salmon have been genetically modified to grow bigger at a faster rate than wild salmon. They are kept in fish-farming facilities and are not released into the wild. Which statement regarding genetically modified salmon is most likely true?

    (1)   Wild salmon reproduce sexually, while genetically modified salmon reproduce asexually.

    (2)   Wild salmon have an altered protein sequence, but genetically modified salmon do not.

    (3)   Genetically modified salmon and wild salmon would have different DNA sequences.

    (4)   Genetically modified salmon and wild salmon would have identical DNA sequences.   4 _____

5. A farmer stopped maintaining a field that was once used to grow crops. Over time, the field eventually became a forest. These changes best illustrate the process of

    (1)   ecological succession   (3)   decomposition

    (2)   nutrient recycling   (4)   competition   5 _____

6. During a woman's menstrual cycle, ovulation occurs and an egg is released. This process is important because it allows for

    (1)   eggs to be produced by mitosis and be fertilized by a sperm

    (2)   sperm to fertilize the egg in the uterus

    (3)   multiple sperm to fertilize one egg, which then forms the placenta

    (4)   the egg to leave the ovary and be fertilized by a sperm   6 _____

7. The primary function of estrogen and progesterone is to

    (1)   regulate growth

    (2)   control heart rate

    (3)   monitor blood sugar levels

    (4)   regulate reproductive cycles   7 _____

8. Which characteristic is common to most types of cancer?

   (1)   production of low levels of ATP

   (2)   inadequate levels of antigens

   (3)   rapid and uncontrolled cell division

   (4)   destruction of red blood cells   8 _____

9. The human body fights an infection when a pathogen is detected. As a result, the pathogen stimulates the production of

   (1)   bacteria   (3)   vaccines

   (2)   antibodies   (4)   antibiotics   9 _____

10. Levels of organization in humans are represented below.

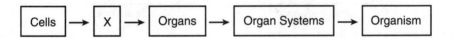

Level *X* most likely represents

   (1)   molecules   (3)   tissues

   (2)   organelles   (4)   ribosomes   10 _____

11. A biological process is represented below.

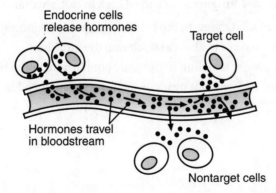

The reason the hormones affect the target cell and *not* the other cells is that the

(1) hormone provides energy only for the target cell

(2) target cell has specific receptors for the hormone

(3) nontarget cells produce antibodies that block the hormone

(4) hormones can only leave the bloodstream near the target cell   11 _____

12. The diagram below represents one of a number of different types of mutations that can occur in DNA.

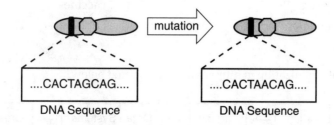

This mutation can best be described as the

(1) pairing of an adenine (A) base with thymine (T)

(2) the insertion of an adenine (A) base into both strands of the DNA molecule

(3) the substitution of an adenine (A) base for guanine (G)

(4) deletion of an adenine (A) base from the DNA molecule   12 _____

13. After digesting the nutrients from a meal high in carbohydrates, the body

(1) releases insulin to return the blood sugar levels to normal

(2) secretes enzymes to absorb starch into the intestines

(3) produces water to maintain dynamic equilibrium in the blood

(4) maintains homeostasis by increasing wastes produced in muscle cells   13 _____

14. The diagram below represents a pair of guard cells changing shape, reducing the size of the stomatal opening in a leaf.

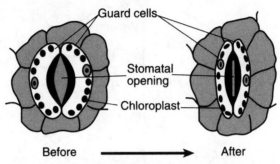

Source: Adapted from: http://o.quizlet.com

This is an adaptation that benefits plants by

(1) increasing the flow of liquid water into leaves, which increases the rate of food and oxygen production

(2) regulating the flow of water vapor out of leaves, preventing excess water loss by the plant

(3) increasing the flow of oxygen molecules into the leaves, which increases the rate of photosynthesis

(4) preventing the flow of carbon dioxide into the leaves, which would reduce the rate of respiration   14 _____

15. Experiments in mice show that a guardian gene that protects against type 1 diabetes can be altered by exposure to antibiotics during development. The exposure alters the gut bacteria, leading to a loss of the guardian gene's protection. The loss of this protection most directly interferes with

(1) homeostasis   (3) reproduction

(2) excretion   (4) respiration   15 _____

16. A food web is represented below.

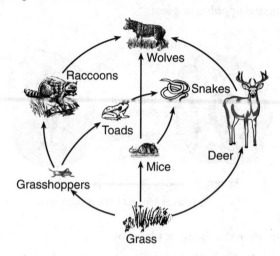

Which organism is correctly paired with its role in the ecosystem?

(1) The grass is both a consumer and a decomposer.

(2) The toads function as consumers and autotrophs.

(3) The grasshoppers function as consumers and heterotrophs.

(4) The snakes are both consumers and herbivores.   16 _____

17. Many adults lack the ability to digest the milk sugar, lactose. Often, this is due to the insufficient production of the enzyme lactase, which breaks down lactose. This is not due to an allergy to milk. Milk allergies are different because they

(1) are often not harmful to the person

(2) result in a build-up of the substance in the body

(3) are the result of the digestive system attacking the substance

(4) result from an overreaction of the immune system to a harmless substance   17 _____

18. One reason energy must be constantly added to a stable ecosystem is because some energy is

(1) lost at each feeding level

(2) incorporated into fossil fuels

(3) destroyed by decomposers

(4) digested by herbivores   18 _____

19. Which statement best illustrates direct competition within a species?

    (1)   A chipmunk is caught and eaten by a hungry fox.

    (2)   A deer attempts to escape a mountain lion that is chasing it.

    (3)   Two muskrats mate and produce a litter of offspring.

    (4)   Several squirrels eat acorns from the oak tree where they live.   19 _____

20. After a kidney transplant, special medications are taken. Otherwise, the patient's immune system might react to the transplanted kidney by

    (1)   producing specialized proteins that would attack the kidney

    (2)   signaling the body to synthesize new DNA molecules

    (3)   stimulating the circulatory system to attack red blood cells

    (4)   stimulating the kidney to produce reproductive hormones   20 _____

21. Information that scientists are able to obtain from the fossil record includes

    (1)   confirmation that Earth is 4.5 million years old

    (2)   data supporting the hypothesis that animal species do not change over time

    (3)   the exact means by which life on Earth began

    (4)   evidence about past environments and the history of life   21 _____

22. Fire ants have a powerful venom that is deadly to the small animals they eat. The deadly venom has reduced the populations of birds who build nests on the ground.

Source: http://www.sbs.utexas.edu/fireant/

The relationship between fire ants and ground-nesting birds is an example of

    (1)   producer/consumer

    (2)   predator/prey

    (3)   scavenger/decomposer

    (4)   parasite/host   22 _____

23. When an altered ecosystem is left undisturbed, the most likely result would be

(1) the gradual evolution of all of the original species

(2) a rapid return to the original ecosystem

(3) the elimination of all of the predator species

(4) a gradual shift toward a stable ecosystem   23 _____

24. Dead zones are areas found in the oceans and some large lakes where there is not enough oxygen to support life. Algae blooms occur when excess nutrients are introduced as pollutants from fertilizers, sewage-treatment plants, and the burning of fossil fuels. When the algae die and undergo decay, bacteria rapidly use up the oxygen in the area. Which human activity would most likely result in a *decrease* in the size and number of dead zones?

(1) irrigating fields and lawns to increase runoff into the ocean and rivers

(2) building more coal-fired electrical generating plants

(3) reducing the use of chemicals on farm fields and golf courses

(4) constructing more sewage-treatment plants on the shores of lakes and rivers   24 _____

25. A recent study found high levels of the toxic industrial pollutant mercury in the feathers of some songbirds. Those birds sang shorter, simpler versions of the songs they use to attract mates. Which statement regarding this finding is supported by the study?

(1) Mercury pollution will result in the extinction of all songbirds.

(2) Mercury prevents songbirds from obtaining required nutrients.

(3) Human activities usually affect the smallest animals in ecosystems.

(4) Human activities can have negative effects on a species.   25 _____

26. Which row in the chart below shows the connection between processes, structures, and hormones involved in the formation of an embryo?

Row	Process	Structure Involved	Hormone Involved
(1)	differentiation	lungs	insulin
(2)	gamete formation	testes	testosterone
(3)	union of gametes	cell nuclei	insulin
(4)	respiration	lungs	estrogen

26 _____

27. A food web and an energy pyramid are represented below.

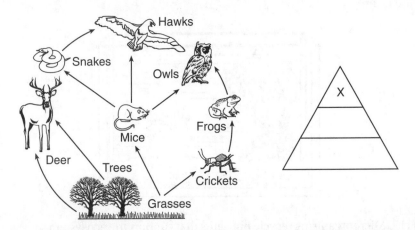

A group of organisms from the food web that would be found at level *X* of the energy pyramid is the

(1)   owls
(2)   deer
(3)   trees
(4)   crickets   27 _____

28. It may be possible to bring back some extinct species using recent advances in genetic technology. Opinions regarding this issue are split within the scientific community. The table below summarizes some of the arguments on both sides.

Pro	Con
■ It would increase the biodiversity of an ecosystem. ■ It would bring back organisms that are extinct.	■ The organisms that are brought back will compete with existing species. ■ The process is very expensive.

The arguments made by both sides provide evidence that

(1)   genetic technology is the best way to correct the damage humans have done to the environment

(2)   the introduction of genetic technology will benefit all organisms equally

(3)   any new technology that increases the biodiversity of the area should be used

(4)   the use of new technology requires decisions based on an assessment of costs, benefits, and risks   28 _____

29. Which statement best explains the purpose of the microorganisms in this aquarium?

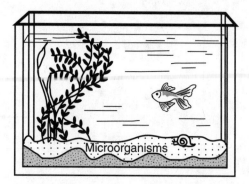

(1)   Microorganisms recycle nutrients that support the ecosystem.

(2)   Microorganisms recycle the energy in this ecosystem.

(3)   Microorganisms are a source of food for the plant.

(4)   Microorganisms are an abiotic factor important for decomposition.   29 _____

30. A new species of floating photosynthetic algae was accidently introduced into a pond ecosystem. It gradually replaced all the original algal species. A possible reason for the replacement could be that the new species

(1)   outcompeted the original algae populations for prey present in the ecosystem

(2)   required more resources than the original algae populations in the pond

(3)   outcompeted the original algae populations for abiotic factors

(4)   is less adapted to the pond ecosystem than the original algae populations   30 _____

PART B-1

Answer all questions in this part. **[13]**

Directions (31–43): For *each* statement or question, record in the space provided the *number* of the word or expression that, of those given, best completes the statement or answers the question.

31. The process of meiotic division in human females is represented below.

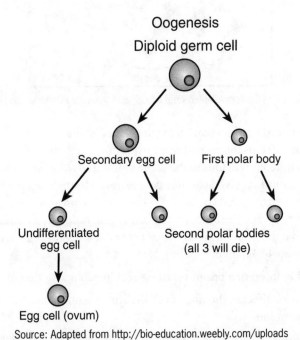

Oogenesis

Diploid germ cell

Secondary egg cell | First polar body

Undifferentiated egg cell | Second polar bodies (all 3 will die)

Egg cell (ovum)

Source: Adapted from http://bio-education.weebly.com/uploads

This process normally produces

(1) one functional gamete with one-quarter of the genetic information found in the diploid germ cell

(2) one functional gamete with one-half of the genetic information found in the diploid germ cell

(3) four functional gametes, each with one-quarter of the genetic information found in the diploid germ cell

(4) four functional gametes, each with one-half of the genetic information found in the diploid germ cell

31 _____

32. A student read that liquid extracted from an *Aloe vera* plant promotes the healing of burned tissue. She decided to investigate the effect of different concentrations of *Aloe vera* extract on the regeneration (regrowth of lost or damaged tissue) rate in planaria. Planaria are small flatworms known for their ability to regenerate.

Planaria Regeneration

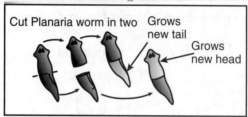

Source: Adapted from https://goo.gl/images/4wfcYv

The student used a sterile scalpel to cut each of 30 planaria in half. This gave her 10 heads and 10 tails for each of three experimental groups. The planaria were kept in separate Petri dishes in the same amount of water and at the same temperature. Group 1 received 0% *Aloe vera* extract, Group 2 received a 20% concentration of the extract, and Group 3 received a 40% concentration. On days 7, 10, and 14, she recorded the amount of tissue regeneration in all three groups. She observed that the group with 20% *Aloe vera* added regenerated more slowly than the group with 40% added.

A reasonable inference based on these results would be that

(1) *Aloe vera* affected the rate of cell division, resulting in an increased rate of regeneration

(2) the control group, which received no *Aloe vera*, did not regenerate

(3) if she applied 30% *Aloe vera* to a group, it would regenerate tissue more rapidly than the 40% group

(4) the application of *Aloe vera* to earthworms would have no effect on tissue regeneration   32 _____

33. The graph below represents the rate of a chemical reaction involving a particular human enzyme that breaks down starch.

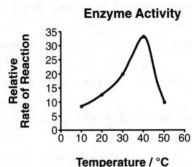

Enzyme Activity

Source: Adapted from http://nygmascience.files.wordpress.
com/2011/11/enzyme-rate-of-reaction1.jpg

The most likely reason the action of the enzyme *decreases* after 40°C is that

(1) the DNA in the enzyme mutates and can no longer break down the starch

(2) enzymes die after working for a long period of constant activity in the body

(3) the shape of the enzyme changes due to environmental conditions

(4) as the temperature of the enzyme rises, the pH of the
environment changes, deactivating the enzyme   33 _____

34. Researchers studied the relationship between lichen nitrogen content and the growth of lichens on trees. They recorded the amount of growth after determining the percentage of the tree that was covered in lichens. Their data are shown in the graph below.

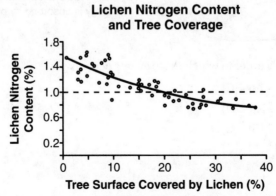

Lichen Nitrogen Content and Tree Coverage

Source: McDermott, Amy, "Sentinels of Forest Health,"
Science News, Nov. 26, 2016, pp.20-23

Which statement best describes the relationship between the nitrogen content and the growth of the lichen?

(1)   As nitrogen content in the lichen increases, the growth of the lichen increases.

(2)   As nitrogen content in the lichen decreases, the growth of the lichen decreases.

(3)   As nitrogen content in the lichen decreases, the growth of the lichen increases.

(4)   There is not a clear relationship between the amount of nitrogen in the lichen and growth.   34 _____

35. Scientific claims should be questioned if

(1)   peer review was used to examine the claims made by scientists

(2)   the experimental results cannot be repeated by other scientists

(3)   conclusions follow logically from the evidence

(4)   the data are based on samples that are very large   35 _____

36. Organisms living in a forest ecosystem rely on the Sun as a source of energy for metabolic processes. The following events occur as energy is captured by a plant and used in the metabolic processes of an herbivore.

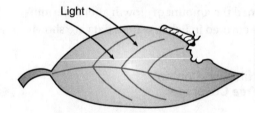
Light

[A] Energy is released from chemical bonds.
[B] Energy is stored in large organic molecules.
[C] Energy is transferred to molecules of ATP.
[D] Energy is absorbed by plant cells.

The most likely order in which these events occur is

(1)   [A] – [D] – [B] – [C]

(2)   [B] – [A] – [C] – [D]

(3)   [D] – [A] – [B] – [C]

(4)   [D] – [B] – [A] – [C]   36 _____

Base your answer to question 37 on the graph below and on your knowledge of biology. The graph shows the carbon dioxide (CO_2) concentration of the atmosphere since the year 1000.

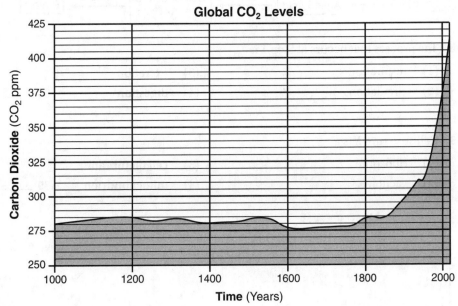

Global CO$_2$ Levels

Source: Adapted from https://www.co2.earth/co2-ice-core-data

37. What was the approximate change in CO_2 level from the year 1000 to the year 2000?

    (1)   an increase of 135 ppm

    (2)   an increase of 95 ppm

    (3)   a decrease of 135 ppm

    (4)   a decrease of 95 ppm   37 _____

Base your answers to questions 38 and 39 on the diagram below and on your knowledge of biology. The diagram represents a series of events that occur within living organisms.

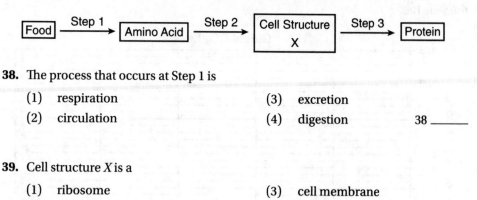

38. The process that occurs at Step 1 is

(1) respiration
(2) circulation
(3) excretion
(4) digestion

38 _____

39. Cell structure X is a

(1) ribosome
(2) vacuole
(3) cell membrane
(4) mitochondrion

39 _____

40. Two types of cells from an individual are represented below.

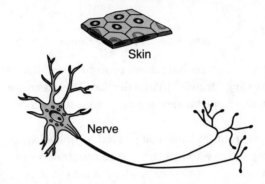

Skin

Nerve

Which model, that shows only some of the chromosomes in each of the two types of cells, best explains why these cells are so different?

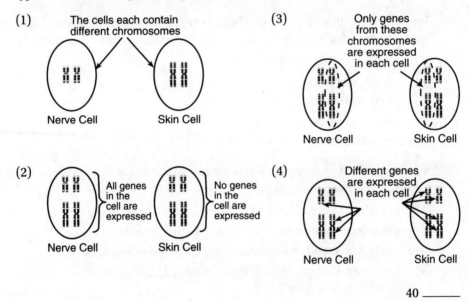

(1) The cells each contain different chromosomes

Nerve Cell   Skin Cell

(3) Only genes from these chromosomes are expressed in each cell

Nerve Cell   Skin Cell

(2) All genes in the cell are expressed / No genes in the cell are expressed

Nerve Cell   Skin Cell

(4) Different genes are expressed in each cell

Nerve Cell   Skin Cell

40 _____

Base your answers to questions 41 and 42 on the passage below and on your knowledge of biology.

Bed Bugs...They're Back!

Bed bugs aren't just a problem from centuries past. Bed bug infestations have been increasing for more than a decade. This has been largely due to the insects' ability to quickly develop resistance to the insecticides used to kill them.

Bed bugs have a tough outer coat, called a cuticle, which helps protect them. Researchers have found that some resistant bed bugs have gene mutations that allow the cuticle to produce substances that break down the insecticides. Others have gene mutations that direct the building of biological pumps, which allow the cuticle to pump the harmful insecticide out of the bug.

41. The substances that allow the bed bugs to break down insecticides and the biological pumps that remove the insecticides from the bed bugs are examples of

    (1)   the failure of homeostasis
    (2)   genetic engineering
    (3)   biological adaptations
    (4)   selective breeding   41 _____

42. A gene mutation resulting in insecticide resistance would most likely increase in the bed bug population because

    (1)   more bed bugs will need to be resistant to the insecticide
    (2)   the insecticide-resistant bed bugs will survive and reproduce
    (3)   the bed bugs with the resistance gene will reproduce asexually
    (4)   spraying an insecticide will allow more bed bugs without mutations to survive   42 _____

43. The graph below summarizes how effective the seasonal flu vaccine has been at preventing infection with the flu virus. The data were collected over a 13-year period.

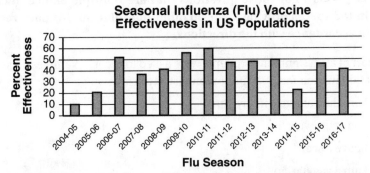

Source: https://www.sciencenews.org/article/
universal-flu-shot-may be-nearing-reality

Based on the data provided, a reasonable interpretation would be that

(1) in 2004–2005, some individuals caught the flu from the vaccine

(2) the virus mutated in 2014–2015, resulting in the vaccine being less effective

(3) people have become immune to the flu vaccine over the 13-year period

(4) the vaccine has become increasingly effective over the 13-year period

43 _____

PART B-2

Answer all questions in this part. [12]

Directions (44–55): **For those questions that are multiple choice, record your answers in the spaces provided. For all other questions in this part, record your answers in accordance with the directions.**

44. As part of an assignment, students were asked to record examples of genetic variation in their family. One student listed the following:

- I am the youngest in my family.
- I have brown eyes.
- I have a scar.
- I am a vegetarian.

Only one of these statements is an example of a genetic trait. Identify the genetic trait and support your answer.   [1]

_____

_____

Base your answers to questions 45 through 49 on the information and data table below and on your knowledge of biology.

Overfishing of Newfoundland Cod

When fishing results in small catches, it is said that the species has been overfished. Over the last 75 years, ocean fish populations have dropped by almost 90%. The data below show the approximate amount, in thousands of tons, of Newfoundland cod caught each year from 1970 to 1995.

Approximate Amount of Newfoundland Cod Catches, 1970–1995

Years	Tons $\times 10^3$ of Newfoundland Cod Caught
1970	1500
1975	1300
1980	600
1983	700
1985	300
1987	400
1990	210
1993	100
1995	50

Directions (45–46): **Using the information in the data table, construct a line graph on the grid provided, following the directions below.**

45. Mark an appropriate scale, without any breaks in the data, on each labeled axis.   [1]

46. Plot the data on the grid. Connect the points and surround each point with a small circle.   [1]

Example:

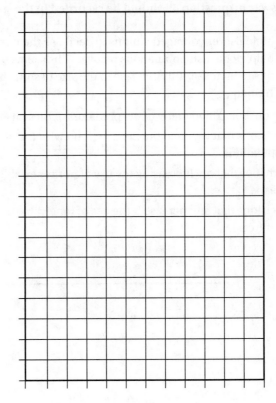

Newfoundland Cod Catches

Catch in Tons X 10³

Years

Note: The answer to question 47 should be recorded in the space provided.

47. During which five-year span did the largest drop in fishing occur?

    (1)   1970 and 1975   (3)   1980 and 1985

    (2)   1975 and 1980   (4)   1990 and 1995   47 _____

48. Other than maintaining an adequate food supply for humans, state *one* other advantage of *not* overfishing the oceans.   [1]

_____

_____

Note: The answer to question 49 should be recorded in the space provided.

49. In 2003, biologists encouraged nations to decrease the number of fish caught in order to help global fish populations recover. This seems to be helping some fish populations to increase. This increase in the size of some fish populations is a result of human

    (1)   actions that killed many of the predators of these fish populations

    (2)   decisions that weighed the need for food with the need to maintain fish populations

    (3)   activities that are increasing the use of nonrenewable resources of the oceans

    (4)   decisions that are increasing the use of renewable ocean resources   49 _____

Base your answers to questions 50 and 51 on the information and diagram below and on your knowledge of biology.

The diagram illustrates the evolution of tetrapods. A tetrapod is a four-footed animal.

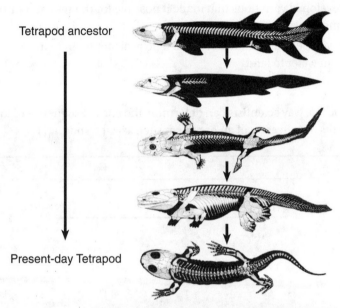

Source: Adapted from Coates, M., *Palaeobiology 2*,
Briggs D. et al., eds., p. 75, © 2001 Blackwell Publishing

Note: The answer to question 50 should be recorded in the space provided.

50. The changes observed over time occurred as the organisms

    (1)   needed to change the habitat where they lived from land to water

    (2)   needed to change the habitat where they lived from water to land

    (3)   developed variations that made it possible for them to move from land to water

    (4)   developed variations that made it possible for them to move from water to land   50 _____

51. Describe *one* way scientists can determine the correct sequence of fossils that represents the ancestry of an organism such as a tetrapod.   [1]

_____

_____

                                _____

Base your answers to questions 52 through 55 on the diagram and graph below and on your knowledge of biology.

The diagram represents some organisms in a pond food web. The graph shows the changes in the size of the bacteria population also present in the food web over time.

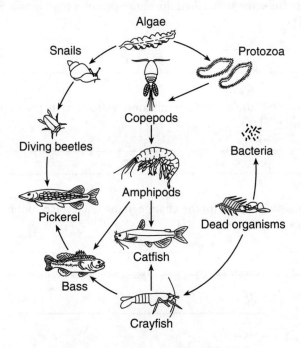

Change in a Bacterial Population

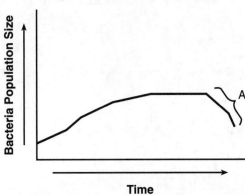

52. Identify the population represented in this food web that has the greatest amount of stored energy. [1]

_____

_____

53. State what would most likely happen to the bass population if a pesticide that was used in this ecosystem killed the entire pickerel population. Support your answer. [1]

_____

_____

54. Identify the role of the bacteria in this food web and state the importance of this particular role. [1]

_____

_____

55. State _one_ possible reason for the change in the bacterial population size in the area labeled _A_ on the graph. [1]

_____

_____

PART C

Answer all questions in this part.   [17]

Directions (56–72): **Record your answers in the spaces provided.**

Base your answers to questions 56 through 58 on the data table below and on your knowledge of biology.

Changes in Size of the Ozone Hole

Year	Ozone Hole Area (million km^2)
1980	3.3
1985	18.8
1990	21.1
1996	26.9
2000	29.9
2005	27.2
2010	22.6
2017	19.6

Source: https://ozonewatch.gsfc.nasa.gov

In 1987 an agreement was reached called the Montreal Protocol, which limited the world's production of chemicals that could damage the ozone shield.

56. Identify *one* risk associated with the destruction of the ozone shield.   [1]

_____

_____

57. Using evidence from the data table, explain whether or not the Montreal Protocol has been effective.   [1]

_____

_____

58. Describe *one* possible *negative* consequence that is important to consider when an international agreement such as the Montreal Protocol is adopted. [1]

_____

_____

Base your answers to questions 59 through 62 on the information below and on your knowledge of biology.

Fungicides and Bumblebees

Source: Adapted from https://polinizador.files.
wordpress.com/2011/03/img670-6-18-07.jpg

Bumblebees are extremely important in agriculture. They pollinate many flowering plants, including food crops such as tomatoes, pumpkins, and blueberries. The bees gather wet, sticky pollen from flowers and take it to their nests. Fungi, present on the pollen, keep it from spoiling. In the nest, bumblebee larvae feed on both the pollen and fungi.

Through his research, Dr. Shawn Steffan discovered that the stored pollen and nectar that bumblebee larvae feed on is rich in yeast, a type of fungus. Based on this observation, he proposed that the application of fungicides, chemicals that kill fungi, on agricultural crops could affect the quality of bumblebee food and ultimately the health of bumblebee colonies. He hypothesized that if the fungi associated with the pollen suffer, then the bumblebee larvae will also suffer.

Dr. Steffan designed an experiment in which five colonies of bumblebees only fed on flowers treated with fungicides. In five other colonies, the bumblebees only fed on flowers that were free of fungicides. At the conclusion of the experiment, the control-bee colonies averaged about 43 individuals. The colonies that fed on flowers with fungicides (and no fungus) averaged only about 12 individuals.

59. Using information from the reading, explain how the results of the experiment support Dr. Steffan's hypothesis.   [1]

_____

_____

_____

60. Dr. Steffan proposed that one way to protect the bees might be to only spray agricultural crops when they were not flowering. Explain how this would prevent harming bumblebee larvae.   [1]

_____

_____

_____

61. In addition to the use of pesticides, studies also show that bee species inhabiting smaller geographic areas are more sensitive to changes in climate. Explain how climate change could have a greater impact on bee species inhabiting smaller geographic areas than those inhabiting larger geographic areas.   [1]

_____

_____

62. Explain why it is important to preserve bumblebee populations.   [1]

_____

_____

63. Scientists build models based on what they know from previous research to develop testable hypotheses. Scientists Watson and Crick first constructed an incorrect triple-helix model of DNA with the bases (A, T, C, G) arranged on the outside of the molecule. Explain why their triple-helix model was valuable even though it was not correct. [1]

_____

_____

64. State *one* reason why a human heart muscle cell would probably contain a higher proportion of mitochondria than a skin cell. [1]

_____

_____

65. Phytoplankton are photosynthetic organisms that live in aquatic environments. Although microscopic, their vast numbers provide a plentiful resource for many aquatic food webs. Explain why populations like phytoplankton are required to sustain an aquatic food web. [1]

_____

_____

Base your answers to questions 66 through 68 on the information below and on your knowledge of biology.

How One Bull Cost the Dairy Industry $420 Million

It all started with a bull named Chief. He had 16,000 daughters, 500,000 granddaughters, and 2 million great-granddaughters. Today, 14% of the genes present in Holstein dairy cows came from Chief.

Chief was popular because his daughters were fantastic milk producers. The problem is, he also had a single copy of a deadly mutation. The mutation spread undetected through the Holstein cow population and was responsible for the spontaneous death of 500,000 fetal calves. The loss of these calves cost the dairy industry $420 million.

Over the past 35 years, using Chief's sperm, instead of sperm from an average bull, resulted in $30 billion in increased milk production. Due to Chief's genetic contribution, the average dairy cow today produces four times more milk than a dairy cow in the 1960s.

Chief embodies the trade-offs associated with selective breeding.

Chief

Source: https://www.progressivedairy.com

66. Explain why using Chief to produce so many offspring is an example of selective breeding.   [1]

_____

_____

67. Explain how the use of Chief to produce offspring had both advantages and disadvantages.   [1]

_____

_____

68. Explain how genetic engineering could be used to improve the chance that more of Chief's offspring would survive.   [1]

_____

_____

_____

Base your answers to questions 69 through 72 on the information below and on your knowledge of biology.

The Tuskless Female Elephants of Gorongosa National Park

Elephants are large mammals that live in parts of Africa and Asia. They typically have tusks, which are a pair of elongated teeth that the animals use to strip bark off of trees and dig holes to obtain water and minerals. Tusks are also used by males when they compete with each other to impress females during the mating season. Males born without tusks are at a high risk of being severely wounded during these competitions.

In several regions in Africa, elephants have been killed for their ivory tusks. The ivory can be sold for large sums of money, even though the sale of ivory is illegal in many parts of the world. During a 15-year civil war in Mozambique, many large-tusked elephants in Gorongosa National Park were killed and their ivory sold to buy arms and ammunition. The elephant population decreased during the war from over 2000 individuals to only a few hundred. Female elephants that had no tusks (an inheritable trait) made up only about 6% of the entire population before the war began.

When the war ended in 1992, the wildlife in the park was better protected against poaching. The elephant population has recovered fairly well, but a significant change has been noted: The tuskless female elephants that survived the civil war now make up more than 50% of the older female population in the park. About 33% of the female offspring that were born after the war are also tuskless. No tuskless males have been seen.

69. Explain how an elephant without the ability to grow tusks could be born into a population of elephants that all have tusks.   [1]

_____

_____

70. At the start of the civil war, only about 6% of the female elephants had no tusks. Explain why over one-half of the females that survived the war had no tusks.   [1]

_____

_____

71. Explain why so many (33%) of the female elephants born in the years after the war have no tusks.   [1]

_____

_____

72. Even without poaching being a factor, explain why tuskless males are rarely seen.   [1]

_____

_____

PART D

Answer all questions in this part.   [13]

Directions (73–85): **For those questions that are multiple choice, record your answers in the spaces provided. For all other questions in this part, record your answers in accordance with the directions.**

Base your answers to questions 73 and 74 on the information below and on your knowledge of biology.

Before watching a scary movie, the members of a theater audience agreed to have their heart rates monitored. They were asked to sit in silence for 10 minutes before the film began. The movie was then shown from beginning to end.

The scatter plot below summarizes the data collected by all of the heart monitors from ten minutes before the start of the movie to the end of the movie.

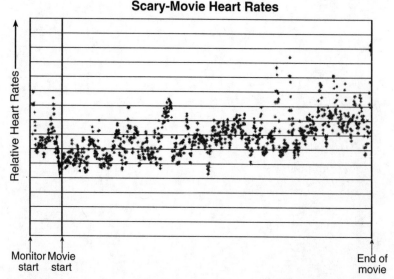

Scary-Movie Heart Rates

Source: http://www.theguardian.com/film/filmblog/2014/sep/01/
watched-horror-film-heart-rate-monitor-as-above-so-below

Note: The answer to question 73 should be recorded in the space provided.

73. In this experiment, the dependent variable is the
  (1) heart rate of the audience members
  (2) scene being viewed by the audience
  (3) amount of time the movie played
  (4) number of viewers with heart-rate monitors   73 _____

Note: The answer to question 74 should be recorded in the space provided.

74. Which is a possible hypothesis most likely being tested in this experiment?
  (1) Silence in a theater increases the heart rates of the audience members.
  (2) The length of a movie causes changes in heart rate.
  (3) Do heart rates increase when watching scary movies?
  (4) Watching scary movies will increase the heart rates of
       audience members.   74 _____

Note: The answer to question 75 should be recorded in the space provided.

75. A student filled two Petri dishes with a clear gel made with corn starch. He was given two unknown solutions (*A* and *B*) and was asked to determine which solution contained a chemical that digests starch.

Using a clean cotton swab, he dipped it into solution *A* and wrote a "?" invisibly onto the gel in one of the Petri dishes. He repeated the same procedure on the second Petri dish with a clean cotton swab he dipped in solution *B*.

Twenty minutes later, he added starch-indicator solution to the surface of both Petri dishes. The surface of the Petri dish with solution *A* added turned completely blue. Most of the surface of the Petri dish to which solution *B* was added was blue, except the "?" was clear. The results are illustrated below.

Petri Dishes With Starch Gel After 20 Minutes

Petri dish swabbed   Petri dish swabbed
with solution A   with solution B

An observation that supports the student's conclusion that solution *B* contained a chemical that digests starch is that the

(1) damp cotton swab absorbed some of the starch where it touched the gel

(2) starch indicator changed the color of the gel to blue

(3) area swabbed with solution *B* remained clear

(4) chemical in the starch indicator reacted with the chemical in *B*   75 _____

Base your answer to question 76 on the information below and on your knowledge of biology. The diagram shows variations in the beaks (bills) of some finches on the Galapagos Islands.

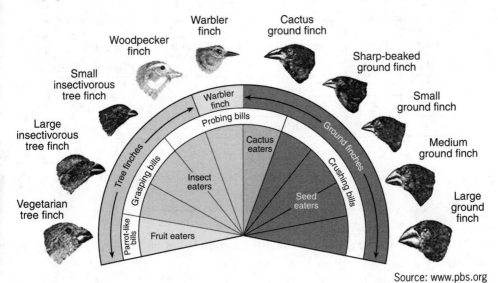

Source: www.pbs.org

The photographs of four different finch species that are found in the Galapagos are shown below.

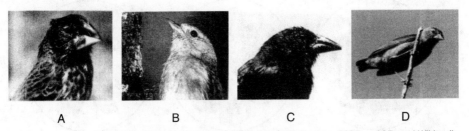

Source: *Biology*, Mader, Sylvia, McGraw-Hill, Boston, 2007, p.287, and Wikipedia

Note: The answer to question 76 should be recorded in the space provided.

76. Which row in the chart below correctly identifies one of these finches?

Row	Finch	Beak Characteristic	Food Source	Species
(1)	A	Probing	Fruit	Large ground finch
(2)	B	Probing	Insects	Warbler
(3)	C	Parrot-like	Seeds	Cactus finch
(4)	D	Crushing	Fruit	Small ground finch

76 _____

Base your answer to question 77 on the diagram below and on your knowledge of biology.

The diagram represents three groups of red blood cells. Groups *A* and *B* were each placed in different solutions for the same period of time.

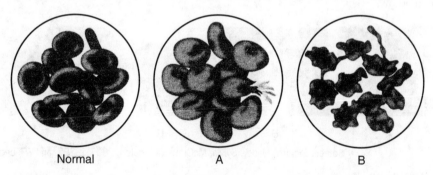

Normal   A   B

77. Identify which group of cells, *A* or *B*, had most likely been placed in distilled water. Support your answer. [1]

_____

_____

Base your answer to question 78 on the information below and on your knowledge of biology.

A student placed artificial cells, each containing a 25% sugar solution, into three different beakers containing sugar solution that varied in concentration from 0% to 25%. The set-ups are shown below.

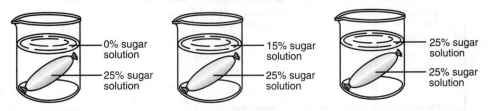

78. The student collected data on the mass of each artificial cell. The student predicted that the cell in the beaker with 25% sugar solution would have the greatest change in mass after 24 hours. Would his prediction be correct? Support your answer. [1]

_____

_____

_____

79. Identify *one* waste product that is more effectively removed from muscle cells as a result of increased pulse rate. [1]

_____

Base your answers to questions 80 and 81 on the information and chart below and on your knowledge of biology.

The chart represents the results of gel electrophoresis of DNA from an unknown individual and four known individuals.

Gel Electrophoresis Results of DNA From Five Individuals

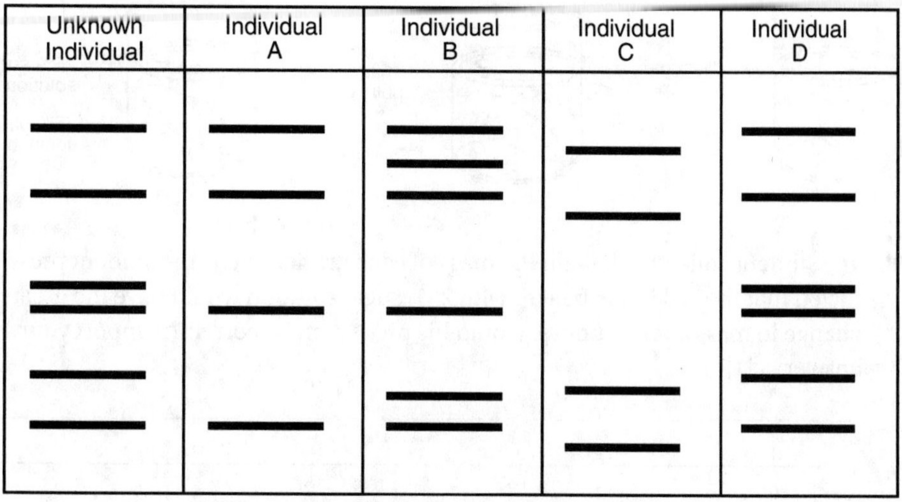

80. Identify the unknown individual as *A, B, C,* or *D* by comparing the gel electrophoresis results. Support your answer.   [1]

_____

_____

Note: The answer to question 81 should be recorded in the space provided.

81. Before conducting an electrophoresis procedure, enzymes are added to DNA in order to

   (1)   convert the DNA into gel
   (2)   cut the DNA into fragments
   (3)   remove smaller DNA fragments from the samples
   (4)   synthesize larger fragments of DNA   81 _____

Note: The answer to question 82 should be recorded in the space provided.

82. Variations in the flying speed of a finch population are represented in the graph below. The top flying speed of a predator of these finches is also indicated on the graph.

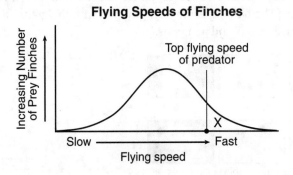

Flying Speeds of Finches

When describing finches with flying speeds in the region indicated by the *X* on the graph, it would be accurate to say that these individuals are more likely to

(1) reproduce and increase the frequency of fast finches in the population

(2) survive and undergo mutations that increase their flying speeds

(3) require less food than the slower finches in the population

(4) produce offspring that fly at average speeds   82 _____

Base your answer to question 83 on the information below and on your knowledge of biology.

RNA Codons and the Amino Acids for Which They Code

AUU, AUC, AUA } ILE (Isoleucine)	ACU, ACC, ACA, ACG } THR (Threonine)	AAU, AAC } ASN (Asparagine)	AGU, AGC } SER (Serine)
AUG MET (Methionine)		AAA, AAG } LYS (Lysine)	AGA, AGG } ARG (Arginine)

83. If a sequence of bases in DNA changes from TGA to TGG, would it result in a new inheritable trait? Support your answer.   [1]

_____

_____

_____

Base your answers to questions 84 and 85 on the information below and on your knowledge of biology.

The diagram below represents a recently developed evolutionary tree for some species of birds. The new tree diagram is based on the analysis of data collected from 169 bird species and includes a change in the placement of flamingos. The flamingos are now grouped with the grebes and pigeons instead of with egrets and penguins.

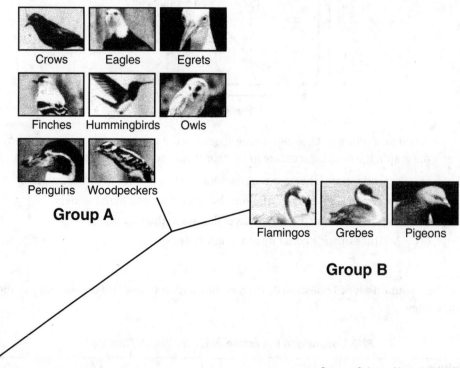

Source: Science News 1/10/15

84. Identify *one* type of molecular evidence that was most likely used to develop this new tree. Explain how this evidence would support the new position of the flamingos. [1]

_____

_____

85. Based on the photos above, select *two* bird species, one from Group *A* and one from Group *B*, and state *one* reason why researchers may have originally thought these two species should be on the same branch of the tree. [1]

Bird species from Group *A*: _____

Bird species from Group *B*: _____

_____

_____

_____

Answers June 2022
Living Environment

Answer Key

PART A

1. 1	6. 4	11. 2	16. 3	21. 4	26. 2
2. 4	7. 4	12. 3	17. 4	22. 2	27. 1
3. 3	8. 3	13. 1	18. 1	23. 4	28. 4
4. 3	9. 2	14. 2	19. 4	24. 3	29. 1
5. 1	10. 3	15. 1	20. 1	25. 4	30. 3

PART B-1

31. 2	34. 3	36. 4	38. 4	40. 4	42. 2
32. 1	35. 2	37. 2	39. 1	41. 3	43. 2
33. 3					

PART B-2

44. *See* Answers Explained.

45. *See* Answers Explained.

46. *See* Answers Explained.

47. 2

48. *See* Answers Explained.

49. 2

50. 4

51. *See* Answers Explained.

52. *See* Answers Explained.

53. *See* Answers Explained.

54. *See* Answers Explained.

55. *See* Answers Explained.

PART C. *See* **Answers Explained**.

PART D

73. 1

74. 4

75. 3

76. 2

77. *See* Answers Explained.

78. *See* Answers Explained.

79. *See* Answers Explained.

80. *See* Answers Explained.

81. 2

82. 1

83. *See* Answers Explained.

84. *See* Answers Explained.

85. *See* Answers Explained.

Answers Explained

PART A

1. 1 The slowing of the rate of growth is most likely due to *limited resources*. The carrying capacity of an environment is the number of individuals of a particular species that may be sustainably supported by the natural resources present in the environment. If this population has leveled off in terms of numbers of individuals, then it is most likely that a limitation of the resources needed by that population has caused this leveling.

WRONG CHOICES EXPLAINED:
(2) The slowing of the rate of growth is *not* most likely due to *renewable energy*. The term renewable energy relates to sources of power (e.g., solar, wind, tidal) used by humans to operate industrial and other technological activities. It is not used to refer to natural processes such as carrying capacity.

(3) The slowing of the rate of growth is *not* most likely due to *an increase in decomposers*. An increase in decomposers (e.g., bacteria, fungi) would accelerate the breakdown of complex organic substances into soluble molecules available to plants for their growth. This would likely increase, not level, the rate of growth of this population in this habitat.

(4) The slowing of the rate of growth is *not* most likely due to *a lack of competition*. A reduction in intraspecies and interspecies competition for available resources would make more resources available for this species. This would likely increase, not level, the rate of growth of this population in this habitat.

2. 4 Both the sweating and the panting are *a response to a stimulus*. As the dog's blood/body temperature rises on a hot day, nerve sensors in the central nervous system detect this increase. The central nervous system then sends commands to the animal's sweat glands and respiratory system to release perspiration and increase the rate of breathing. Both actions would have the effect of cooling the dog's blood/body temperature. In this example, body heat is the stimulus and the sweating and panting are responses to it.

WRONG CHOICES EXPLAINED:
(1) It is *not* true that both the sweating and the panting are *due to the loss of oxygen*. In this instance, the dog's central nervous system may increase the rate of respiration in order to add more oxygen to the dog's blood but would probably not increase the rate of sweating.

(2) It is *not* true that both the sweating and the panting are *a failure of cell communication*. In fact, this stimulus/response mechanism is a good example of a well-working cell communication system involving sensory and motor nerve cells.

(3) It is *not* true that both the sweating and the panting are *due to a lack of adaptation to the environment*. In fact, the sweating/panting responses of the dog are good examples of favorable adaptations to the environment that enhance the survival rate of the dog species.

3. 3 The formation of the many kinds of body cells that make up the embryo begins with *specific genes being activated*. This process, in which some genes are switched on and others are switched off in cells of developing tissues, is known as differentiation. Differentiation of embryonic tissues leads to the development of ectoderm, mesoderm, and endoderm that will further differentiate into the tissues, organs, and organ systems of the fetus and the adult human.

WRONG CHOICES EXPLAINED:
(1) The formation of the many kinds of body cells that make up the embryo does *not* begin with *chemical changes in the cell membranes*. Cell membranes are composed of a bilipid layer and embedded proteins. Although the specific embedded proteins may differ from tissue to tissue, the basic composition of the cell membrane does not change as a result of this process.

(2) The formation of the many kinds of body cells that make up the embryo does *not* begin with *the clumping together of proteins within the cells*. The clumping of proteins is an abnormal process that occurs when proteins are brought into contact with coagulating conditions such as heat, chemical agents, or immune responses. This process is not a normal part of the growth of an embryo.

(4) The formation of the many kinds of body cells that make up the embryo does *not* begin with *the rapid metabolism of sugar molecules*. Sugar molecules are metabolized in mitochondria under the control of respiratory enzymes to release energy for the process of embryonic development. This process is a normal and necessary part of the growth of an embryo and the development of an independent human being.

4. 3 *Genetically modified salmon and wild salmon would have different DNA sequences* is the statement regarding genetically modified salmon that is most likely true. Genetic modification by human scientists, by definition, leads to changes in the DNA of an organism.

WRONG CHOICES EXPLAINED:

(1) *Wild salmon reproduce sexually, while genetically modified salmon reproduce asexually* is *not* the statement regarding genetically modified salmon that is most likely true. Salmon are fish and, like all fish species whether genetically modified or not, reproduce sexually.

(2) *Wild salmon have an altered protein sequence, but genetically modified salmon do not* is *not* the statement regarding genetically modified salmon that is most likely true. Wild salmon have wild-type DNA, whereas genetically modified salmon contain genes inserted by human scientists for the purpose of producing faster- and larger-growing fish for the commercial market.

(4) *Genetically modified salmon and wild salmon would have identical DNA sequences* is *not* the statement regarding genetically modified salmon that is most likely true. Except for possible instances of twinning, no two salmon of any type have identical DNA sequences.

5. 1 These changes best illustrate the process of *ecological succession*. Ecological succession is a term used to describe the replacement of one plant community by another until a stable, self-perpetuating plant community is established. When the farmer stopped using the field for growing crops, pioneer organisms such as grasses soon overtook the field. After several seasons, the grasses were replaced by shrubs and small trees and still later by a series of tree communities. Eventually, the mixed forest community was established and became self-sustaining.

WRONG CHOICES EXPLAINED:

(2), (3) These changes do *not* best illustrate the processes of *nutrient recycling* or *decomposition*. Nutrient recycling is a result of decomposition in which complex biochemicals are broken down by bacteria and fungi into soluble molecules that can be taken up and recycled by plants. These processes operate during ecological succession but are not solely responsible for the changes described.

(4) These changes do *not* best illustrate the process of *competition*. Competition is a term that relates to any conflict among organisms for limited natural resources in the environment such as food, oxygen, water, light, or soil minerals. Such competitions may occur within a species or among different species. In a sense, ecological succession operates through a series of competitions among different plant species for resources such as light, space, and soil minerals.

[NOTE: This response is a potential correct answer to the question as written, but it is not the best answer.]

6. 4 This process is important because it allows for *the egg to leave the ovary and be fertilized by a sperm*. Ovulation is a process that occurs during the menstrual cycle in which a mature egg is released from the ovary and then passes into the oviduct. The oviduct is the normal site of fertilization of one egg by one sperm.

WRONG CHOICES EXPLAINED:
(1) It is *not* true that this process is important because it allows for *eggs to be produced by mitosis and be fertilized by a sperm*. Eggs are produced by the process of meiotic, not mitotic, cell division.

(2) It is *not* true that this process is important because it allows for *sperm to fertilize the egg in the uterus*. Eggs are normally fertilized in the oviduct, not the uterus.

(3) It is *not* true that this process is important because it allows for *multiple sperm to fertilize one egg, which then forms the placenta*. Only one sperm is required to fertilize one egg in normal fertilization. The placenta is formed by a combination of embryonic and uterine tissues, not by a fertilized egg.

7. 4 The primary function of estrogen and progesterone is to *regulate reproductive cycles*. Specifically, these female hormones act to control the menstrual and gestational cycles by establishing and maintaining the conditions that allow for the regular release of mature eggs and controlled development of the fetus.

WRONG CHOICES EXPLAINED:
(1) It is *not* true that the primary function of estrogen and progesterone is to *regulate growth*. Growth is controlled in humans by growth hormone released from the pituitary gland.

(2) It is *not* true that the primary function of estrogen and progesterone is to *control heart rate*. Heart rate in humans is controlled by the nervous system and the hormone adrenaline.

(3) It is *not* true that the primary function of estrogen and progesterone is to *monitor blood sugar levels*. Blood sugar levels in humans are monitored by the nervous system and controlled by the hormones insulin and glucagon.

8. 3 *Rapid and uncontrolled cell division* is the characteristic that is common to most types of cancer. Cancer begins when the cell's normal reproductive mechanism is taken over and rapid, uncontrolled mitotic cell division occurs. This abnormal process often results in a mass of nonfunctioning cells, known as a tumor, that crowds out healthy cells and interferes with normal tissue functions.

WRONG CHOICES EXPLAINED:
(1), (2), (4) *Production of low levels of ATP, inadequate levels of antigens,* and *destruction of red blood cells* are *not* the characteristics that are common to most types of cancer. Any of these outcomes may occur as a result of specific types of cancer, but they are not common to most types of cancer.

9. **2** As a result of this process, the pathogen stimulates the production of *antibodies*. Antibodies are synthesized and released by the immune system in direct response to the presence in the body of a foreign antigen on a pathogen. The antibody acts against the pathogen by attaching to its antigen and neutralizing it.

WRONG CHOICES EXPLAINED:
(1), (3), (4) It is *not* true that, as a result of this process, the pathogen stimulates the production of *bacteria*, *vaccines*, or *antibiotics*. The human body is unable to produce any of these things. Bacteria represent an independent group of living organisms. Vaccines and antibiotics are produced by pharmaceutical laboratories.

10. **3** Level *X* most likely represents *tissues*. Tissues are groupings of similar cells that together perform a specific function in the body, so they are more complex than cells. Several tissue types make up an organ, so tissues are less complex than organs.

WRONG CHOICES EXPLAINED:
(1) Level *X* does *not* most likely represent *molecules*. Molecules are chemical structures that make up the physical components of most matter, including living things.

(2), (4) Level *X* does *not* most likely represent *organelles* or *ribosomes*. The functional components of living cells are known as organelles. Ribosomes are a type of cell organelle that functions to synthesize proteins.

11. **2** The reason the hormones affect the target cell and not the other cells is that the *target cell has specific receptors for the hormone*. During differentiation, the target cell gene responsible for production of the receptor protein was switched on, allowing the hormone to attach to and affect the metabolism of the target cell. At the same time, that gene was switched off in the nontarget cells, so the hormone could not attach to those cells.

WRONG CHOICES EXPLAINED:
(1) The reason the hormones affect the target cell and not the other cells is *not* that the *hormone provides energy only for the target cell*. Hormones are beneficial proteins that exert chemical control over various metabolic processes that aid the maintenance of homeostasis in the body. Hormones are not directly responsible for energy transfer in living things.

(3) The reason the hormones affect the target cell and not the other cells is *not* that the *nontarget cells produce antibodies that block the hormone*. Antibodies are synthesized and released by the immune system in direct response to the presence of a foreign antigen in the body. Hormones released by endocrine cells are not foreign to the body in which they are produced.

(4) The reason the hormones affect the target cell and not the other cells is *not* that the *hormones can only leave the bloodstream near the target cell*. Hormones flow freely in the bloodstream and may move readily through capillary walls at any point to enter the interstitial fluid.

12. **3** This mutation can best be described as *the substitution of an adenine (A) base for guanine (G)*. A close examination of the diagram reveals that the sixth base from the left is G in the original DNA sequence but has changed to A in the mutated DNA sequence. This type of change in the DNA sequence is classified as a substitution because adenine (A) was substituted for guanine (G).

WRONG CHOICES EXPLAINED:
(1) This mutation *cannot* best be described as the *pairing of an adenine (A) base with thymine (T)*. The pairing of bases in a DNA molecule is not considered to be a mutation, since no genetic change has occurred.

(2) This mutation *cannot* best be described as the *insertion of an adenine (A) base into both strands of the DNA molecule*. An insertion mutation involves the addition of one or more base pairs into a DNA sequence. In the example given, no bases were added but only substituted.

(4) This mutation *cannot* best be described as the *deletion of an adenine (A) base from the DNA molecule*. A deletion mutation involves the removal of one or more base pairs from a DNA sequence. In the example given, no bases were removed but only substituted.

13. 1 After digesting the nutrients from a meal high in carbohydrates, the body *releases insulin to return the blood sugar levels to normal.* Insulin is a hormone released from the pancreas that has the effect of storing excess blood sugar in the liver as glycogen, a complex molecule also known as animal starch. Failure to perform this normal bodily function could result in damage to internal organs and the condition known as hyperglycemia or diabetes.

WRONG CHOICES EXPLAINED:

(2) It is *not* true that, after digesting the nutrients from a meal high in carbohydrates, the body *secretes enzymes to absorb starch into the intestines.* Starch is a complex carbohydrate that cannot be absorbed into the intestines until it is digested into simple sugars such as glucose.

(3) It is *not* true that, after digesting the nutrients from a meal high in carbohydrates, the body *produces water to maintain dynamic equilibrium in the blood.* Although certain bodily metabolic activities produce water as a chemical by-product, this water is insufficient to maintain dynamic equilibrium.

(4) It is *not* true that, after digesting the nutrients from a meal high in carbohydrates, the body *maintains homeostasis by increasing wastes produced in muscle cells.* Increasing wastes, such as carbon dioxide or lactic acid, in the muscle cells would have the effect of interfering with, not maintaining, the cells' homeostatic balance.

14. 2 This is an adaptation that benefits plants by *regulating the flow of water vapor out of leaves, preventing excess water loss by the plant.* During daylight hours, the guard cells are rigid. This rigidity forces the stomate to remain open, allowing the free exchange of water vapor and other gases through the opening. During nighttime or in very warm daytime temperatures, the guard cells lose their rigidity and close the stomate opening. This action restricts the loss of water vapor at night or in hot weather, conserving water within the plant.

WRONG CHOICES EXPLAINED:

(1) It is *not* true that this is an adaptation that benefits plants by *increasing the flow of liquid water into leaves, which increases the rate of food and oxygen production.* By closing the stomatal openings, the plant decreases, not increases, the flow of water from the stem into the leaves.

(3) It is *not* true that this is an adaptation that benefits plants by *increasing the flow of oxygen molecules into the leaves, which increases the rate of photosynthesis.* By closing the stomatal openings, the plant decreases, not increases, the flow of oxygen into the leaves from the atmosphere.

(4) It is *not* true that this is an adaptation that benefits plants by *preventing the flow of carbon dioxide into the leaves, which would reduce the rate of respiration.* By closing the stomatal openings, the plant does decrease the flow of carbon dioxide into the leaves from the atmosphere. This might have the effect of reducing the rate of photosynthesis, not the rate of respiration.

15. **1** The loss of this protection most directly interferes with *homeostasis.* Homeostasis is a term that refers to the sum total of all metabolic activities that serve to maintain physical and chemical balance, or steady state, in the body. The disease diabetes reflects a loss of homeostasis because the body cannot maintain acceptable concentrations of sugar in the blood.

WRONG CHOICES EXPLAINED:
(2) The loss of this protection does *not* most directly interfere with *excretion.* Excretion is a life process in which metabolic wastes are eliminated from the body. This particular example would not directly interfere with excretion.

(3) The loss of this protection does *not* most directly interfere with *reproduction.* Reproduction is a life process in which new members of a species are produced. This particular example would not directly interfere with reproduction.

(4) The loss of this protection does *not* most directly interfere with *respiration.* Respiration is a life process in which the energy in the chemical bonds of glucose is converted to the energy in the chemical bonds of ATP. This particular example would not directly interfere with respiration.

16. **3** *The grasshoppers function as consumers and heterotrophs* is the organism that is correctly paired with its role in the ecosystem. The diagram illustrates a food web common to many forest ecosystems. By convention, the arrows in food web diagrams represent the flow of food energy from one trophic (nutritional) level to another. In the diagram, the grasshopper is shown to receive its food energy from grasses, properly illustrating its role as a consumer and a heterotroph. These terms, which both mean other-feeding, are synonymous when discussing the roles of organisms in a food web.

WRONG CHOICES EXPLAINED:
(1) *The grass is both a consumer and a decomposer* is *not* the organism that is correctly paired with its role in the ecosystem. An examination of the food web diagram shows arrows that all lead away from grasses, illustrating its role as a producer, not a consumer, organism. Decomposers such as bacteria and fungi are not indicated in this food web. Grass is not a decomposer.

(2) *The toads function as consumers and autotrophs* is *not* the organism that is correctly paired with its role in the ecosystem. An examination of the food web diagram shows arrows that lead to the toads from grasshoppers, illustrating the toads' role as consumer, not autotrophic (self-feeding), organisms.

(4) *The snakes are both consumers and herbivores* is *not* the organism that is correctly paired with its role in the ecosystem. An examination of the food web diagram shows arrows that lead to the snakes from toads and mice, illustrating the snakes' role as consumers of animal prey and classifying them as carnivores, not herbivores.

17. **4** Milk allergies are different because they *result from an overreaction of the immune system to a harmless substance.* The question states that the cause of so-called milk allergies is the body's inability to produce enough of the enzyme lactase to digest the lactose found in milk; this is the actual difference between regular allergies and milk allergies. The human immune system protects the body from allergens and pathogens by setting up physiological barriers to their entry and/or neutralizing them after they enter. When the body reacts to a harmless substance such as milk, the term allergy is commonly applied because the body's response to it is similar to the response to a foreign invader such as an allergen or pathogen.

WRONG CHOICES EXPLAINED:
(1) It is *not* true that milk allergies are different because they *are often harmful to the person.* Allergies in general are not usually harmful to the people who suffer from them unless the body's allergic reactions (e.g., swelling of tissues, secretion of fluids, elevated body temperature) to them prove harmful.

(2) It is *not* true that milk allergies are different because they *result in a build-up of the substance in the body.* It is unlikely that lactose would build up in the body. Rather, undigested lactose would most likely be passed through the digestive system and egested into the environment with other undigested matter.

(3) It is *not* true that milk allergies are different because they *are the result of the digestive system attacking the substance.* The role of the digestive system is the enzymatic breakdown of complex, digestible foods such as proteins and starches into simple, absorbable foods such as amino acids and glucose. The digestive system does not attack substances such as lactose.

18. 1 One reason energy must be constantly added to a stable ecosystem is because some energy is *lost at each feeding level*. In fact, scientists estimate that as much of 90% of the energy that supports a trophic (feeding) level of a food pyramid is dissipated into the environment as waste heat.

WRONG CHOICES EXPLAINED:
(2) It is *not* true that one reason energy must be constantly added to a stable ecosystem is because some energy is *incorporated into fossil fuels*. Fossil fuels (e.g., coal, oil, natural gas) are substances formed millions of years ago from the decomposed and fossilized remains of ancient plant life.

(3) It is *not* true that one reason energy must be constantly added to a stable ecosystem is because some energy is *destroyed by decomposers*. Energy is neither created nor destroyed in an ecosystem but only transferred from one form to another (e.g., light, chemical bond, heat).

(4) It is *not* true that one reason energy must be constantly added to a stable ecosystem is because some energy is *digested by herbivores*. Herbivores cannot digest energy but can only transform it from one chemical bond to other chemical bonds or release it to the environment as heat.

19. 4 *Several squirrels eat acorns from the oak tree where they live* is the statement that best illustrates direct competition within a species. Competition is a term that relates to any conflict among organisms for limited natural resources in the environment such as food, oxygen, water, or nesting sites. In the example given, the squirrels of the same species compete for a finite number of acorns produced by the oak tree.

WRONG CHOICES EXPLAINED:
(1), (2) *A chipmunk is caught and eaten by a hungry fox* and *a deer attempts to escape a mountain lion that is chasing it* are *not* the statements that best illustrate direct competition within a species. These situations illustrate nutritional (predator/prey) relationships that exist between different species, not within a single species.

(3) *Two muskrats mate and produce a litter of offspring* is *not* the statement that best illustrates direct competition within a species. This situation illustrates a cooperative, not a competitive, relationship between members of the same species that results in continuation of the species.

20. 1 Special medications are taken because, otherwise, the patient's immune system might react to the transplanted kidney by *producing specialized proteins that would attack the kidney*. The medications referenced in the question are probably immunosuppressants, which are drugs designed to decrease the body's immune responses to foreign antigens. Without these drugs, donor antigens on the transplanted kidney would stimulate the production of specialized antibodies in the recipient, a reaction that could result in rejection of the transplanted organ.

WRONG CHOICES EXPLAINED:

(2) It is *not* true that special medications are taken because, otherwise, the patient's immune system might react to the transplanted kidney by *signaling the body to synthesize new DNA molecules*. New DNA molecules are routinely synthesized during both mitotic and meiotic cell division in the body, not in response to foreign antigens.

(3), (4) It is *not* true that special medications are taken because, otherwise, the patient's immune system might react to the transplanted kidney by *stimulating the circulatory system to attack red blood cells* or by *stimulating the kidney to produce reproductive hormones*. These are not reactions that occur in the human body under any normal circumstances.

21. 4 Information that scientists are able to obtain from the fossil record includes *evidence about past environments and the history of life*. Fossils are the preserved remains of, or impressions left by, living things in the far distant past. By studying fossils, scientists often find information that leads them to develop inferences about the nature of ancient ecosystems and the plants and animals that lived within them millions of years ago.

WRONG CHOICES EXPLAINED:

(1) Information that scientists are able to obtain from the fossil record does *not* include *confirmation that Earth is 4.5 million years old*. Information concerning the age of Earth may be obtained by studying the geological, not the fossil, record. Using this information, scientists have inferred that the age of Earth approaches 4.5 billion, not 4.5 million, years.

(2) Information that scientists are able to obtain from the fossil record does *not* include *data supporting the hypothesis that animal species do not change over time*. In fact, the fossil record provides substantial evidence that species do change over time. Scientists have discovered evidence that many species have evolved from common ancestors, maintained a presence on Earth for long periods, and produced new distinct lines or gone extinct over the past 3.5 billion years.

(3) Information that scientists are able to obtain from the fossil record does *not* include *the exact means by which life on Earth began*. Although this question has been asked by human beings for many thousands of years, scientists have developed only broad hypotheses concerning its answer. Unfortunately, the fossil record provides little direct information that would lead to understanding the exact origin of life on Earth.

22. 2 The relationship between fire ants and ground-nesting birds is an example of *predator/prey*. A predator is an animal that hunts, kills, and consumes other animals (prey) for food. Although the fire ant is considerably smaller than any bird, in large numbers they can overwhelm a ground nest and then kill and consume newly hatched birds before the birds have grown protective feathers. If enough young birds are killed in this way, the ability of the ground-nesting bird species to reproduce is diminished. Over time, the population of these ground-nesting birds may be reduced.

WRONG CHOICES EXPLAINED:
(1) The relationship between fire ants and ground-nesting birds is *not* an example of *producer/consumer*. A producer is an organism, usually a green plant, that is capable of producing its own food via the process of photosynthesis. Neither the fire ants nor the ground-nesting birds are producers.

(3) The relationship between fire ants and ground-nesting birds is *not* an example of *scavenger/decomposer*. A decomposer is an organism, usually a bacterium or fungus, that breaks down complex organic molecules into soluble soil nutrients that are taken up by plants. Neither the fire ants nor the ground-nesting birds are decomposers.

(4) The relationship between fire ants and ground-nesting birds is *not* an example of *parasite/host*. A parasite is an organism, such as a flea or tick, that lives in or on the body of another organism (host) and harms that organism as the parasite feeds on its tissues. Neither the fire ants nor the ground-nesting birds are parasites.

23. 4 When an altered ecosystem is left undisturbed, the most likely result would be *a gradual shift toward a stable ecosystem*. Whether the ecosystem has been altered by a natural disaster (e.g., fire, flood, volcanism) or by human activity (e.g., farming, development, logging), when left undisturbed it will tend to revert to its former state. This phenomenon is known as ecological succession and is characterized by the gradual replacement of one plant community by another until a stable, self-perpetuating plant community is established. Ecological succession may require decades or centuries to complete depending on biotic and abiotic factors at work in the environment.

WRONG CHOICES EXPLAINED:

(1) It is *not* true that, when an altered ecosystem is left undisturbed, the most likely result would be *the gradual evolution of all of the original species.* Evolution is a process that depends on mutation, natural selection, and reproduction over many generations to produce new varieties or species of living things. The process described would not directly lead to the evolution of new variations or species.

(2) It is *not* true that, when an altered ecosystem is left undisturbed, the most likely result would be *a rapid return to the original ecosystem.* Ecological succession may require decades or centuries to complete depending on biotic and abiotic factors at work in the environment.

(3) It is *not* true that, when an altered ecosystem is left undisturbed, the most likely result would be *the elimination of all of the predator species.* The presence of predators in an ecosystem is normally dependent on the availability of prey living in that environment. The process described would not directly lead to the elimination of all predators.

24. 3 *Reducing the use of chemicals on farm fields and golf courses* is the human activity that would most likely result in a *decrease* in the size and number of dead zones. Farm fields and golf courses tend to use large amounts of fertilizer to grow crops and grass. By reducing the use of these chemicals, less fertilizer would run off into nearby bodies of water, so fewer algae blooms and resulting bacterial decay would occur.

WRONG CHOICES EXPLAINED:

(1), (4) *Irrigating fields and lawns to increase runoff into the oceans and rivers* or *constructing more sewage-treatment plants on the shores of lakes and rivers* are *not* the human activities that would most likely result in a *decrease* in the size and number of dead zones. These activities would release fertilizers and sewage into the nearby water bodies, causing them to have more algae blooms and bacterial decay.

(2) *Building more coal-fired electrical generating plants* is *not* the human activity that would most likely result in a *decrease* in the size and number of dead zones. This activity would release chemicals into the air that could add to the nutrient load and raise the acidity of any body of water downwind of the power plant.

25. 4 *Human activities can have negative effects on a species* is the statement regarding this finding that is supported by the study. The passage states one negative effect (shortened mating songs) on some songbirds that were found to have been contaminated by the industrial pollutant mercury. Although the passage does not clearly link the mercury contamination to the negative effect on the songbirds, it can be inferred that such a link exists. Further scientific research is needed to establish this link and verify the negative effect of human activities on this species of songbirds.

WRONG CHOICES EXPLAINED:
(1), (2), (3) *Mercury pollution will result in the extinction of all songbirds*, *mercury prevents songbirds from obtaining required nutrients*, and *human activities usually affect the smallest animals in ecosystems* are *not* the statements regarding this finding that are supported by the study. No information is presented in the passage that would lead to any of these outcomes.

26. 2 Row *2* is the row in the chart that shows the connection between processes, structures, and hormones involved in the formation of an embryo. In human beings, males produce a gamete known as sperm in specialized tissues of the physical structures known as the testes. In order to regulate this process, the testes synthesize and secrete the male reproductive hormone testosterone.

WRONG CHOICES EXPLAINED:
(1), (3), (4) Rows *1*, *3*, and *4* are *not* the rows in the chart that show the connection between processes, structures, and hormones involved in the formation of an embryo. None of these combinations of process, structure, and hormone is correct. See the correct answer above.

27. 1 A group of organisms from the food web that would be found at level *X* of the energy pyramid is the *owls*. By convention, the arrows in a food web are drawn to indicate the direction of energy and nutrient flow in the ecosystem. Also by convention, an energy pyramid is drawn to indicate the total relative energy at each trophic (feeding) level, with producers at the bottom (most energy), herbivores in the middle (moderate energy), and predators at the top (least energy). A review of the information in the chart shows that two organisms, the hawk and the owl, have arrows leading to them but not away, representing them as top predators in this food web. Both of these groups of organisms would occupy the top trophic (least energy) level of the energy pyramid.

WRONG CHOICES EXPLAINED:
(2), (4) It is *not* true that a group of organisms from the food web that would be found at level X of the energy pyramid is the *deer* or *crickets*. A review of the information in the chart shows that both the deer and the crickets have arrows leading to them from trees and grasses, representing them as herbivores in this food web. These organisms would occupy the middle (moderate energy) level of the energy pyramid.

(3) It is *not* true that a group of organisms from the food web that would be found at level X of the energy pyramid is the *trees*. A review of the information in the chart shows that trees have arrows leading to them but not away, representing them as producers in this food web. These organisms would occupy the bottom (most energy) level of the energy pyramid.

28. **4** The arguments made by both sides provide evidence that *the use of new technology requires decisions based on an assessment of costs, benefits, and risks*. This concept is often referred to as a cost-benefit analysis or a trade-off analysis. It acknowledges the fact that the introduction of any human process into the natural environment comes with inherent positive and unanticipated negative reactions that must be taken into account before implementing that process. If, after scientific study, a consensus of decision-makers is that the expected positive outcomes outweigh the risks to be taken, the process may be implemented on a limited scale and the negative effects studied further before implementing the process on a broad scale.

WRONG CHOICES EXPLAINED:
(1) The arguments made by both sides do *not* provide evidence that *genetic technology is the best way to correct the damage humans have done to the environment*. The best way to repair damage done to the environment by humans is for humans to stop doing the things that caused the damage in the first place.

(2) The arguments made by both sides do *not* provide evidence that *the introduction of genetic technology will benefit all organisms equally*. In fact, genetic technology primarily benefits the target organism for which the technology was developed, in this case an extinct species that benefits by being brought back into existence.

(3) The arguments made by both sides do *not* provide evidence that *any new technology that increases the biodiversity of the area should be used*. The introduction of invasive species is a technology that technically increases the biodiversity of a native environment but that may have extremely negative effects on the native species with which it competes.

29. **1** *Microorganisms recycle nutrients that support the ecosystem* is the statement that best explains the purpose of the microorganisms in this aquarium. Nutrient recycling is a result of decomposition in which complex biochemicals are broken down by microorganisms (e.g., bacteria and fungi) into soluble molecules that can be taken up and recycled by plants. This is an essential step in any functional food web.

WRONG CHOICES EXPLAINED:

(2) *Microorganisms recycle the energy in the ecosystem* is *not* the statement that best explains the purpose of the microorganisms in this aquarium. Energy is never recycled in a food web but is only converted from one form to another until it dissipates into the environment as waste heat.

(3) *Microorganisms are a source of food for the plant* is *not* the statement that best explains the purpose of the microorganisms in this aquarium. Green plants are producer organisms that manufacture their own food by the process of photosynthesis. Plants do not use microorganisms for food.

(4) *Microorganisms are an abiotic factor important for decomposition* is *not* the statement that best explains the purpose of the microorganisms in this aquarium. Abiotic factors are nonliving components (e.g., moisture, light, temperature, soil minerals, oxygen) that are present in an ecosystem and affect the kinds and numbers of organisms that can inhabit the ecosystem. Microorganisms are biotic (living), not abiotic, factors present in the environment.

30. **3** A possible reason for the replacement could be that the new species *outcompeted the original algae populations for abiotic factors*. Abiotic factors are nonliving components (e.g., moisture, light, temperature, dissolved minerals, oxygen) that are present in an ecosystem and that affect the kinds and numbers of organisms that can inhabit the ecosystem. An introduced algal species might prove better than a native algal species at competing for one or more abiotic factors in their shared ecosystem, thus limiting those factors that may be necessary for the native species' healthy growth.

WRONG CHOICES EXPLAINED:

(1) It is *not* true that the possible reason for the replacement could be that the new species *outcompeted the original algae populations for prey present in the ecosystem*. Algae are producer organisms that manufacture their own food, so they do not consume prey organisms for food.

(2) It is *not* true that the possible reason for the replacement could be that the new species *required more resources than the original algae populations in the pond*. Whether or not the introduced algal species required more resources than the native algal species, the passage makes it clear that the introduced algal species was more successful than the native algal species at competing for those resources.

(4) It is *not* true that the possible reason for the replacement could be that the new species *is less adapted to the pond ecosystem than the original algae populations*. The success of the introduced algal species at outcompeting the native algal species makes it clear that the introduced algal species quickly adapted to the pond ecosystem.

PART B-1

31. 2 This process normally produces *one functional gamete with one-half of the genetic information found in the diploid germ cell*. This process is also known as oogenesis (production of egg cells). In the diagram, the structure labeled "Egg cell (ovum)" is the single gamete (sex cell) produced as a result of this process. The meiotic division of the germ cell separates homologous pairs of chromosomes into four separate cell bodies that each contain exactly one-half of the genetic information (haploid number) needed to create a new human being. Three of these haploid cells are nonfunctional polar bodies, while the fourth haploid cell develops into a mature egg (ovum) ready for fertilization.

WRONG CHOICES EXPLAINED:
(1), (3) This process does *not* normally produce *one functional gamete with one-quarter of the genetic information found in the diploid germ cell* or *four functional gametes, each with one-quarter of the genetic information found in the diploid germ cell*. Normal gametes contain one-half, not one-quarter, of the genetic information found in the diploid germ cell.

(4) This process does *not* normally produce *four functional gametes, each with one-half of the genetic information found in the diploid germ cell*. This outcome is the result of normal spermatogenesis (formation of sperm cells), not oogenesis.

32. 1 A reasonable inference based on these results would be that Aloe vera *affected the rate of cell division, resulting in an increased rate of regeneration*. If we compare only the 20% and the 40% groups, then this inference holds true in relative terms. However, this inference can be drawn definitively only if we assume that the 20% group regenerated more quickly than the control group, a comparison for which no data were reported.

WRONG CHOICES EXPLAINED:
(2) It is *not* true that a reasonable inference based on these results would be that *the control group, which received no Aloe vera, did not regenerate*. No information is provided in the question concerning the rate of regeneration of planaria in the 0% (control) group.

(3) It is *not* true that a reasonable inference based on these results would be that *if she applied 30% Aloe vera to a group, it would regenerate more rapidly than the 40% group*. No information is provided in the question concerning the rate of regeneration of planaria in a hypothetical 30% experimental group.

(4) It is *not* true that a reasonable inference based on these results would be that *the application of* Aloe vera *to earthworms would have no effect on tissue regeneration.* No information is provided in the question concerning the effect of *Aloe vera* on earthworms.

33. 3 The most likely reason the action of the enzyme *decreases* after 40°C is that *the shape of the enzyme changes due to environmental conditions.* Studies of enzymes in humans show that they deform (denature) at temperatures above 40°C (104°F), making them inoperative for catalyzing reactions in human cells and potentially causing tissue damage or death.

WRONG CHOICES EXPLAINED:
(1) It is *not* true that the most likely reason the action of the enzyme *decreases* after 40°C is that *the DNA in the enzyme mutates and can no longer break down the starch.* The function of DNA in humans is to carry the genetic code for cell development and homeostatic balance, not the breakdown of starch.

(2) It is *not* true that the most likely reason the action of the enzyme *decreases* after 40°C is that *enzymes die after working for a long period of constant activity in the body.* Enzymes are protein molecules, not living things. So proteins cannot die but may become less efficient over time and be recycled as new proteins.

(4) It is *not* true that the most likely reason the action of the enzyme *decreases* after 40°C is that *as the temperature of the enzyme rises, the pH of the environment changes, deactivating the enzyme.* Although the pH of the environment is known to affect the operation of enzymes, there is no known mechanism by which increasing temperature could directly change pH in that environment.

34. 3 *As nitrogen content in the lichen decreases, the growth of the lichen increases* is the statement that best describes the relationship between the nitrogen content and the growth of the lichen. A review of the data presented in the graph indicates that at relatively high nitrogen concentrations (1.4% to 1.8%), lichen covers only a small surface area (0% to 10%) of the trees on which they live. At lower nitrogen concentrations (0.6% to 1.0%), the surface area of trees covered by the lichen increases (20% to 40%). This inverse relationship, also known as a negative correlation, is demonstrated by the statistical average line that slopes downward from left to right on the graph.

WRONG CHOICES EXPLAINED:

(1), (2) *As nitrogen content in the lichen increases, the growth of the lichen increases* and *as nitrogen content in the lichen decreases, the growth of the lichen decreases* are *not* the statements that best describe the relationship between the nitrogen content and the growth of the lichen. These statements both infer a positive correlation between the factors (lichen nitrogen content vs. growth rate) in this study. A statistical average line for such a positive correlation would slope upward from left to right, not downward from left to right as shown in the graph.

(4) *There is not a clear relationship between the amount of nitrogen in the lichen and growth* is *not* the statement that best describes the relationship between the nitrogen content and the growth of the lichen. This relationship is quite clear from an examination of the graphed data. It is made still clearer by the addition of the statistical average line superimposed over the individual data points on the graph. See the correct answer above.

35. 2 Scientific claims should be questioned if *the experimental results cannot be repeated by other scientists*. This repetition of experiments by independent scientists is a key element in the process of peer review. It often reveals weaknesses in the methods used by the original experimenters and may refute their claims altogether.

WRONG CHOICES EXPLAINED:

(1), (3), (4) It is *not* true that scientific claims should be questioned if *peer review was used to examine the claims made by scientists, conclusions follow logically from the evidence*, or *the data are based on samples that are very large*. Although each of these actions helps to support the results of a scientific experiment, none by itself is sufficient to preclude the questioning of a scientific claim. Peer review may be performed by biased or incompetent individuals, conclusions may have been developed prior to the experiment, and the samples may appear large but still be too small for a particular study. Only review and repetition of experimental results by competent, independent, and unbiased scientists can verify the claims made about scientific knowledge.

36. 4 The most likely order in which these events occur is *[D] – [B] – [A] – [C]*. The series of events as energy is captured by a plant and used in the metabolic processes of an herbivore begins when the plant absorbs sunlight [D] and uses it in the process of photosynthesis to synthesize molecules of glucose. The plant then uses these molecules of glucose to synthesize complex carbohydrates such as starch and cellulose [B] in its cells. An herbivore such as a caterpillar consumes leaf tissues, digests the complex molecules back into molecules of glucose, and uses them in the process of cellular respiration to release energy [A]. This energy is immediately transferred to molecules of ATP in the caterpillar's cells [C] for use in operating the cells' metabolic processes.

WRONG CHOICES EXPLAINED:

(1), (2), (3) It is *not* true that the most likely order in which these events occur is *[A] – [D] – [B] – [C], [B] – [A] – [C] – [D]*, or *[D] – [A] – [B] – [C]*. None of these sequences presents the correct series of events that occur as energy is captured by a plant and used in the metabolic processes of an herbivore. See the correct answer above.

37. 2 The approximate change in CO_2 level from the year 1000 to the year 2000 was *an increase of 95 ppm*. This can most easily be determined by reading the values for CO_2 level in the years 1000 (280 ppm) and 2000 (375 ppm) and then calculating the difference between these values (375 ppm − 280 ppm = 95 ppm).

WRONG CHOICES EXPLAINED:

(1), (3), (4) It is *not* true that the approximate change in CO_2 level from the year 1000 to the year 2000 was *an increase of 135 ppm, a decrease of 135 ppm*, or *a decrease of 95 ppm*. None of these responses can be correctly determined from the data as presented. Care should be taken to read the CO_2 level for the year 2000 from the vertical line for that year, not the CO_2 level at the graph end line representing approximately the year 2020.

38. 4 The process that occurs at Step 1 is *digestion*. The diagram illustrates key steps in the nutritional process by which complex molecules are broken down into simple components that may be used by the body as the building blocks for new complex molecules. In the first step of this process, foods containing complex proteins are broken down (digested) into amino acids.

WRONG CHOICES EXPLAINED:

(1) It is *not* true that the process that occurs at Step 1 is *respiration*. Respiration is a life process in which the energy in the chemical bonds of glucose is converted to the energy in the chemical bonds of ATP. The process illustrated in the diagram at this step does not fit the definition of respiration.

(2) It is *not* true that the process that occurs at Step 1 is *circulation*. Circulation is a life process by which materials are distributed throughout the body for use by living tissues. The process illustrated in the diagram at this step does not fit the definition of circulation.

(3) It is *not* true that the process that occurs at Step 1 is *excretion*. Excretion is a life process in which metabolic wastes are eliminated from the body. The process illustrated in the diagram at this step does not fit the definition of excretion.

39. 1 Cell structure X is a *ribosome*. Ribosomes are a type of cell organelle that functions to synthesize proteins from free amino acids, as illustrated in the diagram at Steps 2 and 3. This role is essential to the cell's ability to grow, function, and reproduce.

WRONG CHOICES EXPLAINED:
(2) It is *not* true that cell structure X is a *vacuole*. Vacuoles are small, membrane-bound structures in cells that function to store food prior to digestion or wastes prior to excretion.

(3) It is *not* true that cell structure X is a *cell membrane*. Cell membranes form cell boundaries and function to regulate the passage of materials into and out of cells.

(4) It is *not* true that cell structure X is a *mitochondrion*. Mitochondria are cell organelles that function to metabolize sugar molecules under the control of respiratory enzymes in order to release energy for cell processes.

40. 4 Model *4* best explains why these cells are so different. This model illustrates the manner by which specialized cells turn on certain genes and turn off others, allowing the cell to produce the specific proteins it needs to perform its functions. This action results in the differentiation of tissues in the body, each tissue specialized to perform its particular function at a high level of efficiency.

WRONG CHOICES EXPLAINED:
(1) Model *1* does *not* best explain why these cells are so different. All somatic cells of the body contain the diploid chromosome complement typical for the species.

(2) Model *2* does *not* best explain why these cells are so different. If no genes were expressed in a cell, it would be unable to synthesize any proteins, including the basic structural proteins that make up the cell body.

(3) Model *3* does *not* best explain why these cells are so different. Pairs of homologous chromosomes carry gene pairs that work together to control individual traits. If the genes of an entire chromosome were turned off, the genes on its homologous chromosome could not function properly.

41. 3 The substances that allow the bed bugs to break down insecticides and the biological pumps that remove the insecticides from the bed bugs are examples of *biological adaptations*. A biological adaptation is any trait or characteristic of an organism that serves to make it more successful in its environment than other members of the species that lack that trait. The mechanisms described for the bed bug make it more successful by protecting it from insecticides in its environment.

WRONG CHOICES EXPLAINED:
(1) The substances that allow the bed bugs to break down insecticides and the biological pumps that remove the insecticides from the bed bugs are *not* examples of *the failure of homeostasis*. Homeostasis is a term that refers to the sum total of all metabolic activities that serve to maintain physical and chemical balance, or steady state, in the body of an organism. The mechanisms described for the bed bug represent the success, not the failure, of homeostasis for that organism.

(2) The substances that allow the bed bugs to break down insecticides and the biological pumps that remove the insecticides from the bed bugs are *not* examples of *genetic engineering*. Genetic engineering refers to a set of laboratory techniques used by scientists to modify the genetic characteristics of living organisms by inserting genes from other species into the genome of those organisms. The mechanisms described for the bed bug represent the process of natural selection, not artificial gene manipulation.

(4) The substances that allow the bed bugs to break down insecticides and the biological pumps that remove the insecticides from the bed bugs are *not* examples of *selective breeding*. Selective breeding refers to a set of methods used by plant and animal breeders to produce new varieties of plants and animals. They breed parental pairs that express desirable characteristics and hope that the pairs will pass on these characteristics to their offspring. The mechanisms described for the bed bug represent the process of natural selection, not artificial breeding techniques.

42. **2** A gene mutation resulting in insecticide resistance would most likely increase in the bed bug population because *the insecticide-resistant bed bugs will survive and reproduce.* It can be assumed that the original bed bug population had no natural resistance to the insecticide. Over the course of many generations, a mutation for such resistance occurred in the gamete of one individual. However, in the absence of the insecticide, this mutation provided no particular advantage, so the mutation was passed on to only those bed bugs in a direct line of descent from the original mutant. At some point, the insecticide was added to the bed bugs' environment by human action. The insecticide immediately killed susceptible bed bugs but did not kill resistant bed bugs, leaving the resistant bed bugs to breed more resistant offspring. In this scenario, the insecticide serves as a selection pressure on the bed bugs that resulted in increasing the gene frequency for insecticide resistance.

WRONG CHOICES EXPLAINED:

(1) It is *not* true that a gene mutation resulting in insecticide resistance would most likely increase in the bed bug population because *more bed bugs will need to be resistant to the insecticide*. Evolutionary change does not occur because of need but only by the natural processes of mutation and selection described above.

(3) It is *not* true that a gene mutation resulting in insecticide resistance would most likely increase in the bed bug population because *the bed bugs with the resistance gene will reproduce asexually*. Bed bugs are insects and, as do all insects, reproduce sexually, not asexually.

(4) It is *not* true that a gene mutation resulting in insecticide resistance would most likely increase in the bed bug population because *spraying an insecticide will allow more bed bugs without mutations to survive*. Assuming that the mutation referenced is for insecticide resistance, the insecticide will kill more non-mutated bed bugs than it will mutated bed bugs. So fewer, not more. bed bugs without mutations will survive.

43. **2** Based on the data provided, a reasonable interpretation would be that *the virus mutated in 2014–2015, resulting in the vaccine being less effective*. Because the virus is prone to frequent genetic change due to mutation, in any given year, the vaccine development may lag behind the viral mutation rate. When this occurs, the vaccine may be less effective at neutralizing the virus, as happened in 2014–2015.

WRONG CHOICES EXPLAINED:

(1) It is *not* true that, based on the data provided, a reasonable interpretation would be that *in 2004–2005, some individuals caught the flu from the vaccine*. There is no evidence that modern vaccines are capable of transmitting the disease for which they were developed.

(2) It is *not* true that, based on the data provided, a reasonable interpretation would be that *people have become immune to the flu vaccine over the 13-year period*. Immunity is a condition by which the body develops safeguards that block the ability of a virus or other pathogen to invade the body. The flu vaccine is not a pathogen of any type, so the body develops no immunity to it but only to the pathogen itself.

(3) It is *not* true that, based on the data provided, a reasonable interpretation would be that *the vaccine has become increasingly effective over the 13-year period*. A review of the data presented in the graph shows that there has been considerable variation, not steady improvement, in the effectiveness of the flu vaccines developed over the 13-year period.

PART B-2

44. One credit is allowed for correctly identifying the genetic trait and supporting the answer. Acceptable responses include but are not limited to: [1]

- *Eye color is a trait that is inherited from parents. A scar, a birth date, and a decision to be a vegetarian are not inherited.*

- *Brown eye color can be passed on to offspring. An individual does not pass on eating preferences, scars, or birth dates to offspring.*

- *Eye pigmentation is a trait that is determined by genes/alleles, which can be passed on to offspring through normal reproductive mechanisms.*

- *Having brown eyes is a genetic trait, while the others are accidents of birth or personal choices.*

45. One credit is allowed for correctly marking an appropriate scale, without any breaks in the data, on each labeled axis.   [1]

46. One credit is allowed for correctly plotting the data on the grid, connecting the points, and surrounding each point with a small circle.   [1]

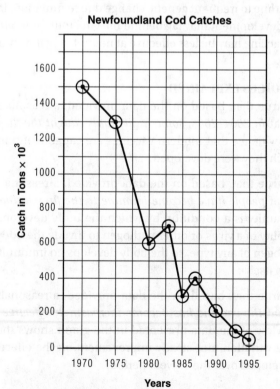

Newfoundland Cod Catches

47. **2** The five-year span between *1975 and 1980* is the span in which the largest drop in fishing occurred. A review of the charted and graphed data shows the steepest decline in cod catches was 700 tons $\times$ 10^3 ($1300 - 600 = 700$) during this period. These data indicate that some biotic or abiotic factor in the cod's marine ecosystem changed dramatically during or just prior to 1975 that had the effect of significantly reducing the carrying capacity of the environment with respect to cod. This phenomenon may be exaggerated by the overfishing of cod in these waters, a human activity that took too many fish out of the ecosystem too rapidly, leaving too few to repopulate the species in that fishing zone.

WRONG CHOICES EXPLAINED:
(1) The five-year span between *1970 and 1975* is *not* the span in which the largest drop in fishing occurred. A review of the data indicates a drop of 200 tons $\times$ 10^3 during this period, which is less than 700 tons $\times$ 10^3.

(3) The five-year span between *1980 and 1985* is *not* the span in which the largest drop in fishing occurred. A review of the data indicates a drop of 300 tons $\times$ 10^3 during this period, which is less than 700 tons $\times$ 10^3.

(4) The five-year span between *1990 and 1995* is *not* the span in which the largest drop in fishing occurred. A review of the data indicates a drop of 160 tons $\times$ 10^3 during this period, which is less than 700 tons $\times$ 10^3.

48. One credit is allowed for correctly stating *one* advantage, other than maintaining an adequate food supply for humans, of *not* overfishing the oceans. Acceptable responses include but are not limited to: [1]

- *Maintaining fish populations in the oceans will help to maintain biodiversity/ stability in the ocean ecosystem.*
- *The organisms the fish feed on will not overpopulate and destroy the ecosystem.*
- *Food chains/webs will not be disrupted.*
- *More food will be available for predators.*
- *Other species may be saved from extinction.*
- *The cod's niche/role in the environment will remain filled, which will help to maintain the balance of nature.*

49. **2** This increase in size of some fish populations is a result of human *decisions that weighed the need for food with the need to maintain fish populations.* This action will leave enough fish in the population to produce offspring each breeding season, which will have the effect of increasing the supply of catchable fish each fishing season. Ideally, it will also have the effect of returning the species to its original abundance in these fishing waters.

WRONG CHOICES EXPLAINED:
(1) This increase in the size of some fish populations is *not* a result of human *actions that killed many of the predators of these fish populations.* Each species that inhabits an ecosystem plays a role that affects other species inhabiting that same environment. This includes predators that take a certain proportion of their prey species for food. Some human programs seek to eliminate predators (e.g., cormorants on Lake Ontario, sharks in Long Island Sound) that fishermen misguidedly see as competitors for the fish they seek for food or profit.

(3) This increase in the size of some fish populations is *not* a result of human *activities that are increasing the use of nonrenewable resources of the oceans.* Nonrenewable resources are minerals that cannot be replaced once they are used. Mining or collecting these resources from the oceans does little to help fish populations to increase and may have the opposite effect.

(4) This increase in the size of some fish populations is *not* a result of human *decisions that are increasing the use of renewable ocean resources.* Renewable ocean resources such as wave, wind, solar, or tidal energy are increasingly being tapped to replace fossil fuels as a source for energy. These decisions may positively affect the environment by slowing global warming and climate change, and they may help to increase fish populations in the bargain.

[NOTE: This response is a potential correct answer to the question as written, but it is not the best answer.]

50. **4** The changes observed over time occurred as the organisms *developed variations that made it possible for them to move from water to land.* Adaptations are changes in an organism's physical or behavioral characteristics that enable the organism to be more successful in its environment than other members of its species. Adaptations occur due to random mutations in the gametes of an ancestral species member that may then be passed on to succeeding generations. Early tetrapod adaptations may have included an enhanced skeletal system that permitted the variants to move fins via muscular control and eventually articulated limbs that allowed some variants to inhabit marginal land environments. The success of such variants would have led to the production of offspring that also exhibited these favorable adaptations.

WRONG CHOICES EXPLAINED:
(1), (2) It is *not* true that the changes observed over time occurred as the organisms *needed to change the habitat where they lived from land to water* or *needed to change the habitat where they lived from water to land.* Generally, organisms adapt to their environments rather than adapting the environment to themselves. The only known species that have evolved adaptations or developed technologies that alter their environments for their benefit are beavers and humans.

(3) It is *not* true that the changes observed over time occurred as the organisms *developed variations that made it possible for them to move from land to water.* The diagram illustrates the gradual changes to tetrapod ancestors that led to their emergence from water to land, not land to water.

51. One credit is allowed for correctly describing *one* way scientists can determine the correct sequence of fossils that represents the ancestry of an organism such as a tetrapod. Acceptable responses include but are not limited to: [1]

- *The fossil organism located in the deepest undisturbed rock layers would most likely be the oldest. Those in layers less deep would be younger.*

- *If DNA for each of the species is available, scientists could determine which ones were the most closely related and which ones seemed to be less similar.*

- *They could use radioactive/carbon dating to determine the actual age of the fossil.*

- *Scientists could determine the ages of the fossils and arrange them in order.*

- *They could compare the bone structures in the fossils.*

52. One credit is allowed correctly identifying the population represented in this food web that has the greatest amount of stored energy. Acceptable responses include: [1]

- *Algae*

[NOTE: Algae represent the producers in this food web. As such, they absorb energy from sunlight and convert it to the chemical bond energy of glucose via the process of photosynthesis. As the basis of the food pyramid, producers contain the most energy available in the food web of any ecosystem.]

53. One credit is allowed for correctly stating what would happen to the bass population if a pesticide that was used in the ecosystem killed the entire pickerel population and supporting the answer. Acceptable responses include but are not limited to: [1]

- *The bass population would initially increase because the pickerel would not be feeding on it.*

- *The bass population would increase because there is no predator. Eventually, the bass population could start to decrease because all the amphipods/crayfish they use for food would be consumed.*

- *The bass population might increase temporarily but might then decrease due to disruptions of the balance of nature in the pond ecosystem caused by the killing of the pickerels by the pesticide.*

[NOTE: The question makes no mention of the susceptibility of bass to the pesticide that killed the pickerels. Without this information, it is not reasonable to assume that the pesticide may have affected the bass population directly.]

54. One credit is allowed for correctly identifying the role of the bacteria in this food web and stating the importance of this particular role. Acceptable responses include but are not limited to: [1]

- *They break down dead organic matter, returning nutrients to the environment.*
- *Decomposers recycle materials in the ecosystem, returning nutrients for the producers to use.*
- *Decomposers: they recycle nutrients.*
- *They break down organic matter/remains of dead organisms and return it to the food web.*

55. One credit is allowed for correctly stating *one* possible reason for the change in the bacterial population size in the area labeled *A* on the graph. Acceptable responses include but are not limited to: [1]

- *There was a decrease in dead organisms.*
- *There was less organic waste available for the bacteria to feed on.*
- *There might have been a disease/antibiotic that affected the bacteria population.*
- *Abiotic factors in the environment may have changed, making the environment less favorable for the growth of bacteria.*
- *This may represent a periodic/seasonal decline in the bacteria population due to temperature/pH or other abiotic factors.*

PART C

56. One credit is allowed for correctly identifying *one* risk associated with the destruction of the ozone shield. Acceptable responses include but are not limited to: [1]

- *The ozone shield protects living things from radiation.*
- *UV radiation can increase the risk of skin cancer.*
- *Destruction of the ozone shield can allow more UV radiation to reach Earth.*
- *Studies have shown that UV rays can cause mutations in living cells exposed to it.*
- *The ozone layer filters out solar UV radiation, so when it's gone, the UV can get through and damage living things.*

57. One credit is allowed for correctly explaining whether or not the Montreal Protocol has been effective. Acceptable responses include but are not limited to: [1]

- *The Montreal Protocol has been effective because the size of the ozone hole has decreased since 2000.*
- *The Montreal Protocol has not been effective because the ozone hole was still 19.6 million km^2 in 2017.*
- *It was not effective since the size of the hole increased several years after it was adopted.*
- *It was effective since the size of the ozone hole eventually started to decrease.*

58. One credit is allowed for correctly describing *one* possible *negative* consequence that it is important to consider when an international agreement such as the Montreal Protocol is adopted. Acceptable responses include but are not limited to: [1]

- *There could be an economic cost of banning ozone-destroying chemicals that may impact the economy of some countries.*
- *Although they destroy the ozone layer, the chemicals may have important uses, and alternatives would have to be developed before banning them.*
- *Limiting the production of certain chemicals may cause economic hardships for some countries.*
- *Developing alternative chemicals could be more expensive.*
- *Taking already-manufactured products off shelves and disposing of them safely could be costly.*

59. One credit is allowed for correctly explaining how the results of the experiment support Dr. Steffan's hypothesis. Acceptable responses include but are not limited to: [1]

- *The control group averaged 43 survivors, while the group exposed to fungicide residues averaged only 12 survivors.*

- *He hypothesized that if the fungi associated with the pollen suffer, then the bumblebee larvae will also suffer, and they did. The colonies exposed to fungicides produced fewer individuals.*

- *Fewer larvae that fed on pollen without the fungus survived.*

- *There were about three times more individuals in the colonies that were not exposed to the fungicide.*

- *Control bee colonies without fungicide/with fungi did better at producing offspring than the experimental group with fungicide/without fungi.*

60. One credit is allowed for correctly explaining how spraying fungicide on crops only when they were not flowering would prevent harming bumblebee larvae. Acceptable responses include but are not limited to: [1]

- *The bees/larvae would not be feeding on pollen at that time.*

- *The pollen the larvae were fed during flowering would still have the beneficial fungus.*

- *If there are no flowers, there will be no pollen for the bumblebees to collect.*

- *Spraying before and/or after flowering would keep fungicide from affecting the yeast in the pollen.*

61. One credit is allowed for correctly explaining how climate change could have a greater impact on bee species inhabiting smaller geographic areas than those inhabiting larger geographic areas. Acceptable responses include but are not limited to: [1]

- *There would be less variation present in a smaller population in a smaller area.*

- *The plants in the smaller area are less diverse and may be affected more easily by temperature changes.*

- *Changes in the climate could result in large changes in the types of pollen-producing plant species present in a smaller area.*

- *The bee species may be able to survive only within a narrow temperature range.*

- *In a smaller area, there are fewer bees present, and their genetic diversity/gene pool may not be sufficiently large to allow them to adapt to the changing climatic conditions.*

62. One credit is allowed for correctly explaining why it is important to preserve bumblebee populations. Acceptable responses include but are not limited to: [1]

- *Without bumblebees, many food crops will not be pollinated and the amount of food available for humans and wildlife will decrease.*

- *If bee populations are lost, biodiversity will decrease and the stability of the ecosystem will decrease.*

- *They are essential for pollinating many flowering plants/food crops on which humans depend for food.*

[NOTE: Bumblebees are not known to produce quantities of honey sufficient for use as human food.]

63. One credit is allowed for correctly explaining why their triple-helix model was valuable even though it was not correct. Acceptable responses include but are not limited to: [1]

- *Their triple-helix model was valuable because it led to further investigation.*

- *It led to discussion and testing by other scientists, which eventually resulted in Watson and Crick revising their model to one that was supported by the data.*

- *Their model was questioned by others and led to the currently accepted model.*

- *Some aspects of the triple-helix model were accurate and so provided a basis for development of an improved model.*

- *Watson and Crick realized their original hypothesis was incorrect and built on their failure to develop a new and more accurate hypothesis.*

64. One credit is allowed for correctly stating *one* reason why a human heart muscle cell would probably contain a higher proportion of mitochondria than a skin cell. Acceptable responses include but are not limited to: [1]

- *Heart muscle cells require more energy to operate than skin cells.*

- *Skin cells are not as active as muscle cells, so they need less energy than muscle cells.*

- *Because heart muscle cells must constantly contract to maintain blood flow, they need tremendous quantities of energy/ATP that the mitochondria produce.*

65. One credit is allowed for correctly explaining why populations like phytoplankton are required to sustain an aquatic food web. Acceptable responses include but are not limited to: [1]

- *Every food web requires organisms that are able to transform solar energy and store it in glucose that can be consumed by other organisms to use for energy.*

- *Other species in the food web are unable to carry on photosynthesis and must rely on species like phytoplankton to trap solar energy in food molecules.*

- *Phytoplankton are required to sustain a food web because they are producers.*

- *Producers like phytoplankton are the foundation of the energy pyramid.*

66. One credit is allowed for correctly explaining why using Chief to produce so many offspring is an example of selective breeding. Acceptable responses include but are not limited to: [1]

- *Chief was selected to breed because he had valuable genetic traits.*

- *Chief was mated with many cows to produce offspring that would be good milk cows.*

- *The dairy farmers wanted cows that would be good milk cows, so Chief was chosen as the sperm donor to produce them.*

- *Chief produced sperm that contained many genes for valuable traits that would be passed on to his offspring.*

- *Chief is credited with making a lot of money for the dairy industry.*

67. One credit is allowed for correctly explaining how the use of Chief to produce offspring had both advantages and disadvantages. Acceptable responses include but are not limited to: [1]

- *Chief's female offspring are great milk producers. However, his overuse as a sire led to the deaths of many fetal calves due to his defective gene being passed to them.*

- *Chief's desirable genetic traits being passed to his daughters came at the cost of the deaths of many fetal calves that inherited a lethal gene from him.*

- *His daughters produced a lot of milk, but there was less diversity in the offspring.*

- *When a single bull is used to breed many cows, the gene pool becomes less diverse, which can lead to the accumulation of bad gene effects in offspring.*

68. One credit is allowed for correctly explaining how genetic engineering could be used to improve the chance that more of Chief's offspring would survive. Acceptable responses include but are not limited to: [1]

- *The lethal gene could be replaced with a healthy one in Chief's gametes.*

- *Genes that increase the risk of death could be detected and removed from Chief's sperm cells.*

- *Geneticists could identify the lethal gene and then take steps to remove it and substitute a normal gene in the DNA of Chief's sperm cells to be used for artificial insemination breeding techniques.*

69. One credit is allowed for correctly explaining how an elephant without the ability to grow tusks could be born into a population of elephants that all have tusks. Acceptable responses include but are not limited to: [1]

- *The baby elephant may have inherited a mutation that prevented it from growing tusks.*

- *Both of the baby elephant's parents contributed a recessive gene for no tusks.*

- *Genetic recombination can accumulate the effects of mutations in populations where crossbreeding of related organisms is common.*

- *A recessive mutation for no tusks that occurred spontaneously in a single gamete of an ancestral elephant was passed on via sexual reproduction to an offspring elephant that produced many gametes containing the mutation. After several generations, crossbreeding of closely related elephant pairs resulted in a baby elephant homozygous recessive for the no tusks trait.*

70. One credit is allowed for correctly explaining why over one-half of the females that survived the war had no tusks. Acceptable responses include but are not limited to: [1]

- *Many tusked elephants died from poaching during the war, leaving an increased proportion of nontusked breeding adults to continue the species. Those nontusked elephants passed on their genes for no tusks to succeeding generations.*

- *Poaching during the war acted as a selection pressure on this isolated population of elephants, eliminating animals carrying the tusk gene and increasing the frequency of the no tusk gene in the elephants' gene pool.*

- *Elephants with tusks were killed by poachers, leaving more tuskless elephants to pass on their genes for no tusks to offspring.*

- *Most of the surviving elephants did not have tusks, so they were able to pass on their genes for tusklessness to their offspring.*

71. One credit is allowed for correctly explaining why so many (33%) of the female elephants born in the years after the war have no tusks. Acceptable responses include, but are not limited to: [1]

- *Many of the mothers had no tusks, so they inherited that trait.*

- *Most of the surviving females had no tusks, so their female offspring were more likely to inherit the trait from them.*

- *Tuskless females made up more than 50% of the females in the population after the war, so the no tusk gene had increased in frequency in the gene pool relative to the tusk gene.*

- *The reduced elephant population made it more probable that closely related elephants that were carriers of the recessive no tusks gene would mate and produced homozygous recessive offspring displaying the phenotype.*

72. One credit is allowed for correctly explaining why tuskless males are rarely seen, even without poaching being a factor. Acceptable responses include but are not limited to: [1]

- *Since they are males, being unable to grow tusks would put them at a disadvantage when competing with tusked males for breeding partners.*

- *Males without tusks would be more likely to be injured or killed by rivals/predators.*

- *Males lacking tusks might not be selected as mates by breeding females, so their gene for no tusks would not be passed on to offspring.*

- *It is possible that the trait for tusklessness is controlled by a recessive sex-linked gene occurring on the X chromosome but not the Y chromosome, causing the tuskless phenotype to appear only in females homozygous recessive for the trait.*

- *It may be that the gene for no tusks is lethal in males, usually killing them as embryos or calves, before they can develop into breeding adults.*

PART D

73. **1** In this experiment, the dependent variable is the *heart rate of the audience members*. In a scientific experiment, the dependent variable is the component that varies in response to an independent variable controlled by the experimenter. A measured value (in this case, relative heart rates) for the dependent variable is recorded by the experimenter as the experiment progresses.

WRONG CHOICES EXPLAINED:
(2) In this experiment, the dependent variable is *not* the *scene being viewed by the audience*. Although it is not explicitly stated in the description, it seems logical to identify this as the independent variable, not the dependent variable, in this experiment since the scenes vary under the control of the experimenters as a function of their selection of the movie shown to the audience.

(3), (4) In this experiment, the dependent variable is *not* the *amount of time the movie played* or the *number of viewers with heart-rate monitors*. These factors remain the same for the duration of the experiment for all experimental subjects. So they represent constants, not variables, for this experiment.

74. **4** *Watching scary movies will increase the heart rates of audience members* is a possible hypothesis being tested in this experiment. A hypothesis is the experimenter's informed prediction (educated guess) as to the expected outcome of a particular experiment. In this case, the experimenter is making an informed prediction that the audience's average heart rates will measure higher during the movie than in the 10-minute silent period prior to it. The data recorded in the scatter plot seem to indicate that this hypothesis was at least partially supported.

WRONG CHOICES EXPLAINED:
(1) *Silence in theaters increases the heart rates of the audience members* is *not* a possible hypothesis being tested in this experiment. The passage states that a 10-minute silent period was imposed on the audience prior to the movie start. The data collected during this period provide a baseline for audience heart rates that can be compared to their rates as measured during the movie. This represents a control, not a hypothesis, for this experiment.

(2) *The length of a movie causes changes in heart rate* is *not* a possible hypothesis being tested in this experiment. The movie length was the same for all audience members. This represents a constant, not a hypothesis, for this experiment.

(3) *Do heart rates increase when watching a scary movie?* is *not* a possible hypothesis being tested in this experiment. This represents a question that might have been asked, not a hypothesis that might have been proposed, by the experimenters prior to developing a hypothesis and designing/executing this experiment.

75. 3 *Area swabbed with solution* B *remained clear* is an observation that supports the student's conclusion that solution *B* contained a chemical that digests starch. The clear gel in the shape of a "?" in the Petri dish swabbed with solution *B* indicates that the starch molecules suspended in the gel at that point had been digested into simple sugars by a chemical in solution *B*. Areas in both Petri dishes that turned blue when tested with starch indicator still contained starch molecules, so they were unaffected by either solution *A* or *B*.

WRONG CHOICES EXPLAINED:

(1) *Damp cotton swab absorbed some of the starch where it touched the gel* is *not* an observation that supports the student's conclusion that solution *B* contained a chemical that digests starch. This observation would not explain why the clear area occurred in Petri dish *B* but not in Petri dish *A*.

(2) *Starch indicator changed the color of the gel to blue* is *not* an observation that supports the student's conclusion that solution *B* contained a chemical that digests starch. This observation indicates only that the starch suspended in both Petri dish gels reacted the same way to the indicator used, by turning blue.

(4) *Chemical in the starch indicator reacted with the chemical in* B is *not* an observation that supports the student's conclusion that solution *B* contained a chemical that digests starch. Starch indicator is specifically designed to react with starch molecules, not with other chemicals, to produce a characteristic blue color. The chemical in solution *B* reacted with the starch molecules by digesting them into simple sugars.

76. 2 Row *2* is the row in the chart that correctly identifies one of these finches. A review of the information provided in the diagram shows that the warbler finch has a probing bill and eats insects, which matches the information provided in the chart in row *2*.

[NOTE: The photographs marked A–D provided in the question represent superfluous and distracting information that is not needed to answer this question correctly.]

WRONG CHOICES EXPLAINED:

(1) Row *1* is *not* the row in the chart that correctly identifies one of these finches. A review of the information provided in the diagram shows that the large ground finch has a crushing bill and eats seeds, which does not match the information provided in the chart in row *1*.

(3) Row *3* is *not* the row in the chart that correctly identifies one of these finches. A review of the information provided in the diagram shows that the cactus finch has a probing bill and eats cactus, which does not match the information provided in the chart in row *3*.

(4) Row *4* is *not* the row in the chart that correctly identifies one of these finches. A review of the information provided in the diagram shows that the small ground finch has a crushing bill and eats seeds, which does not match the information provided in the chart in row 4.

77. One credit is allowed for correctly identifying group *A* as the group that had most likely been placed in distilled water and supporting the answer. Acceptable responses include but are not limited to: [1]

   ■   *Group A: Distilled water moves with the concentration gradient into cells and makes them swell up or even burst.*

   ■   *Group A: Distilled water has a higher water concentration than the cytoplasm of red blood cells, so water will move into these cells, causing them to get larger.*

   ■   *Group A: Osmosis pushed water molecules across the cell membrane into these cells, causing them to swell and burst.*

   ■   *The cells in group A are illustrated as getting larger/bursting, so they must be absorbing water from the surrounding distilled water.*

78. One credit is allowed for correctly predicting that the cell in the beaker with 25% sugar solution would *not* have the greatest increase in mass and supporting the answer. Acceptable responses include but are not limited to: [1]

   ■   *No, the prediction is not correct because the sugar concentrations are equal so no net change in mass would occur.*

   ■   *Since the concentrations are equal, there will be no change in the mass of the cell.*

   ■   *Since the cell and solution have equal concentrations, they are balanced.*

   ■   *The student is wrong. The cell and the solution are both at 25% sugar concentration, so no concentration gradient exists.*

   ■   *No, because the beaker with the 0% solution will have the greatest change as water moves into the cell.*

79. One credit is allowed for correctly identifying *one* waste product that is more effectively removed from muscle cells as a result of increased pulse rate. Acceptable responses include but are not limited to: [1]

   ■   *Carbon dioxide/CO_2*

   ■   *Lactic acid/$C_3H_6O_3$*

   ■   *Water/H_2O*

- *Urea/$CO(NH_2)_2$*
- *Salts*
- *Heat*

80. One credit is allowed for correctly identifying the unknown individual as *D* and supporting the answer. Acceptable responses include but are not limited to: [1]

- *The bars from the DNA of individual* D *all match those of the unknown individual.*

- *Individual* D: *The DNA fragments placed in the well moved through the gel with the electric current and were deposited in the same pattern as the unknown individual.*

- *The unknown individual and individual* D *must be the same person because the DNA fragments were distributed in the gel in exactly the same way.*

81. **2** Before conducting an electrophoresis procedure, enzymes are added to DNA in order to *cut the DNA into fragments*. The enzymes referenced in the question are known as restriction enzymes. They are used to target and break specific chemical bonds at base sequence–specific sites in the DNA molecule. This procedure results in DNA fragments of different sizes with a predictable base sequence at each end.

WRONG CHOICES EXPLAINED:
(1), (3), (4) It is *not* true that, before conducting an electrophoresis procedure, enzymes are added to DNA in order to *convert the DNA into gel*, to *remove smaller DNA fragments from the samples*, or to *synthesize larger fragments of DNA*. None of these things happen at any point in the process of gel electrophoresis, so none is the result of adding enzymes to DNA.

82. **1** When describing finches with flying speeds in the region indicated by the *X* on the graph, it would be accurate to say that these individuals are more likely to *reproduce and increase the frequency of fast finches in the population*. Assuming that there are no negative effects of displaying the trait of fast flying, having the trait would allow the fast-flying finches to more readily escape from fast-flying predators. This ability would allow more of these finches to survive predation and to produce offspring that also display this trait.

WRONG CHOICES EXPLAINED:

(2) When describing finches with flying speeds in the region indicated by the X on the graph, it would *not* be accurate to say that these individuals are more likely to *survive and undergo mutations that increase their flying speeds.* Mutations are random events in which changes occur in the DNA of the cell as a result of exposure to mutagenic agents (e.g., radiation, chemicals, viruses). Mutations are rare, must occur in germ tissue to be passed on, and have unpredictable and often harmful effects on the cell. Finches are unlikely to increase their flying speeds by way of random mutation.

(3) When describing finches with flying speeds in the region indicated by the X on the graph, it would *not* be accurate to say that these individuals are more likely to *require less food than the slower finches in the population.* There is no information provided in the question concerning food requirements of finches.

(4) When describing finches with flying speeds in the region indicated by the X on the graph, it would *not* be accurate to say that these individuals are more likely to *produce offspring that fly at average speeds.* Because of the processes of sexual reproduction and genetic recombination, fast-flying finches are more likely to produce offspring that are also fast flyers.

83. One credit is allowed for correctly indicating that changing a sequence of bases in DNA from TGA to TGG would *not* result in a new inheritable trait and supporting the answer. Acceptable responses include but are not limited to: [1]

- *No, this mutation would not cause any change because they both code for the same amino acid/threonine.*

- *This mutation would still result in threonine being put into the amino acid sequence.*

- *No, because DNA codons TGA and TGG both code for the amino acid THR.*

- *There would be no new trait since mRNA codons ACU and ACC both link to tRNA molecules carrying threonine, so the resulting protein would not be altered compared with the original protein.*

84. One credit is allowed for correctly identifying *one* type of molecular evidence that was most likely used to develop this new tree and explaining how this evidence would support the new position of flamingos. Acceptable responses include but are not limited to: [1]

- *Many similar proteins:*
  - *Organisms that produce many of the same proteins are more closely related.*
  - *Similar proteins are produced by organisms that share a common genetic code.*

- *If flamingos synthesize proteins that are more similar to grebes and pigeons than to crows and owls, then flamingos should be classified in group B in the diagram.*

■ *Similar DNA sequence:*

- *The closer the DNA sequence is of two species, the more closely they are related.*

- *The use of DNA sequencing showed that the flamingos were more closely related to the birds in group B, so flamingos were moved to that group.*

- *If DNA sequencing techniques indicate that flamingo DNA is more similar to the birds in group B than to the birds in group A, then they should be classified in the same group as grebes and pigeons.*

■ *Production of similar enzymes:*

- *Organisms that produce many similar enzymes are more likely to be closely related.*

- *Enzymes catalyze specific chemical reactions in cells that result in genetic traits. When two species produce the same enzymes, they are considered to be closely related.*

85. One credit is allowed for selecting *two* bird species from the photos, one from group *A* and one from group *B*, and correctly stating *one* reason why researchers may have originally thought these two species should be on the same branch of the tree. Acceptable responses include but are not limited to: [1]

■ *Crows and pigeons:*

- *These two species have similar beaks.*

■ *Penguins and grebes:*

- *Both of these species have webbed feet to swim in their water environments.*

■ *Finches and grebes:*

- *These birds have similar feather color patterns.*

■ *Egrets and flamingos*

- *Both of these birds have long necks/legs.*

Standards/Key Ideas	June 2022 Question Numbers	Number of Correct Responses
Standard 1		
Key Idea 1: The central purpose of scientific inquiry is to develop explanations of natural phenomena in a continuing and creative process.	47, 60	
Key Idea 2: Beyond the use of reasoning and consensus, scientific inquiry involves the testing of proposed explanations involving the use of conventional techniques and procedures and usually requiring considerable ingenuity.	59	
Key Idea 3: The observations made while testing proposed explanations, when analyzed using conventional and invented methods, provide new insights into natural phenomena.	35, 37, 57, 63, 72	
Laboratory Checklist	34, 45, 46	
Standard 4		
Key Idea 1: Living things are both similar to and different from each other and from nonliving things.	10, 11, 16, 23, 27, 38, 39, 52, 53, 55, 64	
Key Idea 2: Organisms inherit genetic information in a variety of ways that result in continuity of structure and function between parents and offspring.	3, 4, 12, 40, 44, 66, 67, 68	
Key Idea 3: Individual organisms and species change over time.	21, 41, 42, 50, 51, 69, 70, 71	
Key Idea 4: The continuity of life is sustained through reproduction and development.	6, 7, 26, 31, 32	
Key Idea 5: Organisms maintain a dynamic equilibrium that sustains life.	2, 8, 9, 13, 14, 15, 17, 20, 33, 36, 43	
Key Idea 6: Plants and animals depend on each other and their physical environment.	1, 5, 18, 19, 22, 29, 54, 62, 65	
Key Idea 7: Human decisions and activities have a profound impact on the physical and living environment.	24, 25, 28, 30, 48, 49, 56, 58, 61	
Required Laboratories		
Lab 1: "Relationships and Biodiversity"	80, 81, 83, 84, 85	
Lab 2: "Making Connections"	73, 74, 79	
Lab 3: "The Beaks of Finches"	76, 82	
Lab 5: "Diffusion Through a Membrane"	75, 77, 78	

Examination August 2022
Living Environment

PART A

Answer all questions in this part. [30]

Directions (1–30): For *each* statement or question, record in the space provided the *number* of the word or expression that, of those given, best completes the statement or answers the question.

1. Studying fossils provides evidence for evolution because fossils

    (1) take a long time to form

    (2) can show patterns of biological change over time

    (3) always contain complete DNA sequences

    (4) found in the same area are usually closely related to each other   1 _____

2. Which statement best describes the interactions between the structures found within a single-celled organism?

    (1) They allow the organism to maintain homeostasis.

    (2) They prevent homeostasis from damaging the cell.

    (3) They must act independently of each other and prevent homeostasis.

    (4) They carry out the same life process in order to maintain homeostasis.   2 _____

3. Sexually reproduced offspring have traits similar to their parents because they receive

    (1) all of the proteins from each parent

    (2) some of the proteins from both parents

    (3) all of the genes present in both parents

    (4) some of the genes present in each parent   3 _____

4. Which row in the chart below correctly pairs a group of organisms with the type of nutrition they carry out?

Row	Autotrophic Nutrition	Heterotrophic Nutrition
(1)	carnivores	herbivores
(2)	decomposers	carnivores
(3)	herbivores	producers
(4)	producers	decomposers

4 _____

5. Rubber usually comes from petroleum or from the Asian rubber tree plant. Scientists have modified a single trait in the domestic plant, guayule, to increase its ability to produce rubber for commercial use. Young guayule plants are shown in the photograph below.

Source: http://agresearchmag.ars.usda.gov

The process that was most likely used to modify the plants' trait and increase their natural rubber production was

(1) selective breeding of two similar plant varieties

(2) genetic recombination during sexual reproduction

(3) genetic engineering to alter a specific gene

(4) fertilizing the plants with key substances found in petroleum   5 _____

6. A response of a normally functioning immune system that can be harmful is

(1) being infected by the flu virus

(2) rejecting an organ transplant

(3) recognizing chemical signals

(4) fighting off a bacterial infection   6 _____

7. Which molecules are normally found in single-celled organisms?

(1) organic molecules, only

(2) inorganic molecules, only

(3) both organic and inorganic molecules

(4) neither organic nor inorganic molecules   7 _____

8. Hammerhead sharks are unlike most other shark species. Nearly all shark species either lay eggs or give birth to live young after their eggs hatch internally. Hammerhead sharks form a placenta, a structure more commonly found in mammals, such as humans. One role of the placenta in the development of offspring is normally to

(1) produce blood cells   (3) produce gametes

(2) provide milk   (4) transfer nutrients   8 _____

9. PCR, Polymerase Chain Reaction, is a method for carrying out DNA replication. In order to perform this technique, a scientist would need

(1) a DNA template, ATP, and 20 different amino acid subunits

(2) enzymes, several types of simple sugars, and starch molecules

(3) a DNA template, enzymes, and subunits with A, G, T, and C bases

(4) enzymes, specific receptor molecules, and several hormones   9 _____

10. A student used a microscope to examine some cells. He observed strands located in the nuclei of these cells.

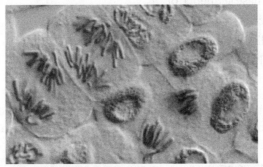

Source: https://www.icr.org/i/wide/mitosis_wide.jpg

These strands are responsible for coding different proteins and are known as

(1)   chromosomes   (3)   ribosomes

(2)   mitochondria   (4)   chloroplasts   10 _____

11. Farmers have been planting crops that express an insecticide gene, so that when pests consume these crops, the pests are poisoned. Unfortunately, since these plants were introduced in 1996, growing numbers of insect pests have developed resistance to the insecticide. The process that led to the insect resistance can best be explained by

(1)   ecological succession

(2)   selective breeding

(3)   asexual reproduction

(4)   natural selection   11 _____

12. Killer whales are an endangered species. The decline in the whales' numbers has been linked to poor nutrition, resulting in the inability to maintain a pregnancy. This risk to developing whale embryos is most likely a result of

(1)   an environmental factor not associated with the embryo's genes

(2)   an infection caused by the embryo's exposure to a pathogen

(3)   faults in the genes of the embryo itself

(4)   toxins that are introduced into the mother from the embryo's blood   12 _____

13. A biotechnology tool, known as CRISPR-Cas9, allows scientists to precisely edit genes. In order to edit genes, CRISPR-Cas9 must be able to

   (1) alter the base sequence of DNA

   (2) prevent cells from differentiating

   (3) block cell receptors from receiving signals

   (4) change the rate at which a cell uses ATP   13 _____

14. By measuring the colors of light reflected by different tree species in a forest, scientists can determine the amount of biodiversity present in different areas. Maintaining biodiversity is important because it

   (1) reduces the carrying capacity of a forest ecosystem

   (2) guarantees that all species within a forest ecosystem will survive

   (3) increases the number of predators that control the population size of prey

   (4) ensures the availability of a variety of genetic material   14 _____

15. The cells in the diagram below were present in the same individual.

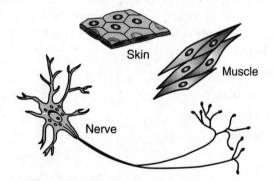

These cells are most similar in the

   (1) amount of energy they release

   (2) type of proteins they synthesize

   (3) rate of their metabolism

   (4) information stored in their DNA   15 _____

16. Large numbers of white-tailed deer on Long Island are infested with ticks that transmit Lyme disease to other mammals. One attempt to control reproduction in these ticks has been the release of large numbers of sterilized male ticks. When compared to using pesticides, this method to control ticks would

(1) cause more environmental pollution

(2) lead to a decrease in the deer population

(3) be less likely to harm the environment

(4) result in an increase in the tick population   16 _____

17. The model below summarizes one pathway of energy transfer in an ocean ecosystem.

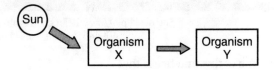

The type of organism represented by box *X* could be

(1) algae

(2) fungi

(3) small fish

(4) sea birds   17 _____

18. Which sequence best represents the correct order of events in the formation of a sexually reproduced individual?

(1) embryo → zygote → gamete → fetus

(2) zygote → embryo → fetus → gamete

(3) gametes → embryo → fetus → zygote

(4) gametes → zygote → embryo → fetus   18 _____

19. Direct harvesting occurs when

(1) pine trees are cut from a forest for use as lumber

(2) corn is planted in a newly plowed field

(3) zebra mussels are accidentally imported to the Great Lakes

(4) roots of plants continually take in water   19 _____

20. In New York State, it is common for farmers to plant large fields of one crop, such as the cornfield shown below.

Source: https://www.123rf.com/photo_
40944515_corn-fields.html

A *negative* outcome of this practice is that

(1) the corn will interbreed with weeds in the area

(2) new predators will be introduced into the ecosystem

(3) the stability of the ecosystem will be reduced

(4) new species of insect-resistant corn will evolve   20 _____

21. The process of differentiation is best described as the

(1) production of a genetically identical copy of an organism

(2) change in shape of a protein due to high temperatures

(3) process by which cells specialize and develop into a specific type of cell

(4) process in which genes are made and transferred into other organisms   21 _____

22. Humans are able to positively or negatively affect their environment in many ways. Which statement accurately describes *one* of these possible effects?

(1) A positive environmental effect is that burning fossil fuels to generate electricity reduces carbon dioxide levels in the atmosphere.

(2) A positive environmental effect is the cutting of trees in rain forests to provide large quantities of lumber to build homes for the increasing world population.

(3) A negative environmental effect is that industrialization provides many jobs and helps the economy grow.

(4) A negative environmental effect is that unregulated fishing in the ocean can disrupt the interactions between organisms in existing food webs.   22 _____

23. Which statement best describes the process of competition?

(1) It may be for abiotic or biotic resources.

(2) It is not affected by changes in the environment.

(3) It always occurs between members of different species.

(4) It allows nutrients in an ecosystem to move from herbivores to autotrophs.   23 _____

24. Changes in external temperatures often result in a person either sweating or shivering, as represented in the diagram below.

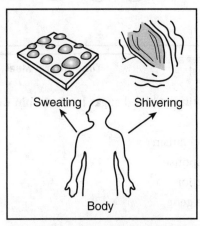

Source: Adapted from http://askabiologist.
asu.edu/sites/default/files/resources/
articles/singing_in_rain/temp_control.gif

These responses are one way

(1) to counteract feedback mechanisms that would otherwise be beneficial

(2) to make the body release insulin to control blood circulation

(3) the body is able to maintain dynamic equilibrium

(4) skin and muscle cells are able to disrupt homeostasis   24 _____

25. Which statement about the response of the body to pathogens is correct?

(1) Red blood cells engulf invaders and produce antibodies that attack invaders.

(2) Vaccinations may contain weakened microbes that stimulate the formation of antibodies.

(3) AIDS is a bacterial disease that strengthens the immune system.

(4) All allergic reactions are caused by an immune response to microorganisms.   25 _____

26. Blood sugar levels increase after a meal and eventually return to normal. This is represented in the diagram below.

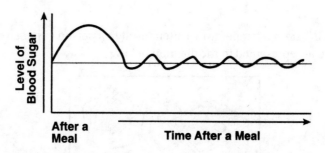

This constant correcting of blood sugar levels within the body is accomplished by

(1) a feedback mechanism

(2) an immune response

(3) an allergic reaction

(4) manipulating a gene   26 _____

27. Sheets of skin are grown in a culture in order to replace the skin of victims with severe burns or frostbite. Undamaged skin cells are obtained from the victim, put in a Petri dish with the proper growth materials, and incubated, as represented in the diagram below.

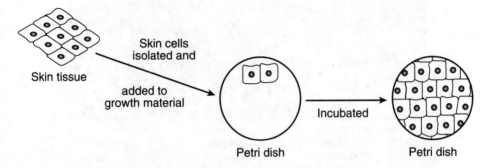

These new skin cells form as a result of

(1) meiotic cell division

(2) sexual reproduction

(3) mitotic cell division

(4) gene recombination   27 _____

28. Which graph below best represents the relationship between the relative number of nuclei, genes, and chromosomes in a typical human cell?

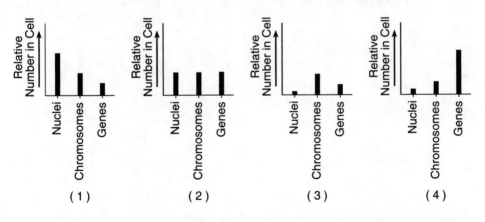

28 _____

29. The diagram below represents a biological process.

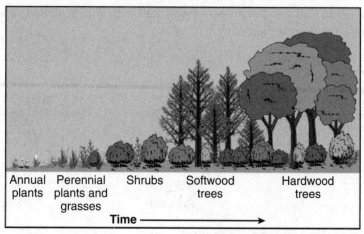

Annual plants   Perennial plants and grasses   Shrubs   Softwood trees   Hardwood trees

Time ⟶

Source: Adapted from http://www.physicalgeography.net/fundamentals/9i.html

Which statement is true about the biological process shown?

(1) This is a short-term process resulting from sudden changes.

(2) This process cannot be altered by humans and other organisms.

(3) If the hardwood trees are destroyed, the altered ecosystem cannot recover.

(4) The shrubs modify the environment, making it more suitable for the softwood trees.

29 _____

30. Cells may divide abnormally and produce cells like some of those shown in the photograph below.

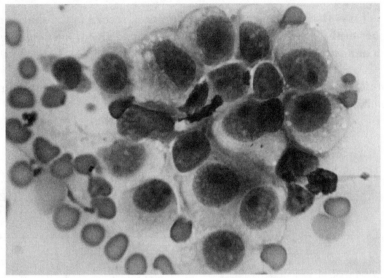

Source: www.popsci.com/July 2018

When cells such as the skin cells shown reproduce abnormally, it could be a sign of

(1) an immune response

(2) dynamic equilibrium

(3) cancerous cell growth

(4) a cellular adaptation

30 _____

PART B-1

Answer all questions in this part.   [13]

Directions (31–43): **For *each* statement or question, record in the space provided the *number* of the word or expression that, of those given, best completes the statement or answers the question.**

Base your answers to questions 31 and 32 on the diagram below and on your knowledge of biology.

Sizes of Various Structures and Ways to View Them

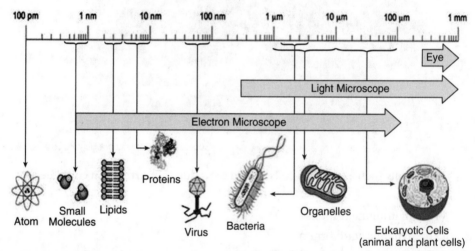

Source: Adapted from https://microbiologyinfo.com/different-size-shape-and-arrangement-of-bacteria-cells/

31. Only an electron microscope can be used to view

(1) bacteria   (3) animal cells

(2) mitochondria   (4) viruses   31 _____

32. A scientist is developing a system to remove harmful bacteria from a contaminated water supply. In order to trap the bacteria and prevent them from going through the filter, she must make sure the pores in the filter are no larger than

(1) 1 nm   (3) 10 μm

(2) 1 μm   (4) 100 μm   32 _____

33. For native human populations in tropical areas, the intensity of ultraviolet (UV) rays from the Sun is strong, and skin color is generally dark. Melanin pigments found in people with darker skin color help block the effects of the UV radiation on skin cells.

In tropical areas, the best explanation for having increased melanin in human skin cells is that it

(1) increases the occurrence of mutations

(2) provides a survival advantage

(3) acts as a feedback mechanism to increase UV exposure

(4) produces antibodies that destroy pathogens   33 _____

34. Smoking increases the risk of certain cancers of the mouth, esophagus, pancreas, kidneys, and uterus. This finding would be most reliable if it were based on

(1) data collected from patients in one cancer-research hospital

(2) research done by scientists in many different countries

(3) reading the information on cigarette cartons

(4) cancer information published on social media sites   34 _____

35. Which diagram below best represents the direction that energy flows through an energy pyramid?

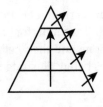

(1)

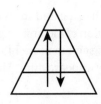

(3)

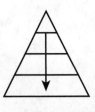

(2)

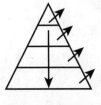

(4)   35 _____

36. An evolutionary tree is shown below.

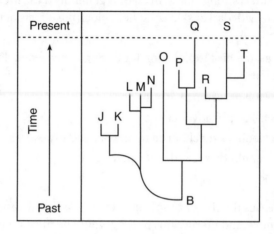

Which conclusion is correct, based on the evolutionary tree?

(1) All of these species have certain DNA sequences in common.

(2) Species *S* is the best adapted of all the species shown.

(3) A common ancestor of species *L* and *M* is species *N*.

(4) Species *O* and *P* are more closely related than species *P* and *Q*.   36 _____

Base your answers to questions 37 and 38 on the information and photographs below and on your knowledge of biology.

An arctic fox has a gland in its brain that secretes a hormone that regulates the production of melanin, a pigment that accounts for brown fur. In the winter, the foxes secrete more of this hormone and their cells stop making melanin, so they appear white. The pictures below illustrate two variations of fur color.

Source: http://www.nationalgeographic.com/animals/mammals/a/arctic-fox/

37. Which *two* rows best support the information provided?

A	winter	increased melanin	white fur
B	summer	increased melanin	brown fur
C	winter	decreased melanin	white fur
D	summer	decreased melanin	brown fur

(1) *A* and *B*

(2) *B* and *C*

(3) *C* and *D*

(4) *D* and *A*   37 _____

38. Which statement is the most likely explanation for the color differences in the fur of the fox at different times of the year?

(1) Mutations can be caused by changes in the number of biotic factors in the environment.

(2) The expression of genes can be modified by the external environment.

(3) Hereditary information is contained in genes located in the chromosomes of each cell.

(4) Random changes in DNA can occur to change the expression of a gene.   38 _____

Base your answer to question 39 on the information below and on your knowledge of biology.

Fighting the Flu

A new technique to attack flu virus antigens is being tested on mice. Normally, antibodies attack the "head" portion of antigens on the surface of the flu virus.

Since the "head" portions mutate frequently, the antibodies do not provide protection for very long. The new technique is to develop antibodies that attack the "stem" portion of the antigen. Since the "stem" regions do not mutate very often, the effectiveness of the vaccine should last longer. This technique is represented below.

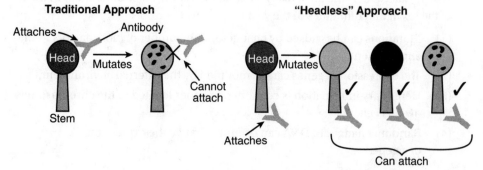

Source: Adapted from www.sciencenews.org/
"A Universal Flu Shot May Be Nearing Reality"/October 28, 2017.

39. Which statement describes an observation that would best support the continued study of using antibodies produced by this new technique against the flu?

(1) A group of 50 mice with flu antibodies formed using the new technique were exposed to mutated forms of the flu. None of the mice became ill.

(2) The use of these antibodies in mice stopped mutations that occur in flu viruses.

(3) Chemical tests showed that the stem antibodies attached to the heads of some flu viruses and destroyed them.

(4) Blood tests showed that only "stem" antibodies attacking the stem of flu antigens can cause the flu in mice. Those attacking the "head" did not.

39 _____

40. Scientists have discovered that pathogenic organisms and the chemicals they produce can cause foodborne illnesses. These illnesses harm the body as a result of interactions between the digestive and immune systems.

Which statement most correctly describes how these two systems interact when an individual comes down with a foodborne illness?

(1) Chemicals produced by pathogens enter the immune system through a cut in the skin. The circulatory system carries the chemical to the digestive system, resulting in foodborne illness.

(2) When specific chemicals produced by pathogens enter the digestive system in contaminated foods, the ability of the immune system to fight off foodborne illness is reduced.

(3) When foods contaminated with pathogens are eaten, the immune system prevents the pathogens from entering the digestive system.

(4) The digestive system breaks down the pathogens in the contaminated foods so that they are harmless. These harmless pathogens are then transferred to the immune system.   40 _____

Base your answers to questions 41 and 42 on the information and diagram below and on your knowledge of biology.

A live plant was placed in a closed container in a lab. Sensors were set up to monitor the levels of oxygen in the container over several hours.

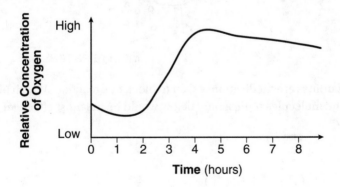

41. At which hour were the lights turned on in the lab?

(1) 8   (3) 0

(2) 2   (4) 4   41 _____

42. During the 8 hours studied, the plant performed

    (1)   photosynthesis, only

    (2)   respiration, only

    (3)   both photosynthesis and respiration

    (4)   neither photosynthesis nor respiration   42 _____

43. Human cells have many molecules attached to their surfaces. Some of these molecules are involved in producing the symptoms associated with allergies. Histamine is a chemical produced by some human cells. When histamine binds to molecules on the surface of cells that line the nose and throat, the cells will swell and leak fluid, causing the characteristic itching, sneezing, and congestion associated with allergies. A model of this mechanism is represented below.

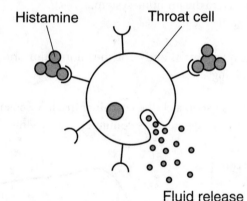

Fluid release

Antihistamines are medications taken to block this reaction. Which of the anti-histamine molecules represented below would be the most effective?

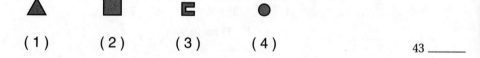

    (1)   (2)   (3)   (4)   43 _____

PART B-2

Answer all questions in this part.   [12]

Directions **(44–55): For those questions that are multiple choice, record your answers in the spaces provided. For all other questions in this part, record your answers in accordance with the directions.**

Base your answers to questions 44 through 48 on the information and data table below and on your knowledge of biology.

West Nile Virus

West Nile virus (WNV) has been detected in a variety of bird species. Crows and jays are known to get sick and die when infected. WNV also infects other animals, including horses, cats, dogs, chipmunks, alligators, and humans.

WNV affects the nervous system. It was first detected in the U.S. in New York City during the summer of 1999, when nearly 5500 crows died within a four-month period. Since then, WNV has spread rapidly throughout the country. Although the virus is widespread, symptoms in humans are usually mild. However, about 1 in 150 people who are infected develop severe, sometimes fatal, symptoms that include the inflammation of the brain and membranes surrounding the brain and spinal cord. There is no human vaccine for WNV. The virus is transmitted to certain bird species when they are bitten by infected mosquitoes. When these bird species are not available, mosquitoes are more likely to bite humans. Humans are a dead-end host, which means that even when infected with the virus, it is not passed on.

The Centers for Disease Control and Prevention (CDC) recorded the number of cases of WNV per 100,000 people in the U.S. from 2002 to 2014. These data are recorded in the table below.

| Incidence of West Nile Virus in the U.S. per 100,000 People ||
Year	Cases per 100,000 People
2002	1.02
2004	0.39
2006	0.50
2008	0.23
2010	0.20
2012	0.91
2014	0.42

Source: https://www.epa.gov/climate-indicators/
climate-change-indicators-west-nile-virus

Directions (44–45): **Using the information in the data table, construct a line graph on the grid provided, following the directions below.**

44. Mark an appropriate scale, without any breaks in the data, on each labeled axis. [1]

45. Plot the data for the incidence of West Nile virus in the U.S. per 100,000 people. Connect the points and surround each point with a small circle. [1]

Example:

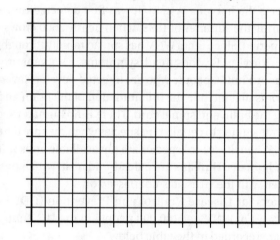

Incidence of West Nile Virus in the U.S. per 100,000 People

Cases per 100,000 People (y-axis)

Year (x-axis)

46. Based on the data, is it possible to predict what the number of cases per 100,000 people will be for the year 2020? Support your answer with data from the graph. [1]

Note: The answer to question 47 should be recorded in the space provided.

47. The two maps below show the number of human cases of West Nile virus per 100,000 people for the years 2007 and 2016.

West Nile Virus Neuroinvasive Disease Incidence Reported to ArboNET, by State, United States

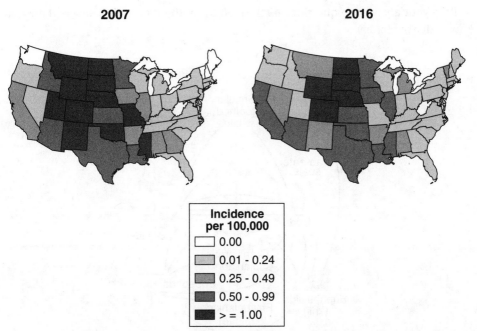

2007   **2016**

Incidence per 100,000
- 0.00
- 0.01 - 0.24
- 0.25 - 0.49
- 0.50 - 0.99
- >= 1.00

Source: http://www.cdc.gov/westnile/resources/pdfs/data/2007StateincidenceMap.pdf
Source: http://www.cdc.gov/westnile/resources/pdfs/data/WNV-Neuro-Incidence-by-State-Map_2016_09292017.pdf

The data represented on the maps best indicate that

(1) birds have spread WNV to every state in the United States

(2) New York State has the highest rate of WNV infection for both of the years shown

(3) once WNV reaches a state, the number of people infected increases every year

(4) for any given year, it is difficult to know which states will have the greatest number of cases

47 _____

48. Explain why some people may be more severely affected by West Nile virus than others.   [1]

_____

_____

_____

Base your answers to questions 49 through 51 on the food web below and on your knowledge of biology.

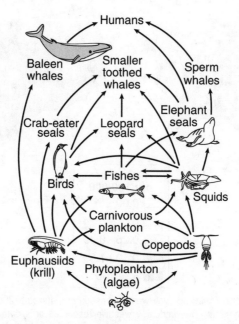

Note: The answer to question 49 should be recorded in the space provided.

49. Based on the food web, the population that contains the greatest amount of available energy would be

(1) seals

(2) fishes

(3) phytoplankton

(4) humans   49 _____

Note: The answer to question 50 should be recorded in the space provided.

50. Which statement best describes what would happen in this ecosystem if the phytoplankton were removed from the food web?
    (1) Copepods and krill would fill the vacant niche.
    (2) The number of heterotrophs would increase.
    (3) The food web would be disrupted, and organisms would die.
    (4) The food web would remain stable.   50 _____

51. Describe the relationship represented by the arrows between squids and fishes.   [1]

_____

_____

—————————————

Base your answer to question 52 on the information below and on your knowledge of biology.

Scientists are interested in studying the effects of a mother's alcohol consumption on the brain development of the fetus during pregnancy. In order to collect data, scientists typically use newborn rats because the rats' brain development after birth is roughly equivalent to that of a human fetus during the third trimester (late in pregnancy). Scientists divided newborn rats into four groups and exposed them to alcohol using the following methods:

Alcohol Exposure in Newborn Rats	
Rat Group	**Alcohol Exposure**
1	No alcohol exposure
2	4.5 g/kg/day given over a 4-hour period
3	4.5 g/kg/day given over an 8-hour period
4	6.6 g/kg/day given over a 24-hour period

At the end of the experiment, scientists measured the total brain weight of the newborn rats, as represented in the graphs below.

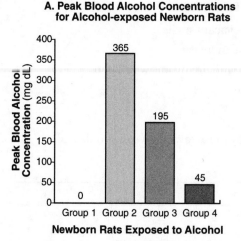

A. Peak Blood Alcohol Concentrations for Alcohol-exposed Newborn Rats

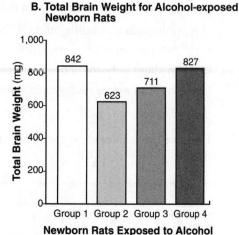

B. Total Brain Weight for Alcohol-exposed Newborn Rats

Source: Adapted from https://pubs.niaaa.nih.gov/publications/arh25-3/168-174.htm

52. State the relationship between peak blood alcohol concentration and total brain weight for alcohol-exposed newborn rats.   [1]

_____

_____

Base your answers to questions 53 and 54 on the diagram below and on your knowledge of biology. The diagram represents a food web in a forest ecosystem.

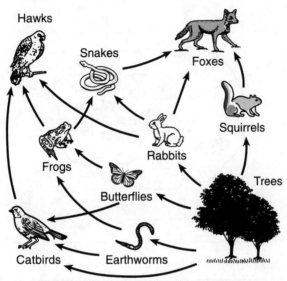

Source: Adapted from The American Biology Teacher
vol. 78, No. 7, September 2016, p. 577

53. A student claims that this food web represents a stable ecosystem. State whether or not her claim is correct. Support your answer. [1]

_____

_____

54. Select *two* organisms from this food web that compete with each other for food, and state *one* reason why they are able to survive in the same ecosystem. [1]

Organisms: _____ and _____

_____

_____

55. In the cell below, identify *both* the number and name of the structure in the cell that produces proteins. [1]

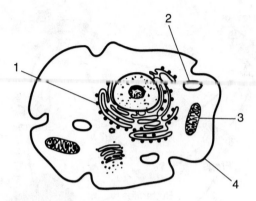

Number of structure: _____

Name of structure: _____

_____

PART C

Answer all questions in this part.   [17]

Directions (56–72): **Record your answers in the spaces provided.**

Base your answer to question 56–57 on the information below and on your knowledge of biology.

Fragile X Syndrome

Fragile *X* syndrome is a genetic disorder associated with a mutation in a gene located on a particular human chromosome called the *X* chromosome. The disorder results in a critical protein, FMR1, not being produced. The normal FMR1 protein helps regulate the production of other proteins that play a role in the development of nerve cells. The situation is summarized below:

*Normal X chromosome → normal FMR1 protein produced →
nerve cell development is regulated*

*Abnormal fragile X chromosome → no FMR1 protein produced →
nerve cell development is unregulated*

56–57. Explain how the mutation in the fragile *X* chromosome affects the body. In your answer, be sure to:

- state *one* specific reason why the mutated gene on the fragile *X* chromosome is unable to produce the FMR1 protein.   [1]

- explain why children with fragile *X* syndrome would often have learning disabilities, including speech and language problems and intellectual disabilities.   [1]

_____

_____

_____

_____

_____

_____

_____

Base your answers to questions 58 and 59 on the information below and on your knowledge of biology.

The chart shows the reproductive characteristics of three species living in an area that has recently undergone a major environmental change.

Species	Method of Reproduction	Frequency of Reproduction	Average Number of Offspring Produced Each Time
A	Asexual	Every two days	2
B	Sexual	Every two years	4
C	Sexual	Every year	20

58. Explain why species *C* might have a greater chance of avoiding extinction in the changed environment than species *B*. Support your answer.   [1]

_____

_____

59. State *one* possible reason why species *A* could be the most successful in surviving an environmental change. Support your answer.   [1]

_____

_____

Base your answers to questions 60 through 63 on the passage below and on your knowledge of biology.

Plastic Bags Everywhere!

As of 2016, Americans used approximately 100 billion plastic bags annually. An average family brought home about 1500 plastic bags a year. Less than 1% of those bags were returned for recycling. Therefore, most of the bags ended up in landfills, where it takes anywhere from months to hundreds of years for them to be broken down. These growing landfills are destroying natural habitats. Many of the bags also make their way into oceans where, if mistaken for food, they can cause animals to choke or starve to death.

A group of researchers in Europe discovered that wax moth caterpillars could break the chemical bonds in polyethylene, a polymer used to produce plastic bags and other products. Though the scientists don't know the exact chemical that the caterpillar is using to break down the plastic, they predict it is an enzyme. Once they isolate the chemical, scientists may be able to mass-produce the chemical in order to break down the plastic bags accumulating in the environment.

60. State *one negative* effect the overuse of plastic bags is having on the environment.   [1]

_____

_____

61. Explain why the researchers suspect it is an enzyme that is enabling wax moth caterpillars to break down the plastic bags.   [1]

_____

_____

62. Explain why using the chemical produced by the caterpillars to break down plastic bags could be considered an ecologically friendly solution to the problem.   [1]

_____

_____

63. Suggest a plan of action that could be carried out in your local community, which would be a step toward solving the plastic bag problem.   [1]

_____

_____

_____

Base your answers to questions 64 and 65 on the information below and on your knowledge of biology.

Plants Clean Up Mining Wastes

The mining of certain metal ores, such as copper and lead, can result in the contamination of soils. Wastes from the mining process can be toxic to plants and animals in the area. It has been discovered that some species of grass are able to grow in these contaminated areas. These grass plants can actually remove some of the toxic wastes from the soil and accumulate them in their tissues.

Growing these resistant grass plants in contaminated soil, then harvesting them to remove the toxic wastes from the environment, has been suggested as a possible way to clean up these areas.

64. Describe *one* positive and *one* negative outcome of mining metal ores.   [1]

_____

_____

65. Explain why importing grasses to clean up mining wastes in areas where those grasses do not normally grow could lead to unexpected environmental problems.   [1]

_____

_____

_____

66. Today, many diseases have been linked to mutations that cause mitochondria to fail. Patients who suffer from mitochondrial diseases may suffer from fatigue and weakness. Explain why patients with a mitochondrial disease would tend to experience these symptoms. [1]

_____

_____

_____

Base your answers to questions 67 through 70 on the information below and on your knowledge of biology.

Lessening Snow Cover Affects Survival of Snowshoe Hare

Snowshoe hares are found in the northern evergreen forests of the United States. The physical characteristics of the hares enable them to hunt for food and hide from their predators during the cold, snow-covered winters. They have large, snowshoe-shaped feet and thick fur. A change in fur color during an annual molt (shedding) occurs before the winter season, causing white fur to replace the brown fur of summer.

The amount of snow cover in these northern forests has decreased in recent years. Research has shown that this decrease has had a significant effect on the snowshoe hare population, even though the carrying capacity of the forests has not changed. Researchers have estimated that for every seven days that snow covers the ground, the snowshoe hare populations are four times more likely to survive.

Since the molt from brown fur to white fur is a response to the decreasing hours of daylight in the fall and not the arrival of snow, the later the snow arrives, the greater the chance that the white hares will be caught by their predators.

The snowshoe hare plays a major role in the stability of these forest ecosystems. Their loss would affect other species such as lynx and great horned owls. If the amount of snow cover continues to decrease, researchers are concerned that it will be harder for the snowshoe hare to survive in their current habitats.

Snowshoe Hare

Source: Science News, April 30, 2016

67. Explain how snow cover affects the population of the snowshoe hare.   [1]

_____

_____

68. Identify the environmental factor that stimulates the fur color of the hare to change from brown to white.   [1]

_____

_____

69. Identify a specific environmental issue that is most likely to affect snowshoe hare populations in northern ecosystems. Support your answer.   [1]

_____

_____

_____

70. Graph *A* below, based on 1985 data, represents the time of the year when the fur color of the snowshoe hares in the northern Rockies does not match the color of their surroundings. The bold line on each graph indicates the period of time that snow covers the ground. The shaded area in the graphs represents this mismatch of color. Graphs *B* and *C* show future projections.

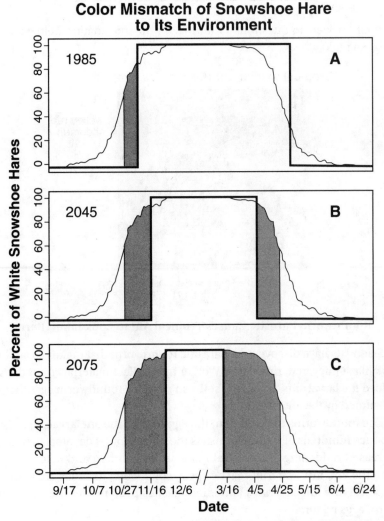

Source: Adapted from L. Scot Mills, et al. PNAS(2013).
DOI: 10.1073/pnas.1222724110

Identify *one* change in the characteristics of the snowshoe hares in this ecosystem that would most likely be selected for if the trend shown in graphs *B* and *C* proves to be true.   [1]

_____

Base your answers to questions 71 and 72 on the information below and on your knowledge of biology.

Tropical worker ant (*Cephalotes atratus*)

_____Swollen red
abdomen

Source: The American Naturalist, 2008,
171:4, 536-544

Roundworm Parasite Causes Tropical Ant to Look Like a Berry

Scientists have discovered a parasitic roundworm that makes its ant host look like a juicy, red, ripe berry. Worker ants collect materials from the soil to feed the larval ants. Often, the soil also contains roundworm eggs that are consumed by the ant larvae.

The roundworms develop from the eggs within the ant larvae, mate, and produce hundreds of roundworms. As the roundworms develop, they cause increased reddening of the developing ant's abdomen and take nutrients from the ant. Just as a fruit reaches peak color when its seeds are ready for dispersal, the infected ant's abdomen reaches peak redness as the roundworm eggs mature.

Birds don't normally eat the foul-tasting ants, but are thought to eat the ants infected with roundworms since they look like red berries. The roundworm eggs move through the bird's digestive system unaffected and pass to the soil in the bird's feces.

71. State *one* reason this roundworm is considered a parasite to this species of tropical ant.   [1]

_____

_____

72. Describe *one* advantage the roundworm has by having birds involved in part of its life cycle.   [1]

_____

_____

PART D

Answer all questions in this part. [13]

Directions (73–85): **For those questions that are multiple choice, record your answers in the spaces provided. For all other questions in this part, record your answers in accordance with the directions.**

Base your answers to questions 73 and 74 on the information below and on your knowledge of biology. The photograph shows two birds on a bird feeder.

Source: http://birdfeederhub.com/best-large-capacity-bird-feeders/

Studies have shown that the length of beaks within a songbird population may be influenced by the presence of bird feeders. When bird feeders were widely used in one area, birds were observed to have longer beaks. In an area where bird feeders were *not* used, the beaks of these species were of average length.

Note: The answer to question 73 should be recorded in the space provided.

73. One possible reason for the increase in beak length is that birds with longer beaks

    (1)   were less likely to have offspring with long beaks

    (2)   had a more successful adaptation for survival in the area

    (3)   needed to reach the seed within the bird feeder, so their beaks grew longer

    (4)   had more competition than other birds at the bird feeders   73 _____

Note: The answer to question 74 should be recorded in the space provided.

74. The presence of bird feeders in an area would represent a

    (1)   selecting agent   (3)   source of mutation

    (2)   feedback mechanism   (4)   biological catalyst   74 _____

Base your answers to questions 75 through 77 on the information below and on your knowledge of biology.

The Elephant Shrew

    The elephant shrew spends its days searching the leaf litter on the forest floor for insect prey. When first discovered, due to structural similarities, the elephant shrew was classified with other shrews and their relatives. However, scientists recently reclassified the elephant shrew, as shown in the evolutionary tree below:

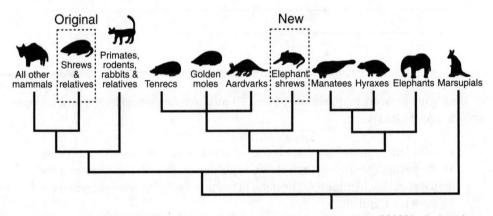

Source: Adapted from http://evolution.berkeley.edu/evolibrary/news/080301_elephantshrew

Note: The answer to question 75 should be recorded in the space provided.

75. The new, more accepted classification of the elephant shrew is most probably based on an analysis of

    (1)   the coloration of the elephant shrew's fur

    (2)   the feeding habits of the elephant shrew compared to other shrews

    (3)   a number of newly found shrew fossils

    (4)   the DNA present in the cells of the elephant shrews   75 _____

Note: The answer to question 76 should be recorded in the space provided.

76. According to the new evolutionary tree, elephant shrews are most closely related to

   (1) manatees and hyraxes

   (2) shrews and their relatives

   (3) tenrecs, golden moles, and aardvarks

   (4) primates, rodents, rabbits, and relatives   76 _____

77. The elephant shrew is at risk for extinction because its habitat is very limited. The elephant shrew can only be found in two forest locations within the country of Tanzania. Even though these locations are protected, they could be harmed by fires and human activity. Explain why it is important to continue to protect the habitat in which the elephant shrew is found.   [1]

   _____

   _____

   _____

   _____

   Base your answers to questions 78 and 79 on the information below and on your knowledge of biology.

   A student hypothesized that the pulse rates of his classmates would increase after walking. The student then obtained pulse rates from five classmates after they walked for 15 minutes. The data, in beats per minute, were recorded as: 78, 68, 84, 88, and 90.

78. Identify the dependent variable in this investigation.   [1]

   _____

79. Identify *one* error in the experimental procedure.   [1]

   _____

   _____

   _____

80. Draw a line on the graph provided that shows the relationship between exercise and oxygen consumption. Support your answer.   [1]

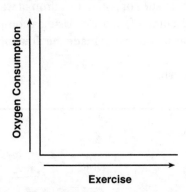

Support: _____

_____

Base your answers to questions 81 and 82 on the information below and on your knowledge of biology.

> A student prepared two potato cubes by cutting 2 cm × 2 cm sections from the same potato. Next, she determined the mass of each of the cubes and recorded the information in her lab notebook.
>
> She then placed one cube in a beaker of distilled water and the other in a beaker with an equal volume of concentrated salt solution. After 20 minutes, she removed both of them from the beakers and again determined the mass of each cube.

Note: The answer to question 81 should be recorded in the space provided.

81. Which statement correctly describes the effect on the mass of one of the cubes after the 20-minute period?

(1) In distilled water, the mass of the potato cube increased due to salt leaving the cells of the potato.

(2) In distilled water, the mass of the potato cube increased due to water moving into the cells from high concentration to low concentration.

(3) In the concentrated salt solution, the mass of the potato cube increased due to salt moving into the cells from low concentration to high concentration.

(4) In the concentrated salt solution, the mass of the potato cube remained the same due to the cell wall preventing the movement of molecules into or out of the cells.   81 _____

Note: The answer to question 82 should be recorded in the space provided.

82. The student placed a thin slice of potato in a drop of water on a glass slide. She added a coverslip and a drop of indicator. Using a compound light microscope, she examined the slide of potato and made the drawing below.

Blue-black-stained structures

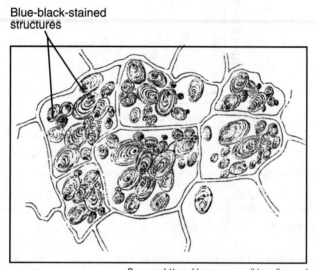

Source: https://commons.wikimedia.org/

The blue-black-stained structures labeled in her drawing are most likely

(1) chloroplasts   (3) ribosomes

(2) starch grains   (4) sugar molecules   82 _____

Base your answers to questions 83 through 85 on the diagram below and on your knowledge of biology.

Tests were performed to help identify the person who committed a crime. Lane *D* contained DNA from evidence found at the crime scene. Lanes *A*, *B*, and *C* contained DNA from each of the three suspects.

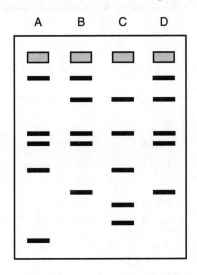

83. Identify the technique used to obtain the results seen in the diagram. [1]

_____

84. Based on the data, which lane most likely contained DNA from the suspect who committed the crime? Support your answer. [1]

_____

_____

85. Identify the lane that contained the band with the shortest fragments of DNA. Support your answer. [1]

_____

_____

Answers August 2022
Living Environment

Answer Key

PART A

1. 2	6. 2	11. 4	16. 3	21. 3	26. 1
2. 1	7. 3	12. 1	17. 1	22. 4	27. 3
3. 4	8. 4	13. 1	18. 4	23. 1	28. 4
4. 4	9. 3	14. 4	19. 1	24. 3	29. 4
5. 3	10. 1	15. 4	20. 3	25. 2	30. 3

PART B-1

31. 4	34. 2	36. 1	38. 2	40. 2	42. 3
32. 2	35. 1	37. 2	39. 1	41. 2	43. 4
33. 2					

PART B-2

44. *See* Answers Explained.

45. *See* Answers Explained.

46. *See* Answers Explained.

47. 4

48. *See* Answers Explained.

49. 3

50. 3

51. *See* Answers Explained.

52. *See* Answers Explained.

53. *See* Answers Explained.

54. *See* Answers Explained.

55. *See* Answers Explained.

PART C. *See* **Answers Explained**.

PART D

73. 2

74. 1

75. 4

76. 3

77. *See* Answers Explained.

78. *See* Answers Explained.

79. *See* Answers Explained.

80. *See* Answers Explained.

81. 2

82. 2

83. *See* Answers Explained.

84. *See* Answers Explained.

85. *See* Answers Explained.

Answers Explained

PART A

1. **2** Studying fossils provides evidence for evolution because fossils *can show patterns of biological change over time.* Fossils are the preserved remains of, or impression left by, living things in the far distant past. By studying fossils, scientists often find information that leads them to develop inferences about the way that present-day living things have evolved over millions of years from ancestral species. They do this by identifying similarities in body structure, such as skeletal elements, preserved as fossils in consecutive layers of sedimentary rock.

WRONG CHOICES EXPLAINED:
(1) It is *not* true that studying fossils provides evidence for evolution because fossils *take a long time to form.* Although it is true that fossils take a long time to form, this fact is not considered evidence of evolution.

(3) It is *not* true that studying fossils provides evidence for evolution because fossils *always contain complete DNA sequences.* Generally, DNA is not well preserved in fossilized remains, so it is not feasible to compare DNA sequences for this purpose.

(4) It is *not* true that studying fossils provides evidence for evolution because fossils *found in the same area are usually closely related to each other.* Close proximity of fossils in a fossil bed can be used as evidence of the diversity of different, unrelated species that inhabited ancient ecosystems at the same point in time. This fact is not considered evidence of evolution.

2. **1** *They allow the organism to maintain homeostasis* is the statement that best describes the interactions between the structures found within a single-celled organism. The structures referenced are known as cell organelles. Each type of organelle in a single-celled organism is specialized to perform the biochemical reactions associated with a different life function necessary for the survival of the cell. The sum total of these life functions allows the cell to maintain homeostatic balance essential to its life.

WRONG CHOICES EXPLAINED:
(2) *They prevent homeostasis from damaging the cell* is *not* the statement that best describes the interactions between the structures found within a single-celled organism. Homeostasis is a term that refers to the sum total of all metabolic activities that serve to maintain physical and chemical balance, or steady state, in the cell. Homeostasis does not damage the cell.

(3) *They must act independently of each other and prevent homeostasis* is *not* the statement that best describes the interactions between the structures found within a single-celled organism. In fact, cell organelles must work together in order to perform all the life functions essential to maintain, not prevent, homeostasis.

(4) *They carry out the same life process in order to maintain homeostasis* is *not* the statement that best describes the interactions between the structures found within a single-celled organism. Each type of cell organelle is responsible for performing different, not the same, life functions.

3. **4** Sexually reproduced offspring have traits similar to their parents because they receive *some of the genes present in each parent*. Sexually reproducing organisms produce male and female gametes (sperm and egg) that each carry one-half of the genetic information found in the parents' body cells. When these gametes fuse in the process of fertilization, complimentary gene pairs unite to create a unique genetic profile in the offspring that is composed equally of genes from the father and the mother. It is for this reason that sexually reproduced offspring resemble their parents.

WRONG CHOICES EXPLAINED:
(1), (2) It is *not* true that sexually reproduced offspring have traits similar to their parents because they receive *all of the proteins from each parent* or *some of the proteins from both parents*. Following fertilization, the embryo immediately begins manufacturing its own proteins within its own cells as determined by its own unique genetic combinations. Proteins are not normally passed from parents to offspring.

(3) It is *not* true that sexually reproduced offspring have traits similar to their parents because they receive *all of the genes present in both parents*. The mechanisms of sexual reproduction ensure the diploid number ($2n$) of the species is maintained from one generation to the next under normal circumstances. See the correct answer above.

4. **4** Row *4* is the row in the chart that correctly pairs a group of organisms with the type of nutrition they carry out. Producers such as green plants carry out autotrophic (self-feeding) nutrition via the process of photosynthesis. Decomposers such as fungi carry out heterotrophic (other-feeding) nutrition by consuming the simple nutrients from plant and animal wastes in the soil.

WRONG CHOICES EXPLAINED:
(1), (2), (3) Rows *1, 2,* and *3* are *not* the rows in the chart that correctly pair a group of organisms with the type of nutrition they carry out. Carnivores, decomposers, and herbivores carry out heterotrophic, not autotrophic, nutrition. Producers carry out autotrophic, not heterotrophic, nutrition.

5. 3 The process that was most likely used to modify the plants' trait and increase their natural rubber production was *genetic engineering to alter a specific gene.* As described in the passage, scientists altered a single trait in the domestic plant. This was most likely accomplished by removing a gene for natural latex (rubber) synthesis from a wild rubber plant and substituting it into the pollen grain(s) of a domestic rubber plant. This altered pollen was then used to fertilize the ova of other domestic rubber plants. The resulting offspring of these fertilizations contained the altered gene in all of their somatic cells.

WRONG CHOICES EXPLAINED:
(1) It is *not* true that the process that was most likely used to modify the plants' trait and increase their natural rubber production was *selective breeding of two similar plant varieties.* Selective breeding refers to a set of methods used by plant and animal breeders to produce new varieties of plants and animals by breeding parental pairs that express desirable characteristics and hoping that they will pass these characteristics on to their offspring. While selective breeding may be used to produce improved varieties of many crop species, the passage does not describe it as being used in this case.

(2) It is *not* true that the process that was most likely used to modify the plants' trait and increase their natural rubber production was *genetic recombination during sexual reproduction.* Genetic recombination refers to the process by which chromosomes and their associated genes are mixed in sexual reproduction to produce offspring that display a set of traits that represent a unique shuffling of the parents' traits. This process is the basis for selective breeding, which the passage does not describe as being used in this case.

(4) It is *not* true that the process that was most likely used to modify the plants' trait and increase their natural rubber production was *fertilizing the plants with key substances found in petroleum.* The term "fertilizing" in this case is most likely used to describe the process of adding nutrients to the soil to enhance the growth of desirable plant species. It is likely that using key substances found in petroleum to fertilize plants in this way would damage or kill the plants.

6. 2 A response of a normally functioning immune system that can be harmful is *rejecting an organ transplant*. Without immunosuppressant drugs, donor antigens on a transplanted organ would stimulate the production of specialized antibodies in the recipient, a reaction that could result in the rejection of the transplanted organ. This can prove harmful to the recipient individual.

WRONG CHOICES EXPLAINED:
(1) It is *not* true that a response of a normally functioning immune system that can be harmful is *being infected by the flu virus*. In humans, the flu is caused by a virus that infects human cells and commandeers them to produce more viral particles that can infect additional cells. This disease-causing mechanism is not a function of the immune system.

(3), (4) It is *not* true that a response of a normally functioning immune system that can be harmful is *recognizing chemical signals* or *fighting off a bacterial infection*. The immune system is able to detect the molecular antigens of foreign invaders such as bacteria and can manufacture chemical substances known as antibodies to neutralize them. These are normal and often lifesaving, not harmful, functions of the human immune system.

7. 3 *Both organic and inorganic molecules* are the molecules that are normally found in single-celled organisms. The cell's membranes and organelles are composed of proteins and fats, which are organic molecules made up primarily of atoms of carbon, hydrogen, and oxygen. Specialized proteins such as enzymes are also found in abundance. Carbohydrates such as starches and sugars are also organic molecules present in cells. In addition, inorganic molecules such as water, salts, carbon dioxide, and oxygen are also found in all living cells.

WRONG CHOICES EXPLAINED:
(1), (2), (4) *Organic molecules, only*; *inorganic molecules, only*; and *neither organic nor inorganic molecules* are *not* the molecules that are normally found in single-celled organisms. Living cells contain both organic and inorganic molecules, as described in the correct answer above.

8. 4 One role of the placenta in the development of offspring is normally to *transfer nutrients*. The placenta is a structure that forms in the uterus from a combination of fetal and maternal tissues. Blood vessels within the placenta serve as the point of exchange of soluble nutrients from the mother's blood to the fetal tissues. By receiving its nutrients in this way, the offspring can survive and grow inside the uterus until birth.

WRONG CHOICES EXPLAINED:
(1), (3) It is *not* true that one role of the placenta in the development of offspring is normally to *produce blood cells* or to *produce gametes*. These differentiated cell types are produced from mesodermal tissues of the developing offspring, not from the placenta, as they are needed.

(2) It is *not* true that one role of the placenta in the development of offspring is normally to *provide milk*. Milk is a nutrient fluid produced and secreted from the mother's mammary glands, not from the placenta, after the birth of the offspring.

9. **3** In order to perform this technique, a scientist would need *a DNA template, enzymes, and subunits with A, G, T, and C bases*. DNA (deoxyribonucleic acid) is a complex biochemical made up of two coiled polymer strands made up of the subunits (nucleotide bases) A, G, T, and C. During the process of normal cell division, DNA is known to self-duplicate in an enzyme-controlled reaction known as replication. In order to perform this process artificially as PCR, a scientist would need at a minimum the components listed above.

WRONG CHOICES EXPLAINED:
(1) It is *not* true that, in order to perform this technique, a scientist would need *a DNA template, ATP, and 20 different amino acid subunits*. ATP is synthesized and used in living cells to transfer energy from the respiratory reactions to power other cellular reactions, including replication. Amino acid subunits are used to synthesize protein molecules, not DNA molecules.

(2), (4) It is *not* true that, in order to perform this technique, a scientist would need *enzymes, several types of simple sugars, and starch molecules* or *enzymes, specific receptor molecules, and several hormones*. These lists are too nonspecific to identify them as being useful in performing any particular laboratory technique.

10. **1** These strands are responsible for coding different proteins and are known as *chromosomes*. The photograph provided shows cells in the process of cell division, during which two sets of replicated chromosomes separate into two daughter cells. Chromosomes are composed primarily of DNA molecules, which are known to carry the genetic information needed to control the synthesis of specific proteins required by cells to enable them to perform their essential life functions.

WRONG CHOICES EXPLAINED:
(2) It is *not* true that these strands are responsible for coding different proteins and are known as *mitochondria*. Mitochondria are cell organelles that function to metabolize sugar molecules under the control of respiratory enzymes in order to release energy for cell processes. Mitochondria are not represented in this photograph.

(3) It is *not* true that these strands are responsible for coding different proteins and are known as *ribosomes*. Ribosomes are a type of cell organelle that functions to synthesize proteins from free amino acids. Ribosomes are not represented in this photograph.

(4) It is *not* true that these strands are responsible for coding different proteins and are known as *chloroplasts*. Chloroplasts are a type of cell organelle found in green plant cells that functions to manufacture glucose from water and carbon dioxide. Chloroplasts are not represented in this photograph.

11. 4 The process that led to the insect resistance can best be explained by *natural selection*. Natural selection, sometimes referred to as survival of the fittest, postulates that certain members of a species are better adapted to their environment than others and that these organisms are more likely to survive and pass on their favorable adaptations to future generations. Selection pressure is any aspect of the natural environment that stresses the species population and favors certain variations over others. In this case, the overuse of insecticide is the selection pressure that results in increased resistance of insect pests to the insecticide.

WRONG CHOICES EXPLAINED:
(1) The process that led to the insect resistance *cannot* best be explained by *ecological selection*. Ecological succession is a phenomenon characterized by the gradual replacement of one plant community by another over time until a stable, self-perpetuating plant community is established. This process would not lead to insect resistance to insecticide.

(2) The process that led to the insect resistance *cannot* best be explained by *selective breeding*. Selective breeding refers to a set of methods used by plant and animal breeders to produce new varieties of plants and animals. They breed parental pairs that express desirable characteristics and hope that the pairs will pass on these characteristics to their offspring. This process would not lead to insect resistance to insecticide.

(3) The process that led to the insect resistance *cannot* best be explained by *asexual reproduction*. Asexual reproduction is a type of reproductive activity in which a single parent organism produces genetically identical offspring by utilizing the process of mitotic cell division. Insects reproduce by sexual, not asexual, means.

12. **1** This risk to developing whale embryos is most likely a result of *an environmental factor not associated with the embryo's genes*. The question explains that the lowered birth rate in killer whales is suspected to be a result of poor nutrition. This indicates that the killer whales' primary food sources, fish and marine mammals, are not abundant enough to support the existing world population of this species. The depletion of prey species represents a biotic environmental factor affecting the survival of this species.

WRONG CHOICES EXPLAINED:
(2) This risk to developing whale embryos is *not* most likely a result of *an infection caused by the embryo's exposure to a pathogen*. There is no mention in the question of a pathogenic cause of this phenomenon. The mother's body serves to protect the embryo from disease-causing organisms during its internal development.

(3) This risk to developing whale embryos is *not* most likely a result of *faults in the genes of the embryo itself*. There is no mention in the question of a genetic cause of this phenomenon. Poor nutrition is a biotic factor in the environment, not a genetic factor.

(4) This risk to developing whale embryos is *not* most likely a result of *toxins that are introduced into the mother from the embryo's blood*. There is no mention in the question of a toxicogenic cause of this phenomenon. The mother's body serves to protect the embryo from bloodborne toxins during its internal development.

13. **1** In order to edit genes, CRISPR-Cas9 must be able to *alter the base sequence of DNA*. Gene editing, also known as genetic engineering, refers to a set of laboratory techniques used by scientists to modify the genetic characteristics of living organisms by inserting genes from other species into the genome of those organisms. This gene insertion process is made possible by restriction enzymes such as Cas9 that catalyze the breaking of chemical bonds at specific sites on a DNA molecule. At the same time, the CRISPR portion of the molecule uses a short bacterial RNA code to alter the base sequence of the DNA strand at those same sites.

WRONG CHOICES EXPLAINED:
(2) It is *not* true that, in order to edit genes, CRISPR-Cas9 must be able to *prevent cells from differentiating*. Differentiation is the process by which the cells of a developing embryo undergo specialization to become specific body tissues. CRISPR-Cas9 cannot perform this function.

(3) It is *not* true that, in order to edit genes, CRISPR-Cas9 must be able to *block cell receptors from receiving signals*. Chemical receptor sites on living cells are specialized to attach to complementary portions of beneficial molecules such as hormones. Receptor sites may be blocked by other molecules having the same chemical signature as the beneficial molecule, thus interfering with the cell's metabolic activity. CRISPR-Cas9 cannot perform this function.

(4) It is *not* true that, in order to edit genes, CRISPR-Cas9 must be able to *change the rate at which a cell uses ATP*. Living cells use ATP at rates required by their metabolic activity and corresponding need for energy. Hormones such as adrenaline may alter the rates at which ATP is used by the cell, but CRISPR-Cas9 cannot perform this function.

14. 4 Maintaining biodiversity is important because it *ensures the availability of a variety of genetic material*. Biodiversity is a measure of the number of different, compatible species that inhabit an area. Each species in a diverse ecosystem is represented by a unique gene pool containing biochemical instructions for millions of traits present in the phenotypes of those species. When a species is eliminated from an ecosystem, its gene pool is eliminated along with it, as are all the traits controlled by them. This represents a loss to the entire ecosystem as well as to human scientific investigation.

WRONG CHOICES EXPLAINED:
(1) It is *not* true that maintaining biodiversity is important because it *reduces the carrying capacity of a forest ecosystem*. Generally, the more biodiverse an ecosystem is, the higher the carrying capacity is for all its member species. However, competition within and between species for limited resources may reduce carrying capacity for some member species at some points in time.

(2) It is *not* true that maintaining biodiversity is important because it *guarantees that all species within a forest ecosystem will survive*. While biodiversity of an ecosystem generally enhances the survival of species living within it, competition within and between species for limited resources may reduce carrying capacity for, or even eliminate, some member species at some points in time.

(3) It is *not* true that maintaining biodiversity is important because it *increases the number of predators that control the population size of prey*. The balance between predator and prey species is a self-governing factor in a biodiverse ecosystem. As prey populations are reduced due to predation, predator populations will decline soon afterward due to lack of food.

15. 4 These cells are most similar in the *information stored in their DNA*. All somatic cells of an individual organism contain the same set of chromosomes and genes. During the process of differentiation, cells take on specialized shapes and functions because some specific genes are switched off and others are switched on. This allows the cells to produce only the enzymes needed to catalyze biochemical reactions related to their particular function.

WRONG CHOICES EXPLAINED:
(1), (3) These cells are *not* most similar in the *amount of energy they release* or in the *rate of their metabolism*. These factors are dictated by the functions of these specialized cells. For example, muscle cells may contract many times a minute and, when doing so, require a higher rate of metabolism and use more energy than a skin cell generally needs to function.

(2) These cells are *not* most similar in the *type of proteins they synthesize*. In specialized cells, some genes are switched off and others are switched on. This allows them to produce only the enzymes (proteins) needed to catalyze biochemical reactions related to their particular function.

16. 3 When compared to using pesticides, this method to control ticks would *be less likely to harm the environment*. Pesticides designed to kill harmful insects are chemical agents that can harm other species as well as the target species. Beneficial insects, such as bees and butterflies that pollinate flowers, may be unintentional victims of pesticides meant to kill ticks. In addition, these toxic chemicals may negatively affect the reproductive cycles of birds, amphibians, mammals, and other vertebrate members of the ecosystem. If these chemicals enter the water supply, they may also prove harmful to human health.

WRONG CHOICES EXPLAINED:
(1) It is *not* true that, when compared to using pesticides, this method to control ticks would *cause more environmental pollution*. The opposite is true. Pesticides have proved to be major pollutants of the natural environment, so the biological control described would be less polluting.

(2) It is *not* true that, when compared to using pesticides, this method to control ticks would *lead to a decrease in the deer population*. The opposite is true. The use of pesticides, not biological controls, would tend to decrease the deer population.

(4) It is *not* true that, when compared to using pesticides, this method to control ticks would *result in an increase in the tick population*. Both methods would have a net effect of reducing the tick population initially, although pesticide-resistant strains of ticks may emerge due to natural selection.

17. 1 The type of organism represented by box *X* would be *algae*. The diagram illustrates a food chain operating in a natural ecosystem. By convention, the arrows in a food chain are drawn to indicate the direction of energy flow in the ecosystem. Since the organism in box *X* directly receives the energy of sunlight and passes it along to a different organism in box *Y*, organism *X* must be a producer. Of the choices given, only the algae are producer organisms.

WRONG CHOICES EXPLAINED:
(2), (3), (4) The type of organism represented by box *X* would *not* be *fungi, small fish,* or *sea birds.* Each of these organisms is a consumer in an ocean ecosystem. None of them is able to receive energy directly from sunlight.

18. 4 *Gametes → zygote → embryo → fetus* is the sequence that best represents the correct order of events in the formation of a sexually reproduced individual. This process begins with the production of gametes (sperm and egg cells) from primary sex cells of male and female parents. One egg fuses with one sperm to create a single zygote (fertilized egg). The zygote undergoes rapid cell division to produce a multicelled embryo lacking highly differentiated tissues. During development, the undifferentiated tissues of the embryo differentiate into the highly specialized tissues and organs of the fetus.

WRONG CHOICES EXPLAINED:
(1), (2), (3) *Embryo → zygote → gamete → fetus, zygote → embryo → fetus → gamete,* and *gametes → embryo → fetus → zygote* are *not* the sequences that best represent the correct order of events in the formation of a sexually reproduced individual. None of these sequences places events in the correct order as described above.

19. 1 Direct harvesting occurs when *pine trees are cut from a forest for use as lumber.* Direct harvesting is a human activity in which plants or animals are removed (harvested) from a natural environment for their economic value. The cutting of forest pine trees for lumber production matches this definition.

WRONG CHOICES EXPLAINED:
(1) It is *not* true that direct harvesting occurs when *corn is planted in a newly plowed field.* This represents the result of a human activity known as monocropping, in which a biodiverse ecosystem is replaced by a single crop species.

(2) It is *not* true that direct harvesting occurs when *zebra mussels are accidently imported to the Great Lakes.* This represents the result of a human activity known as the introduction of invasive species, in which an organism not native to an ecosystem is released into, and causes disruption of, that ecosystem.

(3) It is *not* true that direct harvesting occurs when *roots of plants continually take in water*. This represents a natural process by which plants absorb water from their environment for use in maintaining homeostasis.

20. 3 A *negative* outcome of this practice is that *the stability of the ecosystem will be reduced*. This situation is an example of a human activity known as monocropping, in which a biodiverse ecological community is replaced by a single crop species. The loss of biodiversity (number of different species inhabiting and interacting together) in an ecosystem tends to destabilize that ecosystem.

WRONG CHOICES EXPLAINED:
(1) It is *not* true that a *negative* outcome of this practice is that *the corn will interbreed with the weeds in the area*. Corn and weeds are different species, which by definition cannot interbreed successfully.

(2) It is *not* true that a *negative* outcome of this practice is that *new predators will be introduced into the ecosystem*. When a field is planted in a single crop, new predator species (e.g., foxes, coyotes, owls, hawks) may be attracted to the area to hunt for prey animals that inhabit the open field. This is a positive, not a negative, result of this practice.

(4) It is *not* true that a *negative* outcome of this practice is that *new species of insect-resistant corn will evolve*. Evolution is a process that depends on mutation, natural selection, and reproduction over many generations to produce new varieties or species of living things. It is extremely unlikely that this practice would result in the evolution of corn plants.

21. 3 The process of differentiation is best described as the *process by which cells specialize and develop into a specific type of cell*. Differentiation is the process by which the cells of a developing embryo undergo specialization to become specific body tissues such as muscle, skin, blood, endocrine, and nervous tissues.

WRONG CHOICES EXPLAINED:
(1) The process of differentiation is *not* best described as the *production of a genetically identical copy of an organism*. This describes a laboratory technique known as cloning, in which an egg cell nucleus is replaced with the diploid nucleus taken from an individual donor and then allowed to develop into a clone (genetically identical duplicate) of that donor.

(2) The process of differentiation is *not* best described as the *change in shape of a protein due to high temperatures*. This describes the process of denaturation, in which a protein molecule is deformed such that its function is impaired or destroyed by high temperatures or chemical agents.

(4) The process of differentiation is *not* best described as the *process in which genes are made and transferred into other organisms*. This describes a laboratory technique known as genetic engineering, in which a desirable gene is transferred from the genome of one species and inserted into the genome of a different species.

22. 4 *A negative environmental effect is that unregulated fishing in the ocean can disrupt the interactions between organisms in existing food webs* is the statement that accurately describes *one* of these possible effects. When fish species are hunted and removed in large quantities for processing as human or animal foods, the biodiversity of the ocean ecosystem is diminished as are the food webs that normally exist in that ecosystem.

WRONG CHOICES EXPLAINED:
(1) *A positive environmental effect is that burning fossil fuels to generate electricity reduces carbon dioxide levels in the atmosphere* is *not* the statement that accurately describes *one* of these possible effects. The burning of fossil fuels such as coal and oil produces various pollutants, including carbon dioxide. This activity has a negative, not a positive, effect on the environment.

(2) *A positive environmental effect is the cutting of trees in rain forests to provide large quantities of lumber to build homes for the increasing world population* is *not* the statement that accurately describes *one* of these possible effects. The clear-cutting of rain forests greatly diminishes the quality of both local ecosystems and the global environment. This activity has a negative, not a positive, effect on the environment.

(3) *A negative environmental effect is that industrialization provides many jobs and helps the economy grow* is *not* the statement that accurately describes *one* of these possible effects. The creation of jobs to improve the economy is an economic, not an environmental, effect of this activity.

23. 1 *It may be for abiotic or biotic resources* is the statement that best describes the process of competition. Competition is a term that relates to any conflict between organisms for limited natural resources in the environment. Such competition may exist within or between species and may involve both abiotic (nonliving) and biotic (living) factors. Abiotic factors include such things as oxygen, water, light, soil minerals, or living space. Biotic factors include such things as available food, symbiotic relationships, or breeding partners.

WRONG CHOICES EXPLAINED:
(2) *It is not affected by changes in the environment* is *not* the statement that best describes the process of competition. In fact, changes in the environment such as fire, flood, or drought can drastically affect the level of competition within and between species by altering the availability of abiotic and biotic factors in an ecosystem.

(3) *It always occurs between members of different species* is *not* the statement that best describes the process of competition. In fact, competition may exist either within a single species or between different species, especially when those species' environmental niches overlap.

(4) *It allows nutrients in an ecosystem to move from herbivores to autotrophs* is *not* the statement that best describes the process of competition. In fact, in any natural ecosystem, food energy is transferred from autotrophs (plants) to herbivores (plant eaters), not from herbivores to autotrophs.

24. **3** These responses are one way *the body is able to maintain dynamic equilibrium.* By regulating the body's temperature within an acceptable range, the nervous and endocrine systems work together to allow other tissues and organs to operate at peak efficiency. This is an example of a feedback mechanism that assists in the maintenance of dynamic equilibrium, also known as homeostasis or steady state, in the human body.

WRONG CHOICES EXPLAINED:
(1) It is *not* true that these responses are one way *to counteract feedback mechanisms that would otherwise be beneficial.* In fact, this is an example of a process that commonly functions to assist, not counteract, a feedback mechanism in the human body.

(2) It is *not* true that these responses are one way *to make the body release insulin to control blood circulation.* Insulin is a hormone that regulates blood sugar concentration, not blood circulation, in the human body.

(4) It is *not* true that these responses are one way *skin and muscle cells are able to disrupt homeostasis.* The role of any body tissue is to assist, not disrupt, homeostasis in the human body.

25. **2** *Vaccinations may contain weakened microbes that stimulate the formation of antibodies* is the statement about the response of the body to pathogens that is correct. Traditional vaccines are created by isolating fragments of virus particles or bacterial cells that contain antigens produced by the pathogen. When introduced into the human body, the immune system reacts to these antigens by producing antibodies specifically designed to neutralize them.

WRONG CHOICES EXPLAINED:
(1) *Red blood cells engulf invaders and produce antibodies that attack invaders* is *not* the statement about the response of the body to pathogens that is correct. In the human immune system, this function is performed by white blood cells, not red blood cells.

(3) *AIDS is a bacterial disease that strengthens the immune system* is *not* the statement about the response of the body to pathogens that is correct. AIDS is a viral, not a bacterial, disease that weakens, not strengthens, the human immune system.

(4) *All allergic reactions are caused by an immune response to microorganisms* is *not* the statement about the response of the body to pathogens that is correct. An allergy is an immune response to a usually harmless environmental substance, not to a microorganism. The symptoms caused by the body's immune response to these substances are known collectively as an allergic reaction.

26. 1 This constant correcting of blood sugar levels within the body is accomplished by *a feedback mechanism.* Insulin and glucagon are hormones that regulate blood sugar concentration in the human body. When sensors in the body detect an excess concentration of sugar in the blood, insulin secreted by the pancreas acts by moving sugar out of the blood into the liver for storage as glycogen. When the body detects a blood sugar concentration that is too low, the pancreas slows the secretion of insulin and increases the secretion of glucagon, which releases sugar from glycogen into the blood. This mechanism is an example of a feedback mechanism that assists in the maintenance of dynamic equilibrium, also known as homeostasis or steady state, in the human body.

WRONG CHOICES EXPLAINED:
(2) This constant correcting of blood sugar levels within the body is *not* accomplished by *an immune response.* The immune response refers to the body's reaction to the presence of foreign invaders, especially pathogens. When introduced into the human body, the immune system reacts to these pathogens by producing antibodies specifically designed to neutralize them. This process has no direct effect on blood sugar levels in the body.

(3) This constant correcting of blood sugar levels within the body is *not* accomplished by *an allergic reaction.* An allergy is an immune response to a usually harmless environmental substance. The symptoms caused by the body's immune response to these substances are known collectively as an allergic reaction. This process has no direct effect on blood sugar levels in the body.

(4) This constant correcting of blood sugar levels within the body is *not* accomplished by *manipulating a gene.* Also known as genetic engineering, gene manipulation refers to a set of laboratory techniques used by scientists to modify the genetic characteristics of living organisms by inserting genes from other species into the genome of those organisms. This process has no direct effect on blood sugar levels in the body.

27. 3 These new skin cells form as a result of *mitotic cell division*. Mitotic cell division is the process by which new cells arise from parent cells and in which genetic continuity is assured. The new cells produced by this process are genetically identical to and display all of the characteristics of the parent cell.

WRONG CHOICES EXPLAINED:
(1) These new skin cells do *not* form as a result of *meiotic cell division*. Meiotic cell division is the process by which homologous chromosome pairs and the genes they carry are separated into haploid gametes during gametogenesis. The resulting gametes are not genetically identical to and do not have the characteristics of the parent cell. This process is not at work in this case.

(2) These new skin cells do *not* form as a result of *sexual reproduction*. Sexual reproduction refers to a process by which male and female gametes combine genetic information to produce a new and unique offspring. This process is not at work in this case.

(4) These new skin cells do *not* form as a result of *gene recombination*. Genetic recombination refers to the process by which chromosomes and their associated genes are resorted in sexual reproduction to produce offspring that display a set of traits that represent a unique shuffling of the parents' traits. This process is not at work in this case.

28. 4 Graph *4* is the graph that best represents the relationship between the relative number of nuclei, genes, and chromosomes in a typical human cell. A typical human cell contains only one nucleus. Housed within this nucleus are 46 chromosomes, the diploid ($2n$) number of chromosomes for humans. Each of these chromosomes contains thousands of genes that are responsible for governing the traits of each individual human being. Graph *4* most accurately illustrates these relative numbers.

WRONG CHOICES EXPLAINED:
(1), (2), (3) Graphs *1*, *2*, and *3* are *not* the graphs that best represent the relationship between the relative number of nuclei, genes, and chromosomes in a typical human cell. None of these graphs accurately illustrates the relative numbers of these structures as described in the correct answer above.

29. 4 *The shrubs modify the environment, making it more suitable for the softwood trees* is the statement that is true about the biological process shown. The diagram illustrates the process of ecological succession. Ecological succession is a phenomenon characterized by the gradual replacement of one plant community by another over time until a stable, self-perpetuating plant community is established. The shrubs act to enrich and thicken the soil, creating a condition that favors the growth and dominance of softwood trees.

WRONG CHOICES EXPLAINED:

(1) *This is a short-term process resulting from sudden changes* is *not* the statement that is true about the biological process shown. Ecological succession requires many decades to operate from a pioneer (annual plant) community to a climax (hardwood tree) community.

(2) *This process cannot be altered by humans and other organisms* is *not* the statement that is true about the biological process shown. Ecological succession can be altered by natural events such as fire and flood or by humans and other animal species that have the capability of modifying their own environment.

(3) *If the hardwood trees are destroyed, the altered ecosystem cannot recover* is *not* the statement that is true about the biological process shown. Given enough time and the right biotic and abiotic conditions, ecological succession is usually able to restore any ecosystem to its original state.

30. 3 When cells such as the skin cells shown reproduce abnormally, it could be a sign of *cancerous cell growth*. Cancer begins when the cell's normal reproductive mechanism is disrupted and rapid, uncontrolled mitotic cell division occurs. This abnormal process often results in a mass of nonfunctioning cells, known as a tumor, that crowds out healthy cells and interferes with normal tissue functions.

WRONG CHOICES EXPLAINED:

(1) It is *not* true that, when cells such as the skin cells shown reproduce abnormally, it could be a sign of *an immune response*. The immune response refers to the body's reaction to the presence of foreign invaders, especially pathogens. Abnormal cell growth does not occur as a result of an immune response.

(2) It is *not* true that, when cells such as the skin cells shown reproduce abnormally, it could be a sign of *dynamic equilibrium*. Dynamic equilibrium, also known as homeostasis, is a term that refers to the sum total of all metabolic activities that serve to maintain physical and chemical balance, or steady state, in the cell. Abnormal cell growth does not occur as a result of dynamic equilibrium.

(4) It is *not* true that, when cells such as the skin cells shown reproduce abnormally, it could be a sign of *a cellular adaptation*. Cells may adapt to small changes in their environments by initiating specific responses to them. Abnormal cell growth does not occur as a result of cellular adaptation.

PART B-1

31. 4 Only an electron microscope can be used to view *viruses*. A review of the information presented in the diagram shows that virus particles have sizes that range between 40 nm and 100 nm. The electron microscope can allow scientists to view particles less than 1 nm in size, so viruses are well within their range of resolution. The resolution ranges of both the eye and the light microscope are shown to be incapable of viewing particles this small.

WRONG CHOICES EXPLAINED:
(1), (2), (3) It is *not* true that only an electron microscope can be used to view *bacteria, mitochondria,* and *animal cells*. All of these particles may be viewed through a light microscope since their sizes fall within this instrument's range of resolution.

32. 2 In order to trap the bacteria and prevent them from going through the filter, she must make sure the pores in the filter are no larger than *1 μm*. The information presented in the diagram indicates that bacterial particle sizes range between 1.5 μm and 4.0 μm. In order to prevent any bacteria from passing through the filter, the pores must be smaller than this range, so 1 μm is sufficient.

WRONG CHOICES EXPLAINED:
(1) It is *not* true that, in order to trap the bacteria and prevent them from going through the filter, she must make sure the pores in the filter are no larger than *1 nm*. This filter pore size would be as small as some small molecules, so it would be impractically small for an experiment of this type.

(3), (4) It is *not* true that, in order to trap the bacteria and prevent them from going through the filter, she must make sure the pores in the filter are no larger than *10 μm* or *100 μm*. These filter pore sizes are much bigger than the bacterial particles the scientist hopes to exclude, so they would be impractically large for an experiment of this type.

33. 2 In tropical areas, the best explanation for having increased melanin in human skin is that it *provides a survival advantage*. Natural selection, sometimes referred to as survival of the fittest, postulates that certain members of a species are better adapted to their environment than others and that these organisms are more likely to survive and pass on their favorable adaptations to future generations. Selection pressure is any aspect of the natural environment that stresses the species population and favors certain variations over others. In this case, the intense UV radiation in the tropics is the selection pressure that favors human genes able to synthesize increased melanin in skin cells.

WRONG CHOICES EXPLAINED:

(1) It is *not* true that, in tropical areas, the best explanation for having increased melanin in human skin is that it *increases the occurrence of mutations.* Mutations are random events in which changes occur in the DNA of the cell as a result of exposure to mutagenic agents such as radiation, including UV radiation. The trait of increased melanin may have originally resulted from a mutation, but this trait does not increase mutations.

(3) It is *not* true that, in tropical areas, the best explanation for having increased melanin in human skin is that it *acts as a feedback mechanism to increase UV exposure.* Feedback mechanisms assist in the maintenance of dynamic equilibrium, also known as homeostasis or steady state, in the human body. The trait of increased melanin is not a feedback mechanism and does not increase UV exposure.

(4) It is *not* true that, in tropical areas, the best explanation for having increased melanin in human skin is that it *produces antibodies that destroy pathogens.* Antibodies are synthesized and released by the immune system in direct response to the presence in the body of a foreign antigen on a pathogen. The trait of increased melanin has no direct relationship to the production of antibodies in humans.

34. 2 This finding would be most reliable if it were based on *research done by scientists in many different countries.* This research should include repetition of experiments by independent scientists, which is a key element in the process of peer review. Although peer review often supports the work of other scientists and adds to its reliability, it can also reveal weaknesses in the methods used by the original experimenters and may refute their claims altogether.

WRONG CHOICES EXPLAINED:

(1) It is *not* true that this finding would be most reliable if it were based on *data collected from patients in one cancer-research hospital.* A well-designed study should include a wide range of experimental subjects of many different backgrounds, including healthy individuals. Such a range of subjects cannot be found in a single cancer-research hospital.

(3) It is *not* true that this finding would be most reliable if it were based on *reading the information on cigarette cartons.* Information provided by the manufacturers of tobacco products has proved to be unreliable and misleading. In the past, they even claimed that tobacco use is beneficial to human well-being. So information from this source should not be relied upon in the future.

(4) It is *not* true that this finding would be most reliable if it were based on *cancer information published on social media sites.* Information posted on social media is not subject to review by competent scientists and may be inaccurate or misleading. So it should not be relied on without supporting advice from a competent medical professional.

35. 1 Diagram *1* is the diagram that best represents the direction that energy flows through an energy pyramid. By convention, the arrows in an energy pyramid are drawn to indicate the direction of energy and nutrient flow in the ecosystem, represented by the single arrow drawn from the bottom to the top of the pyramid. Also by convention, an energy pyramid is drawn to indicate the total relative energy at each trophic (feeding) level, with producers at the bottom (most energy), herbivores in the middle (moderate energy), and predators at the top (least energy). As it flows through the ecosystem, as much of 90% of the energy that supports a trophic (feeding) level of a food pyramid is dissipated into the environment as waste heat, a phenomenon that is represented by the arrows exiting the pyramid at each level.

WRONG CHOICES EXPLAINED:
(2), (3), (4) Diagrams *2*, *3*, and *4* are *not* the diagrams that best represent the direction that energy flows through an energy pyramid. None of these diagrams correctly illustrates the energy flow that is known to occur in a naturally occurring ecosystem. See the correct answer above.

36. 1 *All of these species have certain DNA sequences in common* is the conclusion that is correct based on the evolutionary tree. A review of the information presented in the diagram shows that all of the species illustrated share species *B* as a common ancestor. While the process of biological evolution introduces many new variants created over time as a result of small genetic changes, much of the DNA coding present in the ancestral species is passed intact from generation to generation and from species to species. In this way, all species that have ever lived on Earth, no matter how diverse, share certain genetic commonalities.

WRONG CHOICES EXPLAINED:
(2) *Species* S *is the best adapted of all the species shown* is *not* the conclusion that is correct based on the evolutionary tree. All the species shown were successful for a time under the environmental conditions that they confronted. Although species *S* is a variant that is proving to be well adapted for survival in the present, it may or may not be the best adapted of all the species shown.

(3) *A common ancestor of species* L *and* M *is species* N is *not* the conclusion that is correct based on the evolutionary tree. Species *N* cannot be an ancestor of species *L* since species *N* branched off their common line of this evolutionary tree after the appearance of species *L*.

(4) *Species* O *and* P *are more closely related than species* P *and* Q is *not* the conclusion that is correct based on the evolutionary tree. Species *O* branched off the line from common ancestor *B* quite early in this evolutionary tree, while species *P* and *Q* branched off a different line much later. This indicates a closer relationship exists between species *P* and *Q* than exists between species *O* and *P*.

37. **2** Rows B *and* C are the rows that best support the information provided. Row *B* correctly links the summer season with the arctic fox's increased melanin production and resulting brown fur color. Row *C* correctly links the winter season with the arctic fox's decreased melanin production and resulting white fur color.

WRONG CHOICES EXPLAINED:
(1), (3), (4) Rows A *and* B, C *and* D, and D *and* A are *not* the rows that best support the information provided. Rows *A* and *D* both contain inaccurate information, so any pairing that includes one or both must be incorrect. See the correct answer above.

38. **2** *The expression of genes can be modified by the external environment* is the statement that is the most likely explanation for the color differences in the fur of the fox at different times of the year. Geneticists know that certain traits, such as fur color in arctic foxes, are influenced by environmental conditions. In this case, environmental factors such as external temperatures and low light levels that are typical of arctic winters cause the genes responsible for the production of melanin to become inactive. As a result, the fox's winter fur grows in without this pigment and appears white. Warmer summer temperatures allow these genes to function, resulting in melanin-rich, brown summer fur.

WRONG CHOICES EXPLAINED:
(1) *Mutations can be caused by changes in the number of biotic factors in the environment* is *not* the statement that is the most likely explanation for the color differences in the fur of the fox at different times of the year. No information is presented in the passage that discusses biotic factors in the environment that might influence the expression of fur color in arctic foxes.

(3) *Hereditary information is contained in genes located in the chromosomes of each cell* is *not* the statement that is the most likely explanation for the color differences in the fur of the fox at different times of the year. This is a true statement, but it does not explain the seasonal variation in fur color described in the passage.

(3) *Random changes in DNA can occur to change the expression of a gene* is *not* the statement that is the most likely explanation for the color differences in the fur of the fox at different times of the year. There is no information presented in the passage that discusses random changes to DNA (mutations) that might influence the expression of fur color in arctic foxes.

39. 1 *A group of 50 mice with flu antibodies formed using the new technique were exposed to mutated forms of the flu. None of the mice became ill.* This is the statement that describes an observation that would best support the continued study of using antibodies produced by this new technique against the flu. This observation of 50 mice would provide some preliminary evidence in support of the new technique but would be insufficient to provide definitive proof of its success. More testing by qualified, independent scientists would be required to reach a firm conclusion as to the degree of success of this technique.

WRONG CHOICES EXPLAINED:
(2) *The use of these antibodies in mice stopped mutations that occur in flu viruses.* This is *not* the statement that describes an observation that would best support the continued study of using antibodies produced by this new technique against the flu. No information is provided in the passage that would lead to this observation. Antibodies do not function to stop mutations in flu viruses.

(3) *Chemical tests showed that the stem antibodies attached to the heads of some flu viruses and destroyed them.* This is *not* the statement that describes an observation that would best support the continued study of using antibodies produced by this new technique against the flu. No information is provided in the passage that would lead to this observation. Stem antibodies attach only to stem antigens.

(4) *Blood tests showed that only "stem" antibodies attacking the stem of flu antigens can cause the flu in mice. Those attacking the "head" did not.* This is *not* the statement that describes an observation that would best support the continued study of using antibodies produced by this new technique against the flu. No information is provided in the passage that would lead to this observation. Antibodies do not function to cause flu.

40. 2 *When specific chemicals produced by pathogens enter the digestive system in contaminated foods, the ability of the immune system to fight off foodborne illness is reduced.* This statement most correctly describes how these two systems interact when an individual comes down with a foodborne illness. The passage is silent concerning the precise nature of the interactions that occur between the digestive and immune systems under the influence of pathogenic chemical secretions. To answer the question correctly, students must assume that the pathogen enters the body via ingestion into the digestive system and that the pathogenic chemical's effect on the immune system is a reduction in efficiency, as well as through the process of eliminating the incorrect distracters.

[NOTE: Generally, students are not expected to know the details of this inter-action unless pertinent information is provided in the passage or unless instruction has been given on a topic not specifically required by the Living Environment curriculum.]

WRONG CHOICES EXPLAINED:
(1) *Chemicals produced by pathogens enter the immune system through a cut in the skin. The circulatory system carries the chemical to the digestive system, result-ing in a foodborne illness.* This statement does *not* most correctly describe how these two systems interact when an individual comes down with a foodborne illness. Pathogens that cause foodborne illness enter the body via ingestion into the digestive system, not through cuts in the skin.

(3) *When foods contaminated with pathogens are eaten, the immune system prevents the pathogens from entering the digestive system.* This statement does *not* most correctly describe how these two systems interact when an individual comes down with a foodborne illness. The immune system does not function to prevent pathogens from entering the digestive system.

(4) *The digestive system breaks down the pathogens in the contaminated foods so that they are harmless. These harmless pathogens are then transferred to the immune system.* This statement does *not* most correctly describe how these two systems interact when an individual comes down with a foodborne illness. Pathogens are not transferred from the digestive system to the immune system under any known circumstance.

41. **2** The lights were turned on in the lab at hour *2*. A review of the information provided in the graph shows that oxygen concentration had reached its lowest point just prior to hour 2 and then began to rise steadily for three hours before leveling off about hour 5. It is likely that these measurements were caused by the plant absorbing the light energy beginning at hour 2 and using it to carry on the process of photosynthesis, which produces oxygen gas as a by-product, for the remainder of the experiment.

WRONG CHOICES EXPLAINED:
(1), (3), (4) It is *not* true that the lights were turned on in the lab at hour *8*, *0*, or *4*. These times are inconsistent with the information provided in the graph. See the correct answer above.

42. 3 During the 8 hours studied, the plant performed *both photosynthesis and respiration*. Information in the graph indicates that the plant carried out the process of photosynthesis from hour 2 to at least hour 5, which resulted in a net increase in the oxygen concentration in the container during that period. Because it is a living organism, the plant carried out the process of respiration for the entire duration of the experiment, which resulted in a net reduction of measured oxygen concentration in the container from hour 0 to hour 2 and again from hour 5 through hour 8.

WRONG CHOICES EXPLAINED:
(1), (2), (4) It is *not* true that, during the 8 hours studied, the plant performed *photosynthesis, only*; *respiration, only*; or *neither photosynthesis nor respiration*. The plant performed both photosynthesis and respiration, as described in the correct answer above,

43. 4 Molecule *4* is the antihistamine molecule represented that would be most effective. Because its circular shape matches the shape of the chemical receptors on the surface of the throat cell, antihistamine molecule 4 will bind to the histamine receptors and prevent the histamine molecules from attaching and affecting the throat cell.

WRONG CHOICES EXPLAINED:
(1), (2), (3) Molecules *1, 2,* and *3* are *not* the antihistamine molecules represented that would be most effective. The shapes of these molecules are not circular. So they will not bind to the histamine receptors and will not prevent the histamine molecules from attaching to and affecting the throat cell.

PART B-2

44. One credit is allowed for correctly marking an appropriate scale, without any breaks in the data, on each labeled axis.　[1]

45. One credit is allowed for correctly plotting the data on the grid, connecting the points, and surrounding each point with a small circle.　[1]

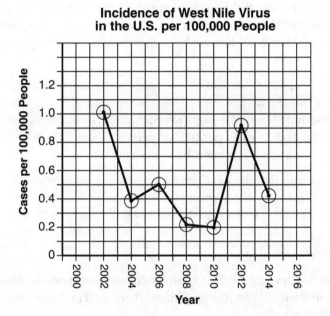

Incidence of West Nile Virus in the U.S. per 100,000 People

46. One credit is allowed for correctly stating whether or not it is possible to predict what the number of cases per 100,000 people will be for the year 2020 and supporting the answer. Acceptable responses include but are not limited to:　[1]

- *It is not possible to make an accurate prediction. In 2002, more than 1 person per 100,000 was infected. It was down to 0.39 people in 2004, back up to 0.91 in 2012, and down again to 0.42 in 2014.*

- *The number of cases per 100,000 people varies widely from one year to the next. There is no consistent trend, making it impossible to predict the number of cases for 2020.*

- *Yes. It is probable that the number of cases will be less than 1.02 per 100,000 in 2020. It has not been that high since 2002.*

- *No. The data vary too much to make a prediction.*

47. **4** The data represented on the maps best indicate that *for any given year, it is difficult to know which states will have the greatest number of cases.* A review of the information presented in the maps shows a great deal of variability in the incidence of West Nile virus (WNV) in the 48 U.S. states illustrated. Although the infection rate of WNV has remained high in four states in the upper Midwest, other states have seen a decrease or a mild increase of WNV from 2007 to 2016.

WRONG CHOICES EXPLAINED.
(1) It is *not* true that the data represented on the maps best indicate that *birds have spread WNV to every state in the United States.* A close examination of the maps reveals that WNV has not been detected in the state of West Virginia in either of the years surveyed.

(2) It is *not* true that the data represented on the maps best indicate that *New York State has the highest rate of WNV infection for both of the years shown.* A close examination of the maps reveals that WNV was measured at an infection rate of 0.01–0.24 cases per 100,000 people in New York in both survey years, which is among the lowest infection rates in the U.S.

(3) It is *not* true that the data represented on the maps best indicate that *once WNV reaches a state, the number of people infected increases every year.* A close examination of the maps reveals that several states (e.g., Utah, New Mexico, Missouri) illustrated have seen decreases, not increases, in their rates of WNV infection from 2007 to 2016.

48. One credit is allowed for correctly explaining why some people may be more severely affected by West Nile virus than others. Acceptable responses include but are not limited to: [1]
- *They may be suffering from a different disease that affects their immune system.*
- *Perhaps their immune system is not able to combat the virus.*
- *They may be very young or very old and not able to combat the virus.*
- *They may have been bitten more times by infected mosquitoes.*
- *They may live in a region where more birds/mosquitoes carry the WNV.*

49. **3** Based on the food web, the population that contains the greatest amount of available energy would be *phytoplankton.* In all naturally occurring, sustainable ecosystems, plants and plantlike (producer) organisms form the basis of the food web. This is possible because producers absorb solar energy and use it to manufacture energy-rich molecules that are used for energy by all the consumer populations that inhabit that ecosystem. As producers at the base of this food web, the phytoplankton (algae) contain the greatest amount of energy of the species populations shown.

WRONG CHOICES EXPLAINED:
(1), (2), (4) Based on the food web, the population that contains the greatest amount of available energy would *not* be *seals*, *fishes*, or *humans*. All of these species are consumers in this food web, so their populations contain less energy than the phytoplankton.

50. **3** *The food web would be disrupted, and organisms would die* is the statement that best describes what would happen in this ecosystem if the phytoplankton were removed from the food web. Because all other populations are either directly or indirectly dependent on the phytoplankton population for food, the ecosystem would suffer total collapse and become completely disrupted under these circumstances.

WRONG CHOICES EXPLAINED:
(1) *Copepods and krill would fill the vacant niche* is *not* the statement that best describes what would happen in this ecosystem if the phytoplankton were removed from the food web. Copepods and krill cannot fill the niche of the phytoplankton because they cannot carry on photosynthesis. As marine consumers that are most directly dependent on the phytoplankton as a food source, the copepod and krill populations would be the first to die out under these circumstances.

(2) *The number of heterotrophs would increase* is *not* the statement that best describes what would happen in this ecosystem if the phytoplankton were removed from the food web. Because no food would be available, the number of heterotrophs (consumers) in this ecosystem would decrease, not increase, under these circumstances.

(4) *The food web would remain stable* is *not* the statement that best describes what would happen in this ecosystem if the phytoplankton were removed from the food web. Because all other populations are either directly or indirectly dependent on the phytoplankton population for food, the ecosystem would become completely destabilized under these circumstances.

51. One credit is allowed for correctly describing the relationship represented by the arrows between squids and fishes. Acceptable responses include but are not limited to: [1]

- *Fish can be predators of the squid, and squid can be predators of the fish.*
- *Each species could be either predator or prey, depending on circumstances.*
- *They can feed on each other.*

52. One credit is allowed for correctly stating the relationship between peak blood alcohol concentration and total brain weight for alcohol-exposed newborn rats. Acceptable responses include but are not limited to: [1]

- *The higher the blood alcohol concentration is, the lower the total brain weight is in newborn rats.*
- *A lower peak blood alcohol results in a higher total brain weight.*
- *As blood alcohol concentration increases, total brain weight decreases.*
- *There is an inverse relationship/correlation between blood alcohol concentration and brain weight in newborn rats.*

53. One credit is allowed for correctly stating whether or not the student's claim that this food web represents a stable ecosystem is correct and supporting the answer. Acceptable responses include but are not limited to: [1]

- *Since there several different species in this ecosystem, the system is probably stable. So she is correct.*
- *No. Because an ecosystem includes all the interacting species in an area and their physical environment, I would say that this ecosystem cannot be stable if it doesn't include abiotic factors like water and soil minerals.*
- *I don't think she is correct. Only biotic factors are shown in this diagram, so the ecosystem can't be considered stable until abiotic factors are added.*
- *I think so. There are producers and consumers in this food web, so their interaction would keep the ecosystem stable.*
- *No. There are no decomposers represented in the food web, so the ecosystem would not be stable because animal and plant wastes would not be recycled.*
- *No, she is not correct. There is only one type of producer (trees) included, which is not biodiverse enough to support a stable ecosystem.*

54. One credit is allowed for selecting *two* organisms from this food web that compete with each other for food and for correctly stating *one* reason why they are able to survive in the same ecosystem. Acceptable responses include but are not limited to: [1]

- *Butterflies and squirrels: They both feed on trees. However, the butterflies eat the leaves, while the squirrels eat the acorns.*
- *Snakes and foxes: They both have other sources of nutrition, so neither is totally dependent on the rabbits for food.*

- *Rabbits and squirrels: They obtain nutrients from different types of plants.*

- *Hawks and foxes: They hunt at different times of the day.*

- *Catbirds and frogs: Frogs catch butterflies when they land close to the ground, and catbirds catch them when they are flying.*

55. One credit is allowed for correctly identifying *both* the number and the name of the structure in the cell that produces proteins. Acceptable responses include: [1]

- *Structure 1: Ribosome*

PART C

56–57. Two credits are allowed for correctly explaining how the mutation in the fragile *X* chromosome affects the body. In your answer, be sure to:

- State *one* specific reason why the mutated gene on the fragile *X* chromosome is unable to produce the FMR1 protein. [1]

- Explain why children with fragile *X* syndrome would often have learning disabilities, including speech and language problems and intellectual disabilities. [1]

Acceptable responses include but are not limited to: [2]

- *The mutated gene on the fragile X chromosome affects the human body by its failure to produce the FMR1 protein. This gene's failure to produce FMR1 is probably the result of a changed base sequence on the DNA molecule of the gene caused by the mutation. [1] Because the FRM1 protein catalyzes the synthesis of other proteins critical to nerve development, it is possible that this condition causes brain damage in the fetus, resulting in the child having learning/speech problems later on. [1]*

- *A mutation could have deleted the gene that codes for the synthesis of the protein. [1] FMR1 protein is associated with nerve cell development. Speech is controlled by the nervous system. [1]*

- *The gene may have been turned off as a result of the mutation. [1] Without this gene, nerve cell development is not well-regulated, so learning disabilities could result. [1]*

- *The mutation messed up the genetic code for the FMR1 protein, so it couldn't be produced. [1] With no FMR1, nerve cells in the brain couldn't work right, so the child has a hard time learning. [1]*

58. One credit is allowed for correctly explaining why species *C* might have a greater chance of avoiding extinction in the changed environment than species *B* and supporting the answer. Acceptable responses include but are not limited to: [1]

- *Species C is more likely to have a greater amount of variation because it has a higher rate of reproduction than species B. So species C has a better chance of surviving the changes.*

- *There is a higher chance of variation in species C because it reproduces more often and has more offspring per year than species B. Some of those offspring may have variations favorable under the changed environment.*

- *Species C has more offspring than species B. The more offspring produced, the higher the chance of variation and the greater the chances of survival.*

- *Some members of species C may already have adaptations that they could pass on to their many offspring, so they're more likely to be able to adapt to the changed environment than species B.*

59. One credit is allowed for correctly stating *one* possible reason why species *A* could be the most successful in surviving an environmental change and supporting the answer. Acceptable responses include but are not limited to: [1]

- *If species A is already adapted to the new environment, it would continue to be successful since the offspring would be identical to the parent.*

- *Its short reproductive cycle would be an advantage since all the offspring would inherit any favorable traits the organism might have.*

- *If a favorable mutation were to occur in one member of species A, the favorable adaptation would be quickly passed on to many offspring under increased selection pressure.*

60. One credit is allowed for correctly stating *one negative* effect the overuse of plastic bags is having on the environment. Acceptable responses include but are not limited to: [1]

- *Plastic bags can kill some animals that mistake them for food, reducing the animals' numbers and interfering with their role in the environment.*

- *The overuse of plastic bags is filling up landfills and increasing litter in the environment.*

- *Natural habitats are being destroyed to make way for new landfills that hold plastic bags.*

- *Plastic bags are produced using petroleum products that pollute the environment.*

- *Plastic bags don't decompose readily, so they will pollute the environment for many years.*

61. One credit is allowed for correctly explaining why the researchers suspect it is an enzyme that is enabling wax moth caterpillars to break down plastic bags. Acceptable responses include but are not limited to: [1]

- *Enzymes are used by living things to digest/hydrolyze complex substances.*
- *One common use of enzymes is to break down big molecules into smaller ones.*
- *It breaks the chemical bonds in polyethylene, so it acts like an enzyme.*
- *Enzymes work by catalyzing chemical reactions such as breaking chemical bonds.*

62. One credit is allowed for correctly explaining why using the chemical produced by the caterpillars to break down plastic bags could be considered an ecologically friendly solution to the problem. Acceptable responses include but are not limited to: [1]

- *It is less likely to be toxic/harmful to the environment than harsh chemical solvents that might otherwise be used to break down the plastic.*
- *Enzymes are proteins that will most likely degrade/break down naturally in the environment.*
- *By synthesizing the chemical in the lab, wax moths will not have to be imported from Europe to the U.S., where they could become an invasive species.*
- *Wax moth caterpillars won't have to be sacrificed to produce the chemical once scientists have figured out how to make it in the laboratory.*

63. One credit is allowed for suggesting a plan of action that could be carried out in your local community that would be a step toward solving the plastic bag problem. Acceptable responses include but are not limited to: [1]

- *Use social media to encourage the use of reusable cloth bags.*
- *Set up plastic bag recycling containers at various locations.*
- *Write letters to local politicians suggesting a ban on the use of plastic bags.*
- *Charge customers a fee for plastic bags so they will be less likely to discard them after one use.*
- *Mount an educational campaign to teach community members about the environmental problems caused by plastic bags.*
- *Petition bag manufacturers to make bags out of decomposable substances such as cellulose.*

64. One credit is allowed for correctly describing *one* positive and *one negative* outcome of mining metal ores. Acceptable responses include but are not limited to: [1]

- *We benefit by obtaining a valuable resource for human use, but contaminated soil can be harmful to people and other organisms.*
- *Positive result: We obtain valuable metal ore to use in manufacturing.*
- *Negative result: Soil contamination is bad for the environment.*
- *The benefit is we get the ore to refine into metals; the risk is that we pollute the environment.*

65. One credit is allowed for correctly explaining why importing grasses to clean up mining wastes in areas where those grasses do not normally grow could lead to unexpected environmental problems. Acceptable responses include but are not limited to: [1]

- *The imported grasses could become invasive in the new area.*
- *The new grasses could compete with native plants for limited resources in the ecosystem.*
- *Herbivores in the area could eat the grasses and accumulate the toxins in their bodies.*
- *Toxic metals could enter the food chain and contaminate cows kept for milk or meat.*
- *There would be a problem of where/how to dispose of the grasses containing the toxic waste.*

66. One credit is allowed for correctly explaining why patients with a mitochondrial disease would tend to experience these symptoms. Acceptable responses include but are not limited to: [1]

- *Mitochondria are the sites of cellular respiration, a process that releases energy in the cell.*
- *If the mitochondria fail, less energy is available. So sufferers may fatigue easily.*
- *The patients would produce less ATP to power other metabolic activities.*
- *No mitochondria, no ATP, no energy—leads to muscle weakness.*

67. One credit is allowed for correctly explaining how snow cover affects the population of the snowshoe hare. Acceptable responses include but are not limited to: [1]

- *Without snow cover, the hares are more visible to predators, so a smaller number will survive.*

■ *Lack of snow cover takes away the hares' winter camouflage, so they can be seen and caught by predators.*

■ *White fur blends in with the snow. If there is no snow, the white hares stand out against the dark ground, and they can be seen and caught by hawks/foxes/owls/lynxes.*

68. One credit is allowed for correctly identifying the environmental factor that stimulates the fur color of the hare to change from brown to white. Acceptable responses include but are not limited to: [1]

■ *The shorter day length is the environmental factor that causes this change.*

■ *The decreasing hours of daylight cause this effect.*

■ *The hare's body senses the shortening day length, causing it to molt from brown to white fur.*

69. One credit is allowed for correctly identifying a specific environmental issue that is most likely to affect snowshoe hare populations in northern ecosystems and supporting the answer. Acceptable responses include but are not limited to: [1]

■ *Environmental issue: global warming. This phenomenon is causing average temperatures to increase. This makes snow cover less common and the hares easier targets for predators.*

■ *I think climate change is the environmental issue that affects the hare population by making it move its range farther north.*

■ *Increasing average temperatures is the environmental change that is putting selection pressure on the snowshoe hare by making white fur color a disadvantageous trait.*

[NOTE: Air pollution and acid rain are not mentioned in the passage as environmental factors affecting the snowshoe hare population.]

70. One credit is allowed for correctly identifying *one* change in the characteristics of the snowshoe hares in this ecosystem that would most likely be selected for if the trend shown in graphs *B* and *C* proves to be true. Acceptable responses include but are not limited to: [1]

■ *The genes that code for white fur color in snowshoe hares will become less common in the hares' gene pool, favoring the genes that code for brown fur color.*

■ *Assuming the hares' gene pool contains white fur variants that will be stimulated by very short day lengths, these genes will be favored over genes stimulated by longer day lengths.*

- *Selection pressures will favor hares that keep their brown fur longer and molt their white fur earlier.*

71. One credit is allowed for correctly stating *one* reason this roundworm is considered a parasite to this species of tropical ant. Acceptable responses include but are not limited to: [1]

- *The roundworm lives in the ants and gets nutrients from the host, harming them in the process.*

- *The roundworm benefits by living in the ant, and the ant is harmed by the infection.*

- *As the roundworms develop, they take nutrients from the ant.*

- *The symbiosis increases the chances that the infected ants will be eaten by birds.*

72. One credit is allowed for correctly describing *one* advantage the roundworm has by having birds involved in part of its life cycle. Acceptable responses include but are not limited to: [1]

- *The birds transport the roundworm eggs to new areas.*

- *Birds disperse the roundworm eggs widely in the environment.*

- *After the birds eat the ants, they deposit the roundworm eggs into the soil of their habitat via their droppings/feces.*

PART D

73. 2 One possible reason for the increase in beak length is that the birds with longer beaks *had a more successful adaptation for survival in the area.* It is common for variations of physical traits to exist among the members of any species. These variations may prove favorable in some situations and less so in others. If beak lengthening in a particular species of bird is observed, it is likely the result of many generations of the species being placed under severe selection pressures for scarce food sources due to changed environmental conditions. It is extremely unlikely that beak length in any naturally occurring bird species would be influenced by the presence of bird feeders in their area over a short span of time.

WRONG CHOICES EXPLAINED:

(1) It is *not* true that one possible reason for the increase in beak length is that the birds with longer beaks *were less likely to have offspring with long beaks.* In fact, the opposite is true. Any trait that is controlled genetically can be passed down to offspring through the process of sexual reproduction. Offspring with long beaks are more likely, not less likely, to be produced by parent birds with long beaks.

(3) It is *not* true that one possible reason for the increase in beak length is that the birds with longer beaks *needed to reach the seed within the bird feeder, so their beaks grew longer*. Evolutionary change in birds does not occur as a function of need but only through the processes of natural selection and sexual reproduction.

(4) It is *not* true that one possible reason for the increase in beak length is that the birds with longer beaks *had more competition than other birds at the bird feeders*. Competition is a term that relates to any conflict between organisms for limited natural resources in the environment. It is extremely unlikely that beak length in any naturally occurring bird species would be influenced by competition at bird feeders in their area over a short span of time.

74. **1** The presence of bird feeders in an area would represent a *selecting agent*. Natural selection, sometimes referred to as survival of the fittest, postulates that certain members of a species are better adapted to their environment than others and that these organisms are more likely to survive and pass on their favorable adaptations to future generations. Selecting agents are any factors present in the natural environment that stress the species population and favor certain variations over others. Human-placed bird feeders would serve as selecting agents only if most or all other food sources in an area became scarce enough to force all birds of a species to feed only at these bird feeders over many generations.

WRONG CHOICES EXPLAINED:
(2) The presence of bird feeders in an area would *not* represent a *feedback mechanism*. Feedback mechanisms assist in the maintenance of dynamic equilibrium, also known as homeostasis or steady state, in living things. Bird feeders do not meet the definition of feedback mechanisms.

(3) The presence of bird feeders in an area would *not* represent a *source of mutation*. Mutations are random events in which changes occur in the DNA of the cell as a result of exposure to mutagenic agents such as radiation and chemical agents. Bird feeders do not meet the definition of sources of mutation.

(4) The presence of bird feeders in an area would *not* represent a *biological catalyst*. Biological catalysts, also known as enzymes, work by speeding up or slowing down chemical reactions in living cells. Bird feeders do not meet the definition of biological catalysts.

75. **4** The new, more accepted classification of the elephant shrew is most probably based on an analysis of *the DNA present in the cells of the elephant shrews*. Prior to the availability of DNA analysis technologies, naturalists and taxonomic scientists used imprecise methods, such as an animal's appearance or behavior, to suggest the relative relatedness of a newly discovered species to other, known species.

These techniques have become outdated in the era of modern science. This reclassification was most likely made possible by modern methods that allow scientists to compare hundreds of DNA samples and assign species to taxonomic groupings based on genetic similarities.

WRONG CHOICES EXPLAINED:
(1), (2) The new, more accepted classification of the elephant shrew is *not* most probably based on an analysis of *the coloration of the elephant shrew's fur* or of *the feeding habits of the elephant shrew compared to other shrews*. These are outdated methods that might have been used by naturalists or taxonomic scientists in the past. These methods are imprecise and prone to error, so their assumptions are being reevaluated using modern technology.

(3) The new, more accepted classification of the elephant shrew is *not* most probably based on an analysis of *a number of newly found shrew fossils*. Older methods such as examination of fossils are still used by taxonomic scientists as a means of developing hypotheses about the classification of newly found species. This method is also used extensively to study and classify extinct organisms whose DNA is no longer available for analysis.

76. **3** According to the new evolutionary tree, elephant shrews are most closely related to *tenrecs, golden moles, and aardvarks*. A close examination of the information presented in the diagram shows that elephant shrews share the nearest common ancestor with these three groups.

    WRONG CHOICES EXPLAINED:
    (1) According to the new evolutionary tree, elephant shrews are *not* most closely related to *manatees and hyraxes*. A close examination of the information presented in the diagram shows that elephant shrews share common ancestor with these groups that is less near.

    (2), (4) According to the new evolutionary tree, elephant shrews are *not* most closely related to *shrews and their relatives* or to *primates, rodents, rabbits, and relatives*. A close examination of the information presented in the diagram shows that elephant shrews share only a distant common ancestor with these groups.

77. One credit is allowed for correctly explaining why it is important to continue to protect the habitat in which the elephant shrew is found. Acceptable responses include but are not limited to: [1]

    - *These animals are important to maintaining biodiversity in their environment.*
    - *The destruction of the habitat could have unintended consequences that could disrupt the entire food web.*

■ *The destruction of their habitat might cause the shrews to become extinct.*

■ *The elephant shrew's habitat probably supports many other plant, animal, and microbial species, all of which are important to the balance of nature.*

78. One credit is allowed for correctly identifying the dependent variable in this investigation. Acceptable responses include but are not limited to: [1]

■ *Pulse rate*

■ *Beats per minute*

■ *Rate of heartbeat*

79. One credit is allowed for correctly identifying *one* error in the experimental procedure. Acceptable responses include but are not limited to: [1]

■ *There is no control group.*

■ *The resting pulse rates of the subjects were not measured at the start of the investigation.*

■ *The student's sample size is too small.*

■ *No information about the age/gender/weight/physical fitness of the subjects was recorded.*

80. One credit is allowed for correctly drawing a line on the graph provided that shows the relationship between exercise and oxygen consumption and supporting the answer. Acceptable responses include but are not limited to: [1]

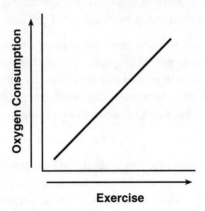

■ *As you exercise, you use more oxygen.*

■ *Exercise causes you to use more oxygen.*

- *There is a direct relationship/positive correlation between these two factors.*
- *During exercise, more energy/ATP is required by the muscle cells, so the cells' respiratory reactions need more oxygen to break down glucose/produce ATP.*

81. **2** *In distilled water, the mass of the potato cube increased due to water moving into the cells from high concentration to low concentration* is the statement that correctly describes the effect on the mass of one of the cubes after a 20-minute period. Distilled water is 100% water (high concentration), whereas the cytoplasm of most cells is approximately 70–80% water (low concentration). When living cells are placed into distilled water, molecules of water move readily into the cell from high concentration to low concentration via the process of osmosis. This causes the cell to swell as the mass of water increases inside the cell membrane/walls.

    WRONG CHOICES EXPLAINED:
    (1) *In distilled water, the mass of the potato cube increased due to salt leaving the cells of the potato* is *not* the statement that correctly describes the effect on the mass of one of the cubes after a 20-minute period. Salt molecules do not pass through the cell membrane in either direction in this process. However, if they were to move out of the cell, this would result in a decrease, not an increase, in cell mass.

    (3) *In the concentrated salt solution, the mass of the potato cube increased due to salt moving into the cells from low concentration to high concentration* is *not* the statement that correctly describes the effect on the mass of one of the cubes after a 20-minute period. Salt molecules do not pass through the cell membrane in either direction in this process. However, if they were to move into the cell, they could do so only by moving molecules from low concentration to high concentration, which would require the expenditure of cell energy.

    (4) *In the concentrated salt solution, the mass of the potato cube remained the same due to the cell wall preventing the movement of molecules into or out of the cells* is *not* the statement that correctly describes the effect on the mass of one of the cubes after a 20-minute period. Cell walls are rigid, porous structures that support plant tissues but do not function to regulate the movement of materials into or out of the cell.

82. **2** The blue-black-stained structures labeled in her drawing are most likely *starch grains*. Lugol's solution, also known as iodine solution, is a common indicator used for detecting the presence of starch molecules in living and nonliving samples. When this tan indicator is placed onto a plant tissue, a blue-black coloration often occurs inside the cells, an observation that indicates the presence of starch molecules in the tissue. The structures in the potato slice turned blue-black, indicating the presence of starch molecules within food vacuoles in the potato cells.